Pathophysiology

made

Incredibly Easy!™

Springhouse Corporation
Springhouse, Pennsylvania

Staff

Executive Director
Matthew Cahill

Clinical Director
Judith A. Schilling McCann, RN, MSN

Art Director
John Hubbard

Senior Editor
Michael Shaw

Clinical Editors
Carla M. Roy, RN, BSN, CCRN (project manager); Joanne M. Bartelmo, RN, MSN, CCRN; Collette Hendler, RN, CCRN; Clare M. Brabson, RN, BSN; Lori Musolf Neri, RN, MSN, CCRN; Deborah Becker, RN, MSN, CCRN; Marybeth Morrell, RN, CCRN

Editors
Donna O. Carpenter (editorial director), Mary Lou Ambrose, Patricia Wittig

Copy Editors
Cynthia C. Breuninger (manager), Karen C. Comerford, Mary T. Durkin, Brenna H. Mayer

Designers
Arlene Putterman (associate art director), Mary Ludwicki (book designer), Joseph Clark

Illustrators
Bot Roda, Jackie Facciolo, Dan Fione, Betty Winnberg, Jean Gardener, Bob Jackson, Judy Newhouse, Julie Devito, John Murphy

Typography
Diane Paluba (manager), Joyce Rossi Biletz, Valerie Rosenberger

Manufacturing
Deborah Meiris (director), Pat Dorshaw (manager), Otto Mezei

Production Coordinator
Stephen Hungerford, Jr.

Editorial Assistant
Beverly Lane

Indexer
Barbara Hodgson

The clinical treatments described and recommended in this publication are based on research and consultation with nursing, medical, and legal authorities. To the best of our knowledge, these procedures reflect currently accepted practice. Nevertheless, they can't be considered absolute and universal recommendations. For individual applications, all recommendations must be considered in light of the patient's clinical condition and, before administration of new or infrequently used drugs, in light of the latest package-insert information. The authors and the publisher disclaim any responsibility for any adverse effects resulting from the suggested procedures, from any undetected errors, or from the reader's misunderstanding of the text.

Printed in the United States of America.

IEP-   D   N   O   S   A
03   02   01   00   10 9 8 7 6 5

Library of Congress Cataloging-in Publication Data

Pathophysiology made incredibly easy.
   p. cm.
   Includes index.
   1. Physiology, Pathological.
   2. Nursing.
   I. Springhouse Corporation.
   [DNLM: 1. PATHOLOGY-NURSES' INSTRUCTION.
QZ 4 P3024 1998]
RB113.P3636 1998
616.07—dc21

DNLM/DLC   97-46681
ISBN 0-87434-935-4   CIP

Contents

Contributors and consultants

Deborah Becker, RN, MSN, CCRN
Lecturer
School of Nursing, University of
Pennsylvania
Philadelphia

Susan J. Brown-Wagner, RN, MSN, CRNP, AOCN
Administrative Director, Oncology Initiative
Anne Arundel Medical Center
Annapolis, Md.

Jacqueline Crocetti, RN, MSN, CRNP
Nursing Professor
Northampton County Community College
Bethlehem, Pa.

Susan B. Dickey, RN,C, PhD
Associate Professor
Department of Nursing CAHP, Temple
University
Philadelphia

DuWayne C. Englert, PhD
Professor
Department of Zoology
Southern Illinois University at Carbondale

Marilyn A. Folcik, RN, MPH, ONC
Assistant Director, Department of Surgery
Hartford (Conn.) Hospital

Ellie Z. Franges, RN, MSN, CNRN, CCRN
Director, Neuroscience Services
Sacred Heart Hospital
Allentown, Pa.

Linda B. Haas, RN, PhC, CDE
Endocrinology Clinical Nurse Specialist
VA Puget Sound Health Care System
Seattle

Nancy A. Haukom, RN, BAN, CIC
Infection Control Nurse
Mayo Medical Center, Saint Mary's
Hospital
Rochester, Minn.

Christine A. Kogel, RN
Clinical Nurse Coordinator, Kidney Stone
Clinic
University Hospital
Denver

Kay L. Luft, MN, CCRN, TNCC
Assistant Professor of Nursing
Saint Luke's College
Kansas City, Mo.

Judith E. Meissner, RN, MSN
Senior Associate Professor
Bucks County Community College
Newtown, Pa.

Teresa Murphy, RN, BSN, CCRN
Nurse Manager, Medical ICU
Allegheny University Hospital, Graduate
Philadelphia

Barbara Barnes Rogers, MN, CRNP, AOCN
Adult Oncology Nurse Practitioner
Fox Chase Cancer Center
Philadelphia

Michael L. Silverman, MD
Clinical Assistant Professor of Medicine
Allegheny University Hospital, Graduate
Philadelphia

Maryann Summers, RN, MS
Nurse-Instructor
James Martin School for LPNs
Philadelphia

Foreword

What memories does the word pathophysiology spark for you? The three things I remember most about studying pathophysiology in nursing school are:

- long lectures

- out-of-focus overheads

- arcane examination questions that tested my ability to read the teacher's mind (unfortunately, a weak skill) but not my understanding of the effects of disease on the human body.

When I entered professional nursing practice, I realized the importance of a strong background in pathophysiology. Understanding the biological basis of disease helps me to better understand a patient's signs and symptoms, prescribed medications and therapies, and response to treatment. What's more, pathophysiology is dynamic. New discoveries are taking place faster than ever. Even nurses with a strong background in pathophysiology need to update their knowledge.

Unfortunately, too many pathophysiology references share some unfortunate characteristics; they're long-winded and (dare I say it) BORING. When seeking to quickly review basic concepts, I find myself weighted down with complex tomes as cumbersome to carry as they are to understand. I wade through a swamp of trivia, bogged down in unimportant details. As I turn the pages, memories of long lectures and out-of-focus overheads come back to haunt me.

Well, folks, I have big news. The problem is solved. In fact, you're holding the solution in your hands. Leave it to the clinical experts at Springhouse to create *Pathophysiology Made Incredibly Easy*, a nursing reference unlike any other on the planet. Stop and think about how much you can benefit from *Pathophysiology Made Incredibly Easy*. If you're like most nurses, you're likely to encounter unfamiliar problems or get a little rusty regarding the science underlying a specific disease. Few of us have time to pour over densely written texts. *Pathophysiology Made Incredibly Easy* helps you update your knowledge quickly so you can provide the best patient care.

This book is unlike any other nursing reference. *Pathophysiology Made Incredibly Easy* is lighthearted, fun-filled, and down-to-earth. The warmth and humor that emanates from these pages is a welcome relief from the day-to-day pressures of today's fast-paced, cost-conscious working environment.

The first chapter of *Pathophysiology Made Incredibly Easy* discusses the basics of pathophysiology. These basics provide an important foundation for understanding specific diseases. The second chapter discusses cancer, the second leading cause of death in the United States. Here you'll find pathophysiology, tests, and treatments for common types of cancer.

Next you'll find chapters covering three hot topics in contemporary health care: infection, immune disorders, and genetics. In each chapter you'll find difficult concepts made crystal clear, along with invaluable clinical information about the pathophysiology of common disorders.

The last 8 chapters focus on body systems. You'll find information on the endocrine, respiratory, cardiovascular, neurologic, hematologic, gastrointestinal, renal, and musculoskeletal systems (try saying that in one breath!). Each chapter begins with a discussion of fundamental concepts and goes on to provide information on pathophysiology, signs and symptoms, tests, and treatments for common disorders.

Within each chapter, you'll find many special features to enhance your understanding and make learning enjoyable. Each chapter begins with an at-a-glance summary of key topics. Numerous checklists make it easy to spot the most important points. There are even cartoon characters that help make learning about pathophysiology a joy. A *Quick quiz* at the end of each chapter helps you assess what you've learned. Special logos throughout each chapter alert you to essential information:

The *Now I get it!* logo alerts you to illustrations, flowcharts, and explanations that make difficult concepts in pathophysiology *incredibly easy.*

Look for the *Battling illness* logo for a quick, clear summary of treatments for disorders discussed in each chapter.

Personally, my favorite feature is the *Incredibly Easy miniguide.* Look on pages 187 to 190 and 223 to 226 and you will find beautiful, full-color illustrations of heart and lung pathophysiology. Here, truly, is pathophysiology like you've never seen it before.

Please take the time to explore this clinically useful and refreshingly original book. I am confident that *Pathophysiology Made Incredibly Easy* will become one of your most frequently utilized and beloved references. Enjoy learning.

Michael A. Carter, RN, DNSc, FAAN
Dean and Professor
College of Nursing
University of Tennessee
Memphis

It all
starts with
me, a
little
tiny cell...

① Pathophysiology basics

Just the facts

This chapter presents basic information about cells and diseases. In this chapter, you'll learn:

♦ the structure of cells and how cells reproduce, age, and die

♦ what homeostasis is and how it affects the body

♦ what causes disease

♦ how diseases develop.

Understanding cells

The cell is the body's basic building block. It's the smallest living component of an organism. Many organisms are made up of one independent, microscopically tiny cell. Others, such as humans, are comprised of millions of cells grouped into highly specialized units that function together throughout the organism's life.

Large groups of individual cells form tissues, such as muscle, blood, and bone. Tissues, in turn, form the organs, such as the brain, heart, and liver, which are integrated into body systems — such as the central nervous system (CNS), cardiovascular system, and digestive system.

Understanding pathophysiology begins with understanding the body's basic building block, the cell.

Cell components

Cells are composed of various structures, or organelles, each with specific functions. (See *Just your average cell.*) The organelles are contained in the cytoplasm — an aqueous mass — which is surrounded by the cell membrane.

The largest organelle, the nucleus, stores deoxyribonucleic acid (DNA), which carries genetic material and is responsible for cellular reproduction or division.

More components

The typical animal cell is characterized by several additional elements:
• Adenosine triphosphate, the energy that fuels cellular activity, is made by the mitochondria.
• Ribosomes and the endoplasmic reticulum synthesize proteins and metabolize fat within the cell.
• The Golgi apparatus holds enzyme systems that assist in completing the cell's metabolic functions.
• Lysosomes contain enzymes that allow cytoplasmic digestion to be completed.

Cell division and reproduction

Individual cells don't live as long as the individual they're part of. They're subject to wear and tear and must reproduce and be replaced. Most cells reproduce as quickly as they die.

Just your average cell

The illustration below shows cell components and structures. Each part has a function in maintaining the cell's life and homeostasis.

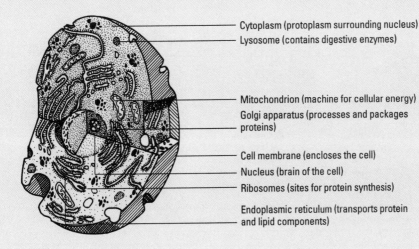

Cytoplasm (protoplasm surrounding nucleus)
Lysosome (contains digestive enzymes)

Mitochondrion (machine for cellular energy)
Golgi apparatus (processes and packages proteins)

Cell membrane (encloses the cell)
Nucleus (brain of the cell)
Ribosomes (sites for protein synthesis)

Endoplasmic reticulum (transports protein and lipid components)

Cell reproduction occurs in two stages. In the first stage, called mitosis, the nucleus and genetic material divide. In the second stage, called cytokinesis, the cytoplasm divides. (See *Replicate and divide*.)

The great divide

Before division, a cell must double its mass and content. This occurs during the growth phase, called interphase. Chromatin, the small, slender rods of the nucleus that give it its granular appearance, begins to form.

Replication and duplication of DNA occur during the four phases of mitosis:

 prophase

 metaphase

 anaphase

 telophase.

Replicate and divide

The illustrations below show the different phases of mitosis, or cell reproduction.

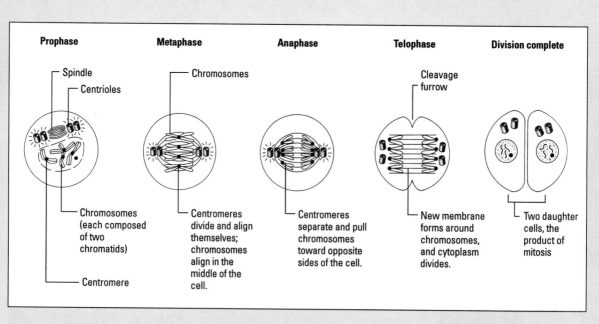

Prophase
- Spindle
- Centrioles
- Chromosomes (each composed of two chromatids)
- Centromere

Metaphase
- Chromosomes
- Centromeres divide and align themselves; chromosomes align in the middle of the cell.

Anaphase
- Centromeres separate and pull chromosomes toward opposite sides of the cell.

Telophase
- Cleavage furrow
- New membrane forms around chromosomes, and cytoplasm divides.

Division complete
- Two daughter cells, the product of mitosis

Prophase

During prophase, the chromosomes coil and shorten and the nuclear membrane dissolves. Each chromosome is made up of a pair of strands called chromatids, which are connected by a spindle of fibers called a centromere.

Metaphase

During metaphase, the centromeres divide, pulling the chromosomes apart. The centromeres then align themselves in the middle of the spindle.

Anaphase

At the onset of anaphase, the centromeres begin to separate and pull the newly replicated chromosomes toward opposite sides of the cell. By the end of anaphase, 46 chromosomes are present on each side of the cell.

Telophase

In the final phase of mitosis — telophase — a new membrane forms around each set of 46 chromosomes. The spindle fibers disappear, and the cytoplasm divides, producing two identical new "daughter" cells.

Cell injury

A person's state of wellness and disease is reflected in the cells. Injury to any of the cell's components can lead to illness.

One of the first indications of cell injury is a biochemical lesion that forms on the cell at the point of injury. This lesion changes the chemistry of metabolic reactions within the cell.

Consider, for example, a patient with chronic alcoholism. The cells of the immune system may be altered, making the patient susceptible to infection. Other cells are also affected, such as the cells of the pancreas and liver. These cells can't be reproduced, which prevents a return to normal functioning.

> **Memory jogger**
>
> To remember the four causes of cell injury, think of how the injury tipped (or *TIPD*) the scale of homeostasis:
>
> T   toxin or other lethal (cytotoxic) substance
> I   infection
> P   physical insult or injury
> D   deficit, or lack of water, oxygen, or nutrients.

> Toxic injuries may result from factors inside or outside the body.

Draw on your reserves, adapt, or die.

When cell integrity is threatened — for example, by hypoxia, anoxia, chemical injury, infection, or temperature extremes — the cell reacts in one of two ways:
- by drawing on its reserves to keep functioning
- by adapting through changes or cellular dysfunction.

If enough cellular reserve is available and the body doesn't detect abnormalities, the cell adapts. If there isn't enough cellular reserve, cell death (necrosis) occurs. Necrosis is usually localized and easily identifiable.

Toxic injury

Toxic injuries may be caused by factors inside the body (called endogenous factors) or outside the body (called exogenous factors). Common endogenous toxins include genetically determined metabolic errors, gross malformations, and hypersensitivity reactions. Exogenous toxins include alcohol, lead, carbon monoxide, and drugs that alter cellular function. Examples of such drugs are chemotherapeutic agents used for cancer and immunosuppressant drugs that prevent rejection in organ transplant recipients.

Infectious injury

Viral, fungal, protozoan, and bacterial infectious agents can cause cell injury or death. These organisms affect cell integrity, usually by interfering with cell synthesis, producing nonviable, mutant cells. For example, human immunodeficiency virus alters the cell when the virus is replicated in the cell's ribonucleic acid.

Physical injury

Physical injury results from a disruption in the cell or in the relationships of the intracellular organelles. Two major types of physical injury are thermal (electrical or radiation) or mechanical (trauma or surgery). Causes of thermal injury include radiation therapy for cancer, X-rays, and ultraviolet radiation. Causes of mechanical injury include motor vehicle accidents, frostbite, and ischemia.

Deficit injury

When a deficit of water, oxygen, or nutrients occurs, or if constant temperature and adequate waste disposal aren't maintained, cellular synthesis can't take place. A lack of just one of these basic requirements can cause cell disruption or death.

Cell degeneration

A type of nonlethal cell damage known as degeneration generally occurs in the cytoplasm of the cell while the nucleus remains unaffected. It usually affects organs with metabolically active cells, such as the liver, heart, and kidneys, and is caused by the following problems:
- increased water in the cell or cellular swelling
- fatty infiltrates
- atrophy
- autophagocytosis, during which the cell absorbs some of its own parts
- pigmentation changes
- calcification
- hyaline infiltration
- hypertrophy
- dysplasia (related to chronic irritation)
- hyperplasia.

Degeneration occurs in the cytoplasm of the cell; the nucleus remains unaffected.

Talkin' 'bout degeneration

When changes within cells are identified, degeneration may be slowed or cell death prevented through prompt health care. An electron microscope makes the identification of changes within cells easier. When a disease is diagnosed before the patient complains of any symptoms, it's termed *subclinical identification*. Unfortunately, many cell changes remain unidentifiable even under a microscope, making early detection impossible.

Down with cell aging.

Cell aging

During the normal process of cell aging, cells lose structure and function. Lost cell structure may cause decrease in size or wasting away, a process called atrophy. Two characteristics of lost cell function are:
• hypertrophy, an abnormal thickening or increase in bulk
• hyperplasia, an increase in the number of cells.
    Cell aging may slow down or speed up depending on the number and extent of injuries and the amount of wear and tear on the cell.

Warning: This cell will self-destruct

Signs of aging occur in all body systems. Examples include decreases in elasticity of blood vessels, bowel motility, muscle mass, and subcutaneous fat. The cell aging process limits the human life span (of course, many people die from disease before they reach the maximum life span of about 110 years). One theory says that cell aging is an inherent self-destructive mechanism that increases with a person's age.
    Cell death may be caused by internal (intrinsic) factors that limit the cells' life span or external (extrinsic) factors that contribute to cell damage and aging. (See *In's and out's of cell aging.*)

The cell aging process is what actually limits the human life span.

Homeostasis

The body is constantly striving to maintain a dynamic, steady state of internal balance called homeostasis. Every cell in the body is involved in maintaining homeostasis, both on the cellular level and as part of an organism.
    Any change or damage at the cellular level can affect the entire body. When homeostasis is disrupted by an external stressor, illness may occur. A few examples of external stressors are injury, lack of nutrients, and invasion by parasites or other organisms. Throughout the course of a person's lifetime, many external stressors affect the body's internal equilibrium.

Maintaining the balance

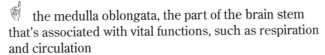

Three structures in the brain are responsible for maintaining homeostasis:

✌ the medulla oblongata, the part of the brain stem that's associated with vital functions, such as respiration and circulation

✌ the pituitary gland, which regulates the function of other glands and thereby a person's growth, maturation, and reproduction

✌ the reticular formation, a group of nerve cells or nuclei that form a large network of connected tissues that help control vital reflexes, such as cardiovascular function and respiration.

Feedback mechanisms

Homeostasis is maintained by self-regulating feedback mechanisms. These mechanisms have three components:

✌ a sensor mechanism that senses disruptions in homeostasis

✌ a control center that regulates the body's response to disruptions in homeostasis

✌ an effector mechanism that acts to restore homeostasis.

The sensor mechanism is usually controlled by an endocrine, or hormone-secreting, gland. A signal is sent to the control center in the CNS, which initiates the effector mechanism.

There are two types of feedback mechanisms:

✌ a negative feedback mechanism, which works to restore homeostasis by correcting a deficit within the system

✌ a positive feedback mechanism, which occurs when hormone secretion triggers additional hormone secretion. It indicates a trend away from homeostasis.

Accentuate the negative

For negative feedback mechanisms to be effective, they must sense a change in the body — such as a high blood glucose level — and attempt to return body functions to

---

In's and out's of cell aging

Factors that affect cell aging may be intrinsic or extrinsic, as outlined below.

Intrinsic factors
- Psychogenic
- Inherited
- Congenital
- Metabolic
- Degenerative
- Neoplastic
- Immunologic
- Nutritional

Extrinsic factors
Physical agents
- Force
- Temperature
- Humidity
- Radiation
- Electricity
- Chemicals

Infectious agents
- Viruses
- Bacteria
- Fungi
- Protozoa
- Insects
- Worms

Negative feedback, positive result

This flow chart shows how a negative feedback mechanism works to restore homeostasis in a patient with a high blood glucose level.

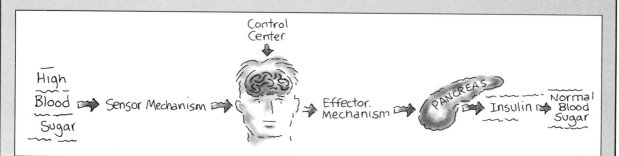

normal. In the case of a high blood glucose level, the effector mechanism triggers increased insulin production by the pancreas, returning blood glucose levels to normal and restoring homeostasis. (See *Negative feedback, positive result*.)

Disease and illness

Although disease and illness are not synonymous, they're often used interchangeably. Disease occurs when homeostasis isn't maintained. Illness occurs when a person is no longer in a state of "normal" health.

For example, a person may have coronary artery disease, diabetes, or asthma but not be ill all the time because his body has adapted to the disease. In this situation, a person can perform necessary activities of daily living.

Genetic factors plus

The course and outcome of a disease are influenced by genetic factors (such as a tendency toward obesity), unhealthy behaviors (such as smoking), attitudes (such as being a "Type A" personality), and even the person's perception of the disease (such as acceptance or denial). Diseases are dynamic and may be manifested in a variety of ways depending on the patient and his environment.

Genetic factors, plus unhealthy behaviors, plus personality type, plus attitude, plus perception may add up to disease.

Cause

One aspect of disease is its cause (the fancy term is etiology). The cause of disease may be intrinsic or extrinsic. Inheritance, age, the person's sex, infectious agents, or behaviors (such as inactivity, smoking, or using drugs) can all cause disease. Diseases with no known cause are called idiopathic.

Development

A disease's development is called its pathogenesis. Unless identified and successfully treated, most diseases progress according to a typical pattern of symptoms. Some diseases are self-limiting or resolve quickly with limited or no intervention; others are chronic and never resolve. Patients with chronic diseases may undergo remission and exacerbation.

Telltale signs

Usually, a disease is uncovered because of an increase or decrease in metabolism or cell division. Other signs and symptoms may include hypofunction (such as constipation), hyperfunction (such as increased mucus production), or increased mechanical function (such as a seizure).

How the cells respond to disease depends on the causative agent and the affected cells, tissues, and organs. Resolution of disease depends on many factors functioning over a period of time.

Disease stages

Typically, diseases progress through the following stages:

exposure or injury. Target tissue is exposed to a causative agent or is injured.

latent or incubation period. No signs or symptoms are evident.

prodromal period. Signs and symptoms are usually mild and nonspecific.

acute phase. The disease reaches its full intensity and complications often arise. If the patient can still function in a pre-illness fashion during this phase, it's called the sub-clinical acute phase.

remission. This second latent phase occurs in some diseases, and is often followed by another acute phase.

convalescence. In this stage of rehabilitation, the patient progresses toward recovery after the termination of a disease.

recovery. In this stage the patient regains health or normal functioning. No signs or symptoms of disease occur.

People with certain diseases never fully recover; periodic remission is their best outcome.

Stress and disease

When a stressor such as a life change occurs, a person can respond in one of two ways: by adapting successfully or by failing to adapt. A maladaptive response to stress may result in disease. The underlying stressor may be real or perceived.

Stressful stages

Hans Selye, a pioneer in the study of stress and disease, describes stages of adaptation to a stressful event: alarm, resistance, and exhaustion or recovery. (See *When stress strikes.*) In the alarm stage, the body senses stress and the CNS is aroused. The body releases chemicals to mobilize the fight-or-flight response. This release is the adrenaline rush associated with panic or aggression. In the resistance stage, the body either adapts and achieves homeostasis, or it fails to adapt and enters the exhaustion stage, resulting in disease.

Everything is under control

The stress response is controlled by actions taking place in the nervous and endocrine systems. These actions try to redirect energy to the organ — such as the heart, lungs, or brain — that is most affected by the stress.

Mind—body connection

Physiologic stressors may elicit a harmful response leading to an identifiable illness or set of symptoms. Psychological stressors, such as the death of a loved one, may also cause a maladaptive response. Stressful events can exacerbate some chronic diseases, such as diabetes and multiple sclerosis. Effective coping strategies can prevent or reduce harmful effects of stress.

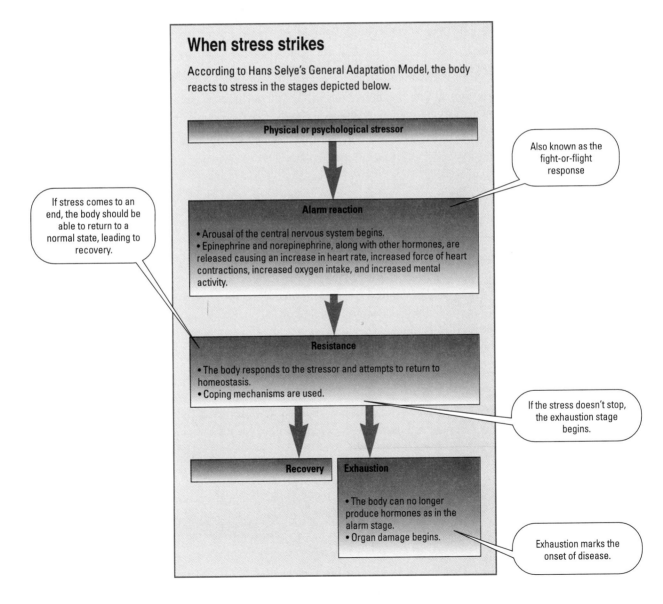

When stress strikes

According to Hans Selye's General Adaptation Model, the body reacts to stress in the stages depicted below.

Physical or psychological stressor

Alarm reaction

• Arousal of the central nervous system begins.
• Epinephrine and norepinephrine, along with other hormones, are released causing an increase in heart rate, increased force of heart contractions, increased oxygen intake, and increased mental activity.

Also known as the fight-or-flight response

If stress comes to an end, the body should be able to return to a normal state, leading to recovery.

Resistance

• The body responds to the stressor and attempts to return to homeostasis.
• Coping mechanisms are used.

If the stress doesn't stop, the exhaustion stage begins.

Recovery   **Exhaustion**

• The body can no longer produce hormones as in the alarm stage.
• Organ damage begins.

Exhaustion marks the onset of disease.

Quick quiz

Now that you've finished the first chapter, take our quick quiz. We promise it will be low stress.

1. The organelle that contains the cell's DNA is:
  A. the mitochondria.
  B. the nucleus.
  C. the ribosome.

Answer: B. The nucleus, the largest organelle, stores DNA and is responsible for cellular reproduction.

2. When a cell gets injured, the first sign is:
  A. a biochemical lesion.
  B. an area of hyperplasia.
  C. a chromatid.

Answer: A. Chemical reactions in the cell occur as a result of injury and form a biochemical lesion.

3. An extrinsic factor that can cause cell aging and death is:
  A. Down syndrome.
  B. sickle cell anemia.
  C. ultraviolet radiation.

Answer: C. An extrinsic factor comes from an outside source.

4. Homeostasis can be defined as:
  A. a steady, dynamic state.
  B. a state of flux.
  C. an unbalanced state.

Answer: A. Homeostasis may also be defined as a balancing act performed by the body to prevent illness.

Scoring

 If you answered all four items correctly, fantastic! Your intrinsic and extrinsic understanding of pathophysiology basics is excellent.

If you answered two or three correctly, wonderful! Your test-taking skills have achieved a nice homeostasis!

 If you answered fewer than two correctly, no sweat! Twelve more quick quizzes to go!

Cancer

2

Just the facts

In this chapter you'll learn:

♦ what causes abnormal cell growth

♦ the seven warning signs of cancer

♦ how cancers are classified

♦ how cancer metastasizes

♦ pathophysiology, tests, and treatments for common types of cancer.

Understanding cancer

Cancer ranks second to cardiovascular disease as the leading cause of death in the United States. Some epidemiologists predict that it will outrank cardiovascular disease by the year 2010. Every year, more than 1 million cancer cases are diagnosed in the United States and 500,000 people die of cancer-related causes.

The importance of early detection

In most cases, early detection of cancer enables more effective treatment and a better prognosis for the patient. A careful assessment, beginning with an exhaustive patient history, is critical. In gathering assessment information, you need to ask the patient about risk factors, such as cigarette smoking, a family history of cancer, and exposure to potential hazards.

In most cases, early detection of cancer is key to achieving an optimal outcome.

Memory jogger

When asking assessment questions, remember the American Cancer Society's mnemonic device, CAUTION:

C Change in bowel or bladder habits

A sore that doesn't heal

U Unusual bleeding or discharge

T Thickening or lump

I Indigestion or difficulty swallowing

O Obvious changes in a wart or mole

N Nagging cough or hoarseness.

Abnormal cell growth

Cancer is classified by the tissues or blood cells in which it originates. Most cancers derive from epithelial tissues and are called carcinomas. Others arise from the following tissues and cells:
- glandular tissues (adenocarcinomas)
- connective, muscle, and bone tissues (sarcomas)
- supportive tissue of the brain and spinal cord (gliomas)
- pigmented cells (melanomas)
- plasma cells (myelomas)
- lymphatic tissue (lymphomas)
- leukocytes (leukemia)
- erythrocytes (erythroleukemia).

Uncontrolled growth

Cancer cells grow without the control that characterizes normal cell growth. At a certain stage of development, the cancer cells fail to mature. Cancer cells have the capability to spread from the site of origin, a process called metastasis.

Cancer cells metastasize three ways:
- by circulation through the blood and lymphatic system
- by accidental transplantation during surgery
- by spreading to adjacent organs and tissues.

What causes cancer?

A cell's transformation from normal to cancerous is called carcinogenesis. Carcinogenesis has no single cause, but probably results from complex interactions between viruses, physical and chemical carcinogens, and genetic, dietary, immunologic, and hormonal factors.

The virus factor

Animal studies show that viruses can transform cells. The Epstein-Barr virus that causes infectious mononucleosis is associated with Burkett's lymphoma and nasopharyngeal cancer. Certain types of human papilloma virus are linked to cancer of the cervix. The hepatitis B virus causes hepatocellular carcinoma. And the human T-cell lymphotropic (HTLV-1) virus is suspected of causing adult T-cell leukemia.

Overexposed cells

The relationship between excessive exposure to the sun's ultraviolet rays and skin cancer is well established. Damage caused by ultraviolet light and subsequent sunburn may also be linked to melanoma.

Radiation exposure may induce tumor development. Other factors, such as the patient's tissue type, age, and hormonal status, also contribute to the carcinogenic effect.

Something in the environment

Substances in the environment can cause cancer by damaging deoxyribonucleic acid in the cells. Examples of common carcinogens and related cancers are:
• tobacco (lung, pancreatic, kidney, bladder, and esophageal cancer)
• asbestos and airborne aromatic hydrocarbons (lung cancer)
• alkylating agents (leukemia).

Danger at the diner

Colorectal cancer is associated with high-protein and high-fat diets. Food additives such as nitrates and food preparation methods such as charbroiling may also contribute to cancer.

The genetic factor

Some cancers and precancerous lesions are genetically linked. They may be autosomal recessive, X-linked, or autosomal dominant disorders (see Chapter 5, Genetics). Such cancers share the following characteristics:
• early onset
• increased incidence of bilateral cancer in paired organs (breasts, adrenal glands, and kidneys)
• increased incidence of multiple primary cancers in nonpaired organs
• abnormal chromosome complement in tumor cells
• unique tumor site combinations
• two or more family members in the same generation with the same cancer.

Hormones: Helping or hurting?

The role hormones play in cancer is controversial. Excessive hormone use, especially of estrogen, may contribute to certain forms of cancer while reducing the risk of other forms.

The best defense

Recent theory suggests that the body develops cancer cells continuously but that the immune system recognizes them as foreign and destroys them. According to the theory, this defense mechanism, called immunosurveillance, promotes antibody production, cellular immunity, and immunologic memory. Therefore, an interruption in immunosurveillance could lead to the overproduction of cancer cells and, possibly, a tumor.

Through immunosurveillance, the immune system recognizes cancer cells as foreign and destroys them.

Disorders

Types of cancer discussed below include:
• breast cancer
• colorectal cancer
• Hodgkin's disease
• leukemia
• lung cancer
• malignant melanoma
• multiple myeloma
• prostate cancer.

Breast cancer

Breast cancer is the most common cancer in women. Although the disease may develop any time after puberty, 70% of cases occur in women over age 50.

Pathophysiology

The exact causes of breast cancer remain elusive. Scientists have discovered specific genes linked to breast cancer (called BRCA1 and BRCA2), which confirms that the disease can be inherited from a person's mother or father. Other significant risk factors have been identified. These include:
- a family history of breast cancer
- radiation exposure
- being a premenopausal woman over age 45
- obesity
- early onset of menses or late menopause
- nulligravida (never pregnant)
- first pregnancy after age 35
- high-fat diet
- endometrial or ovarian cancer
- estrogen therapy
- antihypertensive therapy
- alcohol and tobacco use
- benign breast disease.

About half of all breast cancers develop in the upper outer quadrant. (See *Breast tumor sites,* page 20.)

Numerous risk factors for breast cancer have been identified, but the exact cause remains unknown.

Understanding classification

Breast cancer is generally classified by the tissue of origin and the location of the lesion; for example:
- Adenocarcinoma, the most common form of breast cancer, arises from the epithelial tissues.
- Intraductal cancer develops within the ducts.
- Infiltrating cancer arises in the parenchymal tissue.

Breast tumor sites

This illustration shows the location and frequency of breast tumors.

50% —————

15% —————

18% —————

11% —————

6% —————

• Inflammatory cancer (rare) grows rapidly and causes the overlying skin to become edematous, inflamed, and indurated.
• Lobular cancer involves the lobes of glandular tissue.
• Medullary or circumscribed cancer is a tumor that grows rapidly.

What to look for

Breast cancer is usually detected on a mammogram before a lesion becomes palpable. Otherwise, the patient discovers a painless lump or mass in her breast or a thickening of breast tissue. Inspection may reveal nipple retraction, scaly skin around the nipple, skin changes, erythema, and clear, milky, or bloody discharge. Edema in the arm indicates advanced nodal involvement.

Palpation may identify a hard lump, mass, or thickening of breast tissue. Palpation of the cervical supraclavicular and axillary nodes may reveal lumps or enlargement.

Although growth rates vary, a lump may take up to 8 years to become palpable at ⅜″ (1 cm). Breast cancers can spread via the lymphatic system and bloodstream, through the right side of the heart to the lungs and, eventually, to the other breast, chest wall, liver, bone, and brain. (See *Treating breast cancer,* page 22.)

What tests tell you

The following tests are used to diagnose breast cancer:
• Breast self-examination, if done regularly, is the most reliable method for detecting breast lumps early.
• Mammography — the primary test for breast cancer — can be used to detect tumors that are too small to palpate.
• Fine-needle aspiration and excisional biopsy provide cells for histologic examination to confirm diagnosis.
• Hormonal receptor assay can be used to pinpoint whether the tumor is estrogen- or progesterone-dependent so that appropriate therapy can be chosen.
• Ultrasonography can be used to distinguish between a fluid-filled cyst and a solid mass.
• Chest X-rays can be used to pinpoint chest metastases.
• Scans of the bone, brain, liver, and other organs can be used to detect distant metastases.
• Laboratory tests, such as alkaline phosphatase levels and liver function tests, can uncover distant metastases.

Colorectal cancer

Colorectal cancer accounts for 10% to 15% of all new cancer cases in the United States. It occurs equally in both sexes and is the third most frequent cause of death in men and the second in women. It strikes most often in people over age 60.

The importance of a thorough assessment

Because colorectal cancer progresses slowly and remains localized for a long time, early detection is key to recovery. Unless the tumor metastasizes, the 5-year survival rate is 80% for rectal cancer and 85% for colon cancer. (See *Clues to colorectal cancer*, page 23.)

Battling illness

Treating breast cancer

Treatment for breast cancer may include a combination of surgery, radiation, chemotherapy, and hormonal therapy, depending on the disease stage and type, the woman's age and menopausal status, and the disfiguring effects of the surgery. Each treatment is explained below.

Lumpectomy

Through a small incision, the surgeon removes the tumor, surrounding tissue and, possibly, nearby lymph nodes. The patient usually undergoes radiation therapy afterward.

Lumpectomy is used for small, well-defined lesions. Studies show that lumpectomy and radiation are as effective as mastectomy in early-stage breast cancer.

Partial mastectomy

The surgeon removes the tumor along with a wedge of normal tissue, skin, fascia and, possibly, axillary lymph nodes. Radiation therapy or chemotherapy is usually used after surgery to destroy undetected disease in other breast areas.

Total mastectomy

Also called simple mastectomy, a total mastectomy involves removal of the entire breast. This procedure is used if the cancer is confined to breast tissue and no lymph node involvement is detected. Chemotherapy or radiation therapy may follow. If the patient doesn't have advanced disease, reconstructive surgery can be used to create a breast mound.

Modified radical mastectomy

The surgeon removes the entire breast, axillary lymph nodes, and the lining that covers the chest muscles. If the lymph nodes contain cancer cells, radiation therapy and chemotherapy follow. This procedure has replaced radical mastectomy as the one most widely used. The difference is that it preserves the pectoral muscles.

Before or after tumor removal, radiation therapy may be used to destroy a small, early-stage tumor without distant metastases. It can also be used to prevent or treat local recurrence. Preoperative radiation therapy to the breast also "sterilizes" the area, making the tumor more manageable surgically, especially in inflammatory breast cancer.

Chemotherapy

Cytotoxic drugs may be used either as adjuvant therapy or primary therapy. Decisions to start chemotherapy are based on several factors, such as the stage of the cancer and hormonal receptor assay results.

Chemotherapy may be administered in a hospital, doctor's office, clinic, or patient's home. The drugs may be given orally, by I.M. or S.C. injection, or I.V.

Chemotherapy commonly involves use of a combination of drugs. A typical regimen makes use of cyclophosphamide, methotrexate, and fluorouracil.

Hormonal therapy

Hormonal therapy lowers the levels of estrogen and other hormones suspected of nourishing breast cancer cells. Antiestrogen therapy, especially with tamoxifen, is used in postmenopausal women. Other commonly used drugs include the antiandrogen aminoglutethimide, the androgen fluoxymesterone, the estrogen diethylstilbestrol, and the progestin megestrol.

> Breast cancer treatment is controversial. In choosing therapy, the stage of the disease, age, menopausal status, and possible body image alteration should be considered.

Pathophysiology

Several disorders and preexisting conditions are linked to colorectal cancer. They include:
• familial adenomatous polyposis, such as Gardner's syndrome and Peutz-Jeghers syndrome
• ulcerative colitis
• Crohn's disease
• Turcot's syndrome
• hereditary nonpolyposis colorectal carcinoma
• previous colorectal cancer
• other pelvic cancers treated with abdominal radiation.

Polyp wallop

Colorectal polyps are closely tied to colon cancer. The larger the polyp, the greater the risk.

The genetic angle

The cause of colorectal cancer may be related to genetic factors — deletions on chromosomes 5, 17, and 18 — which may promote mutation and transition of the mucosal cells to a malignant state.

High fat, low fiber

A high-fat, low-fiber diet is thought to contribute to colorectal cancer by slowing fecal movement through the bowel. This results in prolonged exposure of the bowel mucosa to digested materials and may encourage mucosal cells to mutate.

What to look for

Signs and symptoms of colorectal cancer depend on the tumor location.

Right side tumor

If the tumor develops on the right side of the ascending colon, where stool is still in liquid form, signs and symptoms are usually not present in the early stages. However, abdominal aching, cramps, and black, tarry stools sometimes occur. Palpation of the abdomen often reveals a mass.

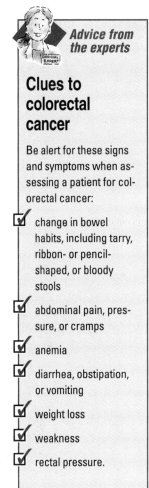

Advice from the experts

Clues to colorectal cancer

Be alert for these signs and symptoms when assessing a patient for colorectal cancer:

☑ change in bowel habits, including tarry, ribbon- or pencil-shaped, or bloody stools

☑ abdominal pain, pressure, or cramps

☑ anemia

☑ diarrhea, obstipation, or vomiting

☑ weight loss

☑ weakness

☑ rectal pressure.

Left side tumor

A tumor on the left side of the descending colon causes early symptoms of obstruction because stools are more completely formed there.

The patient may report a change in bowel habits, such as diarrhea or constipation. The patient may also report rectal bleeding, bloody stools, nausea, vomiting, intermittent abdominal fullness, cramping, and rectal pressure. (See *Treating colorectal cancer.*)

Because colorectal cancer progresses slowly, early detection is the key to recovery.

Battling illness

Treating colorectal cancer

Chemotherapy is necessary if metastasis has occurred or if the patient has residual disease or a recurrent inoperable tumor. Commonly used drugs include fluorouracil combined with levamisole or leucovorin. Radiation therapy before or after surgery causes tumor regression.

The most effective treatment for colorectal cancer is surgery to remove the malignant tumor, adjacent tissues, and cancerous lymph nodes. Afterward, treatment includes chemotherapy or radiation therapy. The surgical site depends on the location of the tumor.

The illustration below shows the different locations in the colon that may be resected for tumor removal.

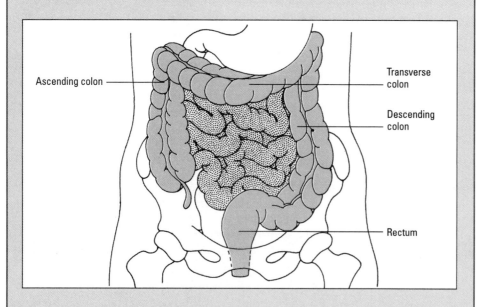

Ascending colon

Transverse colon

Descending colon

Rectum

What tests tell you

Several tests are used to diagnose colorectal cancer:
• Digital rectal examination (DRE) is used to detect 15% of colorectal cancers, specifically rectal and perianal lesions.
• Fecal occult blood test is used to detect blood in stools.
• Proctoscopy or sigmoidoscopy is used to visualize the lower GI tract and aids detection of two-thirds of all colorectal cancers.
• Colonoscopy is used to visualize and photograph the colon up to the ileocecal valve and provides access for polypectomies and biopsies.
• Barium enema is used to locate lesions that aren't visible or palpable.
• Computed tomography (CT) scan helps to detect cancer spread.
• Carcinoembryonic antigen, a tumor marker, is used to monitor the patient before and after treatment to detect metastasis or recurrence.

Hodgkin's disease

This type of cancer causes painless, progressive enlargement of the lymph nodes, spleen, and other lymphoid tissue.

Hodgkin's disease has a higher incidence in men, is slightly more common in whites, and occurs most often in two age-groups: 15 to 38 and over 50. A family history increases the likelihood of acquiring the disease.

Although the disease is fatal if untreated, recent advances have made Hodgkin's disease potentially curable, even in advanced stages. With appropriate treatment, about 90% of patients live at least 5 years. (See *Treating Hodgkin's disease*.)

An indirect relationship may exist between the Epstein-Barr virus and Hodgkin's disease.

Pathophysiology

The cause of Hodgkin's disease is unknown. It probably involves a virus. Many patients with Hodgkin's disease have had infectious mononucleosis, so an indirect relationship may exist between the Epstein-Barr virus and Hodgkin's disease.

Enlargement of the lymph nodes, spleen, and other lymphoid tissues results from proliferation of lymphocytes, histiocytes, and eosinophils. Patients also have dis-

Battling illness

Treating Hodgkin's disease

Depending on the stage of the disease, the patient may receive chemotherapy, radiation therapy, or both. Correct treatment leads to longer survival and even cures many patients.

Other treatments include autologous bone marrow transplantation and immunotherapy, used in conjunction with chemotherapy and radiation.

Chemotherapy combinations
Chemotherapy consists of various combinations of drugs. The well-known MOPP protocol (mechlorethamine, Oncovin [vincristine], procarbazine, and prednisone) was the first to result in significant cures. Another useful combination is doxorubicin, bleomycin, vinblastine, and dacarbazine. Antiemetics, sedatives, and antidiarrheals may be given with these drugs to prevent adverse GI effects.

tinct chromosome abnormalities in their lymph node cells.
(See *Looking at Reed-Sternberg cells.*)

What to look for

Sooner or later, most people with Hodgkin's disease develop signs of whole-body involvement, including the following:

- painless swelling of the lymph nodes in the stomach
- cancerous masses in the spleen, liver, and bones
- painless swelling of the face and neck
- progressive anemia
- possible jaundice
- nerve pain
- increased susceptibility to infection.

Looking at Reed-Sternberg cells

This enlarged, abnormal histiocyte—or Reed-Sternberg cell—from an excised lymph node suggests Hodgkin's disease. Note the large, distinct nucleoli. Reed-Sternberg cells must be present in a patient's blood and lymph tissue to confirm a diagnosis of Hodgkin's disease.

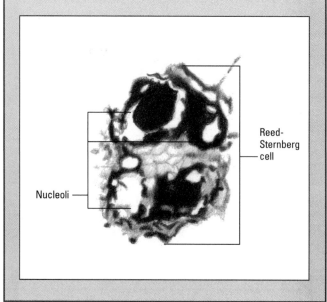

What tests tell you

The following tests may be used to rule out other disorders that enlarge the lymph nodes:

• Lymph node biopsy confirms the presence of Reed-Sternberg cells, abnormal histiocyte (macrophage) proliferation, and nodular fibrosis and necrosis. It's also used to determine the extent of lymph node involvement.

• Bone marrow, liver, mediastinal, and spleen biopsy are used to determine the extent of lymph node involvement.

• Chest X-ray, abdominal CT scan, lung and bone scans, lymphangiography, and laparoscopy are used to determine the extent and stage of the disease.

• Hematologic tests may show mild to severe normocytic anemia, normochromic anemia (in 50% of patients), and elevated, normal, or reduced white blood cell (WBC) count and differential with neutrophilia, lymphocytopenia, monocytosis, or eosinophilia (in any combination). Elevated serum alkaline phosphatase levels indicate liver or bone involvement.

• Staging laparotomy may be performed for developing a therapeutic plan of care.

> **Facts about leukemia**
>
> **Classification**
> Leukemias are classified as:
> • chronic or acute
> • myeloid or lymphoid.
>
> **Statistics**
> • In the United States, about 27,800 people develop leukemia annually.
> • Leukemia is more common in men, whites (especially of Jewish ancestry), and people living in urban and industrialized areas.

Acute leukemia

Acute leukemia is one form of leukemia. Leukemia refers to a group of malignant disorders characterized by abnormal proliferation and maturation of lymphocytes and nonlymphocytic cells, leading to the suppression of normal cells. (See *Facts about leukemia*.)

If untreated, acute leukemia is fatal, usually due to complications from leukemic cell infiltration of bone marrow or vital organs. With treatment, the prognosis varies.

Two types of acute leukemia are acute lymphocytic leukemia and acute myelogenous leukemia (also known as acute nonlymphocytic leukemia or acute myeloblastic leukemia).

Acute lymphocytic leukemia

Acute lymphocytic leukemia accounts for 80% of all childhood leukemias. Treatment leads to remission in 90% of children, who survive an average of 5 years, and in 65% of adults, who survive an average of 2 years. Children between ages 2 and 8 who receive intensive therapy have the best survival rate.

Acute myelogenous leukemia

This is one of the most common leukemias in adults. Average survival time is only 1 year after diagnosis, even with aggressive treatment. Remissions lasting 2 to 10 months occur in 50% of children.

The patient's history usually reveals a sudden onset of high fever and abnormal bleeding, such as bruising after minor trauma, nosebleeds, gingival bleeding, and purpura. Fatigue and night sweats may also occur.

Pathophysiology

The exact cause of acute leukemia isn't known; however, 40% to 50% of patients have mutations in their chromosomes. Further, individuals with certain chromosomal disorders have a higher risk. People with Down syndrome, trisomy 13, and other heredity disorders have a higher incidence of leukemia.

Other risk factors include exposure to large doses of ionizing radiation or drugs that depress the bone marrow. Certain viruses, such as HTLV-1, are also associated with an increased risk for leukemia.

Leukemic cells may crowd out other types of cells, thereby stopping production of normal red and white blood cells and platelets.

Stop pushing!

What happens

Here's a description of what probably happens when a patient develops leukemia:
• Immature cells undergo an abnormal transformation, giving rise to leukemic cells.
• Leukemic cells multiply and accumulate, crowding out other types of cells.
• Crowding prevents production of normal red and white blood cells and platelets, leading to pancytopenia — a reduction in the number of blood cells being produced by all cell lines.

What to look for

Signs and symptoms of acute lymphocytic and acute myelogenous leukemia are similar. Effects of both leukemias are related to suppression of elements of the bone marrow and include:

- infection
- bleeding
- anemia
- malaise, fever, lethargy
- weight loss
- night sweats. (See *Treating acute leukemia*.)

What tests tell you

The following tests are used to diagnose acute leukemia:
- Blood counts show thrombocytopenia and neutropenia; WBC differential determines cell type.
- Lumbar puncture is used to detect meningeal involvement; cerebrospinal fluid analysis reveals abnormal WBC invasion of the central nervous system (CNS).
- Bone marrow aspiration and biopsy confirms the disease by showing a proliferation of immature WBCs. It's also used to determine whether the leukemia is lymphocytic or myelogenous (important because the treatments and prognoses are very different).
- CT scan shows which organs are affected.

Battling illness

Treating acute leukemia

Systemic chemotherapy is used to eradicate leukemic cells and induce remission. It's used when fewer than 5% of blast cells in the bone marrow and peripheral blood are normal. Treatment may also include the following:
- antibiotic, antifungal, and antiviral drugs
- colony-stimulating factors, such as filgrastim, to spur the growth of granuloctyes, red blood cells, and platelets
- transfusions of platelets to prevent bleeding and red blood cells to prevent anemia.

Three-stage treatment for ALL
The treatment of acute lymphocytic leukemia (ALL) is divided into three stages:
1. induction therapy—usually vincristine, prednisone, and L-asparaginase, plus an anthracycline for adults
2. central nervous system prophylaxis—intrathecal methotrexate and intracranial radiation
3. postremission therapy—high-dose chemotherapy (intensification) followed by 2 to 3 years of maintenance chemotherapy.

Two-stage treatment for AML
Treatment of acute myelogenous leukemia (AML) occurs in two stages:
1. induction therapy —cytarabine and an anthracycline
2. postremission therapy—intensification, maintenance chemotherapy, or bone marrow transplant.

Chronic myelogenous leukemia

Also called chronic granulocytic leukemia, this type of leukemia is characterized by myeloproliferation in bone marrow, peripheral blood, and body tissues. It's most common in middle age, but may affect any age-group and affects both sexes equally. In the United States, between 3,000 and 4,000 cases are diagnosed annually, accounting for about 20% of all cases of leukemia.

Pathophysiology

The exact causes of chronic myelogenous leukemia aren't known. Almost all patients have the Philadelphia chromosome, in which the long arm of chromosome 22 translocates to chromosome 9. Radiation exposure may cause this abnormality.

Other suspected causes include myeloproliferative diseases or an unidentified virus.

An unyielding course

The clinical course of chronic myelogenous leukemia proceeds in two distinct phases:
• the chronic phase, characterized by anemia and bleeding abnormalities
• the terminal phase (blast crisis), in which myeloblasts, the most primitive granulocytic precursors, proliferate rapidly.

Chronic myelogenous leukemia is a deadly disease. The average survival time is 3 to 4 years after onset of the chronic phase and 3 to 6 months after onset of the terminal phase. (See *Treating chronic myelogenous leukemia.*)

What to look for

Common signs and symptoms of chronic myelogenous leukemia include:
• fatigue
• weight loss
• heat intolerance
• increased diaphoresis
• splenomegaly with abdominal fullness
• bruising due to low platelet count.

Battling illness

Treating chronic myelogenous leukemia

In chronic myelogenous leukemia, the goal of treatment is to control leukocytosis and thrombocytosis. Commonly used drugs are busulfan and hydroxyurea. Aspirin may help prevent a cerebrovascular accident if the patient's platelet count exceeds 1 million/μl. Bone marrow transplantation leads to remission in more than 60% of patients.

Supplemental therapy
Other treatments include the following:
• local splenic radiation or splenectomy to increase the platelet count and decrease effects of splenomegaly
• leukapheresis (selective leukocyte removal) to reduce the white blood cell count
• allopurinol to prevent secondary hyperuricemia and colchicine to relieve gouty attacks
• antibiotic treatment of infections resulting from bone marrow suppression.

What tests tell you

The following tests are used to diagnose chronic myelogenous leukemia:
- Chromosomal studies of peripheral blood or bone marrow confirm the disease.
- Serum analysis shows WBC abnormalities, such as leukocytosis (WBC count over 50,000/µl, to as high as 250,000/µl), leukopenia (WBC count under 5,000/µl), neutropenia (neutrophil count under 1,500/µl) despite high WBC count, and increased circulating myeloblasts. Other blood abnormalities include a decreased hemoglobin level (below 10 g/dl), low hematocrit (less than 30%), and thrombocytosis (more than 1 million thrombocytes/µl). The serum uric acid level may exceed 8 mg/dl.
- Leukocyte alkaline phosphatase levels may be elevated.
- Bone marrow biopsy may be hypercellular, showing bone marrow infiltration by a high number of myeloid elements. In the acute phase, myeloblasts predominate.
- CT scan may be used to identify the affected organs.

Chronic lymphocytic leukemia

This type of chronic leukemia occurs most often in the elderly, and more than half of the cases are men. According to the American Cancer Society, this type of leukemia accounts for about one-third of new adult cases annually.

Chronic lymphocytic leukemia is the most benign and slowest progressing form of leukemia. However, the prognosis is poor if anemia, thrombocytopenia, neutropenia, bulky lymphadenopathy, or severe lymphocytosis develops. Gross bone marrow replacement by abnormal lymphocytes is the most common cause of death, usually 4 to 5 years after diagnosis. (See *Treating chronic lymphocytic leukemia*.)

Pathophysiology

Chronic lymphocytic leukemia is a generalized, progressive disease. It causes a proliferation and accumulation of relatively mature looking but immunologically inefficient lymphocytes. Once these cells infiltrate, clinical signs appear.

Although the cause of chronic lymphocytic leukemia is unknown, hereditary factors are suspected.

Battling illness

Treating chronic lymphocytic leukemia

If the patient is asymptomatic, treatment may begin with close monitoring. When autoimmune hemolytic anemia or thrombocytopenia is present, systemic chemotherapy is administered, using the alkylating agents chlorambucil or cyclophosphamide or the corticosteroid prednisone.

When obstruction or organ impairment or enlargement occurs, local radiation reduces organ size and splenectomy helps relieve symptoms. Radiation therapy is also used for enlarged lymph nodes, painful bony lesions, or massive splenomegaly. Allopurinol can prevent hyperuricemia.

What to look for

Clinical signs may include fever and frequent infections, fatigue and enlargement of lymph nodes and splenomegaly.

What tests tell you

The following tests are used to diagnose chronic lymphocytic leukemia:
• Lymph node biopsy distinguishes between benign and malignant tumors.
• Routine blood tests usually uncover the disease. In the early stages, the lymphocyte count is slightly elevated (greater than 20,000/µl). Although granulocytopenia is generally seen first, the lymphocyte count climbs (greater than 100,000/µl) as the disease progresses. A hemoglobin level under 11 g/dl, hypogammaglobulinemia, and depressed serum globulin levels are also evident.
• Bone marrow aspiration and biopsy shows lymphocytic invasion.

Lung cancer

Although lung cancer is largely preventable, it remains the most common cause of cancer death in men. In women, lung cancer is a more common cause of death than breast cancer.

The most common forms of lung cancer are:
• adenocarcinoma
• squamous cell (epidermoid)
• large-cell (anaplastic)
• small-cell (oat cell).

The prognosis for lung cancer is usually poor, depending on the extent of the cancer and the cells' growth rate. Only about 13% of patients survive 5 years after diagnosis.

Pathophysiology

Lung cancer most commonly results from repeated tissue trauma from inhalation of irritants or carcinogens. These substances include tobacco smoke, air pollution, arsenic, asbestos, nickel, and radon.

Almost all lung cancers start in the epithelium. Here's what happens:

Find out what happens to lung cells when we're exposed to irritants or carcinogens.

• In normal lungs, the epithelium lines and protects the tissue below it. But when exposed to irritants or carcinogens, the epithelium continually replaces itself until the

Common lung cancers

The table below describes the growth rate, metastasis sites, diagnostic tests, signs and symptoms, and treatments for four common cancers.

Type of cancer	Rate of growth	Metastasis	Signs and symptoms	Diagnostic tests
Adenocarcinoma	Moderate	Early metastasis	Pleural effusion	Fiberoptic bronchoscopy, radiography, electron microscopy
Squamous cell (epidermoid)	Slow	Late metastasis mainly to hilar lymph nodes, chest wall and mediastinum	Airway obstruction, cough, sputum production	Sputum analysis, biopsy, immunohistochemistry, electron microscopy, bronchoscopy
Large-cell (anaplastic)	Fast	Early, extensive metastasis	Cough, hemoptysis, chest wall pain, pleural effusion, sputum production, pneumonia-induced airway obstruction	Bronchoscopy, sputum analysis, electron microscopy
Small-cell (oat cell)	Very fast	Very early metastasis to mediastinum, hilar lymph nodes, and other organ sites	Chest pain, cough, hemoptysis, dyspnea, localized wheezing, pneumonia-induced airway obstruction, muscle weakness, facial edema, hypokalemia, hyperglycemia, hypertension, and other signs and symptoms related to excessive hormone secretion	Sputum analysis, immunohistochemistry, electron microscopy, radiography, bronchoscopy

---

cells develop chromosomal changes and become dysplastic (altered in size, shape, and organization).
• Dysplastic cells don't function well as protectors, so underlying tissue gets exposed to irritants and carcinogens.
• Eventually, the dysplastic cells turn into neoplastic carcinoma and start invading deeper tissues.

What to look for

For information on the signs and symptoms of different types of lung cancers, see *Common lung cancers,* page 33.

What tests tell you

The following tests are used to diagnose lung cancer:
• Chest X-ray usually shows an advanced lesion but can reveal damage 2 years before signs and symptoms appear. This test may be used to determine tumor size and location.

Battling illness

Treating lung cancer

Various combinations of surgery, radiation therapy, and chemotherapy may improve the patient's prognosis and prolong survival. Treatment depends on the stage of illness. Unfortunately, lung cancer is usually advanced at diagnosis.

Surgery

Surgery may involve partial lung removal (wedge resection, segmental resection, lobectomy, or radical lobectomy) or total lung removal (pneumonectomy or radical pneumonectomy). Complete surgical resection is the only chance for a cure, but fewer than 25% of patients have disease that is responsive to surgery.

Radiation

Preoperative radiation therapy may reduce the tumor's bulk, allowing surgical resection and improving the patient's response.

Radiation therapy is also generally recommended for stage I and stage II lesions if surgery is contraindicated and for stage III disease confined to the involved hemithorax and the ipsilateral supraclavicular lymph nodes.

Chemotherapy

Chemotherapy has caused dramatic, although temporary, responses in patients with small-cell carcinoma. The other types of lung cancer are fairly resistant to it. Drug combinations include cyclophosphamide, doxorubicin, and vincristine; cyclophosphamide, doxorubicin, vincristine, and etoposide; and etoposide, cisplatin, cyclophosphamide, and doxorubicin. Unfortunately, patients usually relapse in 7 to 14 months.

Other treatments

Investigational therapies include immunotherapy and laser therapy. Immunotherapy with bacille Calmette-Guérin vaccine or *Corynebacterium parvum* offers the most promise. Laser therapy uses a beam directed through a bronchoscope to destroy local tumors, primarily those causing airway obstruction.

• Cytologic sputum analysis is 75% reliable and requires a specimen expectorated from the lungs and tracheobronchial tree.
• Bronchoscopy may reveal the tumor site; bronchoscopic washings provide material for cytologic and histologic study.
• Needle biopsy is used to locate peripheral tumors in lungs and to collect tissue specimens for analysis; it confirms diagnosis in 80% of patients.
• Tissue biopsy of metastatic sites is used to assess the extent (stage) of the disease and determine prognosis and treatment.
• Thoracentesis allows chemical and cytologic examination of pleural fluid.
• CT scan may help to evaluate mediastinal and hilar lymph node involvement and the extent of the disease.
• Bone scan, CT brain scan, liver function studies, and gallium scans of the liver and spleen are used to detect metastasis. (See *Treating lung cancer.*)

Malignant melanoma

Malignant melanoma is the most lethal skin cancer. It accounts for 1% to 2% of all malignant tumors, is slightly more common in women, is unusual in children, and occurs most often between ages 50 and 70. The incidence in younger age-groups is increasing because of increased sun exposure or, possibly, a decrease in the ozone layer.

Pathophysiology

This disease arises from melanocytes (cells that synthesize the pigment melanin). Besides the skin, melanocytes are also found in the meninges, alimentary canal, respiratory tract, and lymph nodes.

Melanoma spreads through the lymphatic and vascular systems and metastasizes to the regional lymph nodes, skin, liver, lungs, and CNS. In most patients, superficial lesions are curable, but deeper lesions are more likely to metastasize.

Common sites are the head and neck in men, the legs in women, and the backs of people exposed to excessive sunlight. Up to 70% of malignant melanomas arise from a preexisting nevus (circumscribed malformation of the skin), or mole. (See *Treating malignant melanoma, page 36.*)

In malignant melanoma, superficial lesions are usually curable, but deeper lesions are more likely to metastasize.

Battling illness

Treating malignant melanoma

Treatment always involves surgical resection of the tumor and a 3- to 5-cm margin. The extent of resection depends on the size and location of the primary lesion. If a skin graft is needed to close a wide resection, plastic surgery provides excellent cosmetic repair. Surgical treatment may also include regional lymphadenectomy.

Other treatments

In addition to surgery, four other treatments are common:

☝ Deep primary lesions may require chemotherapy, usually with dacarbazine and carmustine. After surgery, intra-arterial isolation perfusions can prevent recurrence and metastatic spread.

✌ Immunotherapy with bacille Calmette-Guérin vaccine is used in advanced melanoma. In theory, this treatment combats cancer by boosting the body's disease-fighting systems.

🤟 For metastatic disease, chemotherapy with dacarbazine and nitrosoureas has been used with some success.

🖐 Radiation therapy, usually reserved for metastatic disease, doesn't prolong survival but may reduce pain and tumor size.

> Regardless of the treatment used, melanoma requires long-term follow-up care to detect metastases and recurrences.

Run down on risk factors

Risk factors for developing melanoma are:
• excessive exposure to sunlight
• increased nevi
• tendency to freckle from the sun
• red hair, fair skin, blue eyes, susceptibility to sunburn, and Celtic or Scandinavian ancestry (Melanoma is rare in blacks.)
• hormonal factors such as pregnancy
• a family history of melanoma
• a past history of melanoma.

An organelle in the melanocyte called a melanosome is responsible for producing melanin. One theory proposes that melanoma arises because the melanosome is abnormal or absent.

What to look for

Suspect malignant melanoma when any preexisting skin lesion or nevus enlarges, changes color, becomes in-

> **Memory jogger**
>
> Use the ABCD rule to assess a mole's malignant potential:
>
> **A** Asymmetry: Is the mole irregular in shape?
>
> **B** Border: Is the border irregular, notched, or poorly defined?
>
> **C** Color: Does the color vary, for example, between shades of brown, red, white, or black?
>
> **D** Diameter: Is the diameter more than 6 mm?

flamed or sore, itches, ulcerates, bleeds, changes texture, or shows signs of surrounding pigment regression. When seeking to assess the malignant potential of a mole, look for asymmetry, an irregular border, color variation, and a diameter greater than 6 mm.

What tests tell you

Diagnostic tests for malignant melanoma include the following:
• Excisional biopsy and full-depth punch biopsy with histologic examination are used to distinguish malignant melanoma from a benign nevus, seborrheic keratosis, or pigmented basal cell epithelioma; they are also used to determine tumor thickness and disease stage.
• Chest X-ray, gallium scan, bone scan, magnetic resonance imaging (MRI), and CT scans of the chest, abdomen, or brain may be used to diagnose or detect metastasis, depending on the depth of tumor invasion.

Multiple myeloma

Multiple myeloma is a disseminated cancer of marrow plasma cells that infiltrates bone to produce lesions throughout the flat bones, vertebrae, skull, pelvis, and ribs. In late stages, it infiltrates the liver, spleen, lymph nodes, lungs, adrenal glands, kidneys, skin, and GI tract.

The best hope: An early diagnosis

Multiple myeloma strikes about 12,300 people yearly, mostly men between the ages of 50 and 80. The prognosis is poor because, by the time it's diagnosed, the vertebrae, pelvis, skull, ribs, clavicles, and sternum are already infiltrated and skeletal destruction is widespread. More than half of patients die within 3 months of diagnosis; 90% die within 2 years. However, if the disease is diagnosed early, treatment may prolong life by 3 to 5 years. (See *Treating multiple myeloma*.)

Pathophysiology

The cause of multiple myeloma isn't known. The disorder has been linked to genetic factors and occupational exposure to certain chemicals and radiation.

In normal cell activity, stem cells within the bone marrow can self-replicate or differentiate (mature into specific cell types). The lymphoid stem cell can differentiate into T lymphocytes (which participate in cell-mediated immunity) or B lymphocytes (which participate in humoral immunity). B lymphocytes eventually become plasma cells that produce and release immunoglobulins. Immunoglobulin G (IgG) is the most common immunoglobulin in humans.

In multiple myeloma, an unknown factor stimulates the B lymphocytes to turn into malignant plasma cells,

Here's a description of normal cell activity and how things go awry in multiple myeloma.

Battling illness

Treating multiple myeloma

Long-term treatment of multiple myeloma consists mainly of chemotherapy to suppress plasma cell growth and control pain. Combinations of melphalan and prednisone or cyclophosphamide and prednisone are used. Local radiation reduces acute lesions and relieves the pain of collapsed vertebrae.

Other treatment includes analgesics for pain, laminectomy for vertebral compression, and dialysis for renal complications. Interferon may prolong the plateau phase once the initial chemotherapy is completed.

Calcium surplus

Because patients may have bone demineralization and may lose large amounts of calcium into blood and urine, they're prime candidates for renal calculi and renal failure from hypercalcemia. Hypercalcemia is treated with hydration, diuretics, corticosteroids, oral phosphate, and I.V. gallium. Plasmapheresis temporarily removes the Bence Jones protein from withdrawn blood and retransfuses the cells to the patient.

which produce huge amounts of IgG (Bence Jones or myeloma protein). This leads to a hyperviscosity syndrome commonly seen in myeloma patients.

What to look for

Infection is a common complication of multiple myeloma. However, patients may exhibit a variety of symptoms, including:
- visual disturbances
- headaches
- somnolence
- irritability
- confusion
- intolerance to cold.

What tests tell you

The following tests are used to diagnose multiple myeloma:
- Complete blood count reveals moderate to severe anemia, with 40% to 50% lymphocytes but seldom more than 3% plasma cells. A differential smear also shows rouleaux formation (blood cells that stick together, resembling stacks of coins). Often the first clue, this results from elevation of the erythrocyte sedimentation rate.
- Urine studies may show proteinuria, Bence Jones protein, and hypercalciuria. Absence of Bence Jones protein doesn't rule out multiple myeloma, but its presence almost always confirms the disease.
- Bone marrow aspiration is used to detect an abnormal number of immature plasma cells (10% to 95% instead of 3% to 5%).
- Serum electrophoresis shows an elevated globulin spike.
- X-rays during the early stages may reveal only diffuse osteoporosis; later, they show multiple, sharply circumscribed osteolytic (punched out) lesions, particularly on the skull, pelvis, and spine.
- Serum calcium levels are elevated because calcium lost from the bone is reabsorbed into the serum.

Prostate cancer

Prostate cancer is a leading cause of death in men. About 85% of these cancers originate in the posterior prostate gland; the rest grow near the urethra. Adenocarcinoma is the most common form. (See *Location of the prostate gland*.)

    Prostate cancer accounts for about 22% of all cancers, with the highest incidence among Blacks and lowest

Location of the prostate gland

This drawing shows the location of the prostate gland. Note that enlargement may occlude the urethra, causing urine retention.

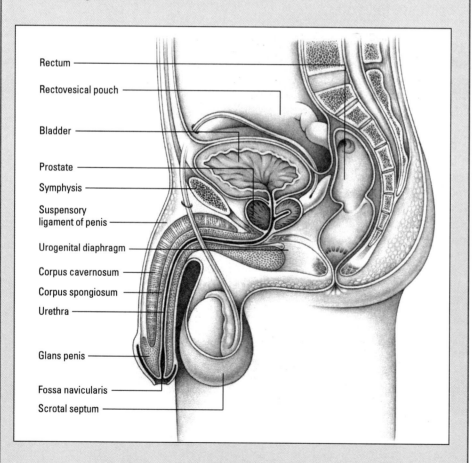

Rectum

Rectovesical pouch

Bladder

Prostate

Symphysis

Suspensory ligament of penis

Urogenital diaphragm

Corpus cavernosum

Corpus spongiosum

Urethra

Glans penis

Fossa navicularis

Scrotal septum

among Asians. Socioeconomic status and fertility don't appear to be factors.

Pathophysiology

Risk factors for prostate cancer include the following:
• age over 40
• exposure to infectious agents, such as herpesvirus, simian vacuolating virus 40, cytomegalovirus, and ribonucleic acid viruses (Viruses haven't been proven to be a cause of cancer.)
• hormonal factors. (Testosterone may initiate or promote prostate cancer.)

A slow progress

Prostate cancer grows slowly. When primary lesions spread beyond the prostate gland, they invade the prostatic capsule and then spread along the ejaculatory ducts in the space between the seminal vesicles or perivesicular fascia.

Again, early detection is key

When prostate cancer is treated in its localized form, the 5-year survival rate is 70%; after metastasis, it's under 35%. Fatal prostate cancer usually results from widespread bone metastases. (See *Treating prostate cancer,* page 42.)

To date, no link has been found between benign prostatic hyperplasia and prostate cancer, although they may occur simultaneously.

What to look for

Prostate cancer seldom produces signs and symptoms until it's well advanced. Signs of advanced disease include a slow urinary stream, urinary hesitancy, incomplete bladder emptying, and dysuria. These symptoms are due to obstruction caused by tumor progression.

What tests tell you

The following tests and results are common in prostate cancer:
• Digital rectal examination is performed to determine prostate location and size.
• Biopsy of the prostate may distinguish a benign or malignant mass.
• Cytology examination of urine and prostate fluid may indicate an abnormality.

Battling illness

Treating prostate cancer

Therapy varies depending on the stage of the cancer and may include radiation, prostatectomy, and hormone therapy. Because most prostate cancers are androgen- or hormone-dependent, the main treatments are antiandrogens to suppress adrenal function or medical castration with estrogen or gonadotropin-releasing hormone analogs.

Radical prostatectomy usually works for localized lesions without metastasis. Transurethral resection of the prostate may relieve an obstruction.

Nonsurgical treatments

Radiation therapy may cure locally invasive lesions in early disease and may relieve bone pain from metastatic skeletal involvement. It's also used prophylactically to prevent tumor growth for patients with tumors in regional lymph nodes. Radioactive "seeds" may also be implanted into the prostate. This treatment increases radiation to the area while minimizing exposure to surrounding tissues.

Chemotherapy with combinations of cyclophosphamide, doxorubicin, fluorouracil, and cisplatin offers limited benefits.

- Blood tests may show elevated levels of prostate-specific antigen. Although an elevated prostate-specific antigen level occurs with metastasis, it also occurs with other prostate diseases.
- Transrectal prostatic ultrasonography may be used for patients with abnormal DRE and prostate-specific antigen findings.
- Bone scan and excretory urography determine the extent of the disease.
- MRI and CT scans help define the tumor's extent.

Quick quiz

1. Lymphedema occurs in the patient with breast cancer because:

    A. the breast tissue causes the patient to retain fluid.

    B. the lymphatic system is swollen with immune-response factors to fight the cancer.

    C. the area is traumatized, which causes it to swell.

Answer: B. The lymphatic system brings immune response factors to the area to fight the cancer, which causes lymphedema.

2. The presence of Reed-Sternberg cells is associated with:
  A. prostate cancer.
  B. malignant melanoma.
  C. Hodgkin's disease.

Answer: C. Reed-Sternberg cells must be present before a diagnosis of Hodgkin's disease can be made.

3. Bone marrow aspiration is performed on a patient with leukemia in order to:
  A. examine the platelets.
  B. determine if the cancer is lymphocytic or myelogenous.
  C. measure the amount of marrow in the bone.

Answer: B. Bone marrow aspiration determines what type of immature white blood cell is involved, which directs the type of treatment.

4. A change in the shape, color, and texture of a nevus may indicate:
  A. multiple myeloma.
  B. malignant melanoma.
  C. healing of the nevus.

Answer: B. An obvious change in a wart or mole is a sign of malignant melanoma and is one of the seven warning signs of cancer identified by the American Cancer Society.

5. Multiple myeloma is related to the immunity factor:
  A. histamine.
  B. serotonin.
  C. IgG.

Answer: C. In multiple myeloma, plasma cells secrete an unusually large amount of IgG.

6. A risk factor for prostate cancer is:
  A. history of infertility.
  B. being over age 40.
  C. poverty.

Answer: B. Prostate cancer seldom develops in men under age 40. Socioeconomic status and infertility don't appear to affect risk for prostate cancer.

Scoring

☆☆☆   If you answered all six items correctly, amazing! Your knowledge of pathophysiology is proliferating with remarkable speed and efficiency.

☆☆   If you answered four or five correctly, good work! Your skill at learning is due to a remarkable combination of environmental, genetic, dietary, and other unspecified factors.

☆   If you answered fewer than four correctly, never fear. Remember that early detection is key, and this is only your second quick quiz.

Infection

Just the facts

In this chapter you'll learn:

♦ how the body naturally prevents infection

♦ the four types of infective microorganisms and how they invade the body

♦ the causes, pathophysiology, diagnostic tests, and treatments for several common infectious diseases.

Understanding infection

An infection is a host organism's response to a pathogen, or disease-causing substance. It results when tissue-destroying microorganisms enter and multiply in the body. Some infections take the form of minor illnesses, such as colds and ear infections. Others result in a life-threatening condition called sepsis, which causes wide-spread vasodilation and multiorgan system failure.

Infection-causing microbes

Four types of microorganisms can enter the body and cause infection:

 viruses

 bacteria

 fungi

 parasites.

An infection is a host organism's response to a pathogen.

Viruses

Viruses are microscopic genetic parasites that may contain genetic material such as deoxyribonucleic acid (DNA) or ribonucleic acid. They have no metabolic capability and need a host cell to replicate.

Viral hide-and-seek

Viral infections occur when normal inflammatory and immune responses fail. The virus develops in the cell and "hides" there. Once it's introduced into the host cell, the inner capsule releases genetic material, causing the infection. Some viruses surround the host cell and preserve it; others kill the host cell on contact.

Viruses contain genetic material and need a host cell to replicate.

Bacteria

Bacteria are one-celled microorganisms that break down dead tissue. They have no true nucleus and reproduce by cell division. Pathogenic bacteria contain cell-damaging proteins that cause infection. These proteins come in two forms:
• endotoxins — released when the bacterial cell wall decomposes. These endotoxins cause fever and aren't affected by antibiotics.
• exotoxins — released during cell growth.

Bacteria are classified several other ways, such as by their shape, growth requirements, motility, and whether they're aerobic (requiring oxygen) or anaerobic.

Bacteria are single-celled organisms that reproduce by cell division and may contain harmful proteins.

Fungi

Fungi are nonphotosynthetic microorganisms that reproduce asexually (by division). They're large compared to other microorganisms and contain a true nucleus. Fungi are classified as:
• yeasts — round, single-celled, facultative anaerobes (which can live with or without oxygen)
• molds — filament-like, multinucleated, aerobic microorganisms.

There's a fungus among us

Although fungi are part of the human body's normal flora, they can overproduce, especially when the normal flora is compromised. For example, vaginal yeast infections can occur with antibiotic treatment because normal flora are killed by the antibiotic, allowing yeast to reproduce. Infections caused by fungi are called mycotic infections because pathogenic fungi release mycotoxin. Most of these infections are mild unless they become systemic or the patient's immune system is compromised.

Parasites

Parasitic infections are uncommon except in hot, moist climates. Parasites are single-celled or multi-celled organisms that depend on a host for food and a protective environment. Most common parasitic infections, such as tapeworm infestation, occur in the intestines.

Barriers to infection

A healthy person can usually ward off infections with the body's own defense mechanisms. The body has many built-in infection barriers, such as the skin and secretions from the eyes, nasal passages, prostate gland, testicles, stomach, and vagina. Most of these secretions contain bacteria-killing particles called lysozymes. Other body structures, such as cilia in the pulmonary airways that sweep foreign material from the breathing passages, also offer infection protection.

Trillions of harmless inhabitants

Normal flora are harmless microorganisms that reside on and in the body. They're found on the skin and in the nose, mouth, pharynx, distal intestine, colon, distal urethra, and vagina. The skin contains 100,0000 microorganisms per square centimeter. And trillions of microorganisms are secreted from the GI tract daily.

Many of these microorganisms provide useful, protective functions. For example, the intestinal flora help synthesize vitamin K, which is an important part of the body's blood clotting mechanism.

The body has built-in defense mechanisms against pathogens.

The infection process

Infection occurs when the body's defense mechanisms break down or when certain properties of microorganisms, such as virulence or toxin production, override the defense system.

Other factors that create a climate for infection include:
- poor nutrition
- stress
- humidity
- poor sanitation
- crowded living conditions
- pollution
- dust.

Enter, attach, and spread

Infection results when a pathogen enters the body through direct contact, inhalation, ingestion, or an insect or animal bite. The pathogen then attaches itself to a cell and releases enzymes that destroy the cell's protective membrane. Next, it spreads through the bloodstream and lymphatic system, finally multiplying and causing infection in the target tissue or organ.

Striking while there's opportunity

Infections that strike people with altered, weak immune systems are called opportunistic infections. For example, patients with acquired immunodeficiency syndrome (AIDS) are plagued by opportunistic infections such as *Pneumocystis carinii* pneumonia.

A pathogen may attach itself to a cell and release enzymes that destroy the cell's protective membrane.

Infectious disorders

The infections discussed below include herpes simplex types 1 and 2, herpes zoster, infectious mononucleosis, Lyme disease, rabies, respiratory syncytial virus, rubella, salmonellosis, and toxoplasmosis.

Herpes simplex

A recurrent viral infection, herpes simplex occurs as two types:

Type 1 primarily affects the skin and mucous membranes and commonly produces cold sores and fever blisters.

Type 2 primarily affects the genital area, causing painful clusters of small ulcerations.

Both types of herpes simplex virus can infect the eyes and other organs in the body. In addition, both types can result in localized or generalized infection.

Herpes simplex occurs worldwide and with equal frequency in men and women. It's most prevalent in lower socioeconomic groups, probably because of crowded living conditions.

The risk of recurrent attacks

Although herpes simplex may be latent for years, initial infection makes the patient a carrier susceptible to recurrent attacks. Outbreaks may be provoked by fever, menses, stress, heat, cold, lack of sleep, or sun exposure.

Potentially serious

Pregnant women should avoid exposure to herpes simplex because it can cause severe congenital anomalies in newborns. These range from localized skin lesions to disseminated infection of major organs. Some examples of common complications in neonates are seizures, mental retardation, blindness, and deafness.

Herpes may also cause severe illness in immunocompromised patients; examples include pneumonias, hepatitis, and neurologic complications. Women with herpes simplex Type 2 may have an increased risk of cervical cancer.

Pathophysiology

Herpesvirus hominis, a widespread infectious agent, causes both types of herpes simplex. Type 1 is transmitted by oral and respiratory secretions, and Type 2 is transmitted by sexual contact. However, cross-infection may result from orogenital sex. The average incubation for generalized infection is 2 to 12 days; for localized genital infection, 3 to 7 days.

The herpes simplex virus is a linear, double-stranded DNA molecule with an outer coating of lipid-type membrane. Here's what happens during exposure:
• The virus fuses to the host cell membrane.
• The virus releases proteins, turning off the host cells' protein production or synthesis.
• The virus replicates and synthesizes structural proteins.

• The virus pushes its nucleocapsid (protein coat and nucleic acid) into the cytoplasm of the host cell and releases the viral DNA.

• Complete virus particles capable of surviving and infecting a living cell (called virion) are transported to the cell's surface.

Herpes simplex infection doesn't end in cell death. Instead, the virus enters a latent state during which it's maintained by the cell. Viral replication and redevelopment of herpetic lesions is called reactivation.

Herpes simplex Type 1, spread by oral and respiratory secretions, causes cold sores and fever blisters in the skin and mucous membranes.

What to look for

Type 1 herpes simplex may cause generalized or localized infection as the virus invades the cells around the mouth. Generalized infection begins with fever and a sore, red, swollen throat. Along with characteristic vesicles, the patient may develop submaxillary lymphadenopathy, increased salivation, halitosis, and anorexia. The patient usually reports severe mouth pain. After a brief prodromal period, primary lesions erupt.

Examination of the mouth may reveal edema and small vesicles (blisters) on a red base. These vesicles eventually rupture, leaving a painful ulcer and then yellow crusting. Common sites for vesicles are the tongue, gingiva, and cheeks, but they may occur anywhere in or around the mouth. The cervical glands may also be swollen. A generalized infection usually lasts 4 to 10 days. (See *What's a vesicle?*)

With primary genital, or Type 2, herpes simplex, the patient usually complains first of tingling in the area involved, malaise, dysuria, dyspareunia (painful intercourse) and, in females, leukorrhea (white vaginal discharge containing mucus and pus cells).

Next, localized, fluid-filled vesicles appear and may last for weeks. In women, they occur on the cervix, labia, perianal skin, vulva, and vagina. In men, they develop on the glans penis, foreskin, and penile shaft. Lesions may also occur on the mouth or anus. After rupture, vesicles become shallow, painful ulcers with redness, edema, and oozing yellow centers. Inguinal swelling may also be present.

Patients should be taught how to care for themselves during a herpes outbreak and how to avoid infecting others. (See *Treating herpes simplex.*)

Type 2 mainly affects the genital area and is spread by sexual contact.

What's a vesicle?

A vesicle is a raised, circumscribed, fluid-filled lesion, less than ¼″ (0.6 cm) in diameter. It's the typical lesion of chickenpox and herpes simplex.

What tests tell you

Confirmation of herpes simplex requires isolating the virus from local lesions and performing a tissue biopsy. In primary infection, an increase in antibodies and a moderate increase in white blood cell (WBC) count support the diagnosis.

Herpes zoster

Also called shingles, herpes zoster is an acute inflammation of dorsal root ganglia — nerve cell clusters found on the dorsal root of each spinal nerve.

Herpes zoster occurs mainly in people over age 50. The prognosis is good and most patients recover completely unless the infection spreads to the brain on nerve roots that originate there.

Pathophysiology

This disorder is caused by a virus called the herpesvirus varicella-zoster, the same virus that causes chickenpox. Because the disease affects nerves at their roots, localized, vesicular skin lesions are usually confined to an area of skin supplied by branches from a single nerve (called a dermatome). The patient may have severe pain in peripheral areas innervated by the inflamed nerve root. Chronic pain — called postherpetic neuralgia — is the most common persisting adverse effect.

Reactivation of the varicella-zoster virus is what causes shingles, but what triggers the reactivation is unknown.

The varicella-zoster virus reactivates after lying dormant in the cerebral ganglia or the ganglia of posterior nerve roots. Some believe that the virus multiplies as it reactivates and that antibodies from the initial chickenpox infection usually neutralize it. However, without opposition from effective antibodies, the virus will continue to multiply in the ganglia, destroying neurons and spreading down the sensory nerves to the skin.

What to look for

The first symptom of herpes zoster is pain within the dermatome. Patients may also have fever and malaise.

Battling illness

Treating herpes simplex

A generalized primary herpes infection usually requires drugs to reduce fever and pain. Anesthetic mouthwashes, such as viscous lidocaine, may help the patient with oral lesions eat and drink with less pain. Drying agents, such as calamine lotion, may soothe labial and skin lesions. (Petroleum-based salves or dressings can cause viral spread and slow healing.)

Acyclovir, the drug most often used to treat herpes, may reduce symptoms, viral shedding, and healing time. It's available in topical, oral, and I.V. forms. Valacyclovir or famciclovir may also be given.

Don't pass it on
Patients with genital herpes should avoid sexual intercourse until lesions completely heal. They should also inform sexual partners of their condition.

What happens next

After 2 to 4 days, severe, intermittent, or continuous deep pain may occur. Pruritus, paresthesia (unusual skin sensations such as prickling), or hyperesthesia (heightened skin sensitivity) in the trunk, arms, or legs may also occur as more nerves are affected.

48 to 72 hours later

After the pain starts, small, red, nodular skin lesions erupt on painful areas and spread unilaterally around the thorax or vertically over the arms or legs. They change rapidly into pus- or fluid-filled vesicles, which may become infected or even gangrenous. Bacterial infection of the skin is usually caused by *Staphylococcus aureus,* or *S. pyogenes,* and may result from scratching.

10 to 21 days later

About 10 to 21 days after the rash appears, the vesicles dry and form scabs.

Nerve specific

When branches of the trigeminal nerve are involved, lesions appear on the face, in the mouth, or in the eyes. When the sensory branch of the facial nerve is involved, lesions appear in the ear canal and on the tongue. (See *Treating herpes zoster.*)

The exact cause of shingles is a mystery, but antibodies from a childhood chickenpox infection usually neutralize it.

Battling illness

Treating herpes zoster

Primary treatment includes antipruritics, such as calamine lotion, to relieve itching and analgesics, such as aspirin, acetaminophen, or codeine, to relieve pain. A systemic corticosteroid, such as cortisone or corticotropin, may also be used to relieve pain and reduce inflammation.

Tincture of benzoin applied to unbroken lesions helps prevent a secondary infection. If lesions rupture and become infected with bacteria, systemic antibiotics are given. Herpes zoster affecting the trigeminal nerve and cornea calls for idoxuridine ointment or other antiviral agent. Other medications used are tranquilizers, sedatives, and tricyclic antidepressants with phenothiazines.

Acyclovir may be prescribed for immunocompromised patients and those with infections of the ophthalmic branch of the trigeminal nerve. The drug stops the rash from spreading, reduces the duration of viral shedding and acute pain, and prevents visceral complications. Valacyclovir and famciclovir also are used for the treatment of most herpes infections.

If other pain relief measures fail, transcutaneous peripheral nerve stimulation, patient-controlled analgesia, or a small dose of radiotherapy may be effective.

What tests tell you

Vesicular fluid analysis is used to diagnose herpes zoster. This test differentiates herpes zoster from localized herpes simplex.

Infectious mononucleosis

An acute infectious disease, infectious mononucleosis has three hallmarks:

 fever

 sore throat

 swollen cervical lymph nodes.

It also may cause liver dysfunction, increased numbers of lymphocytes and monocytes, and development and persistence of heterophil antibodies.

Infectious mononucleosis mainly affects young adults and children, although it's usually so mild in children that it's often overlooked. It's fairly prevalent in the United States, Canada, and Europe, and both sexes are affected equally. Here's the good news:
• Major complications are uncommon.
• About 90% of children over age 4 have acquired antibodies.

Other types of mononucleosis are caused by cytomegalovirus and are benign and self-limiting.

Pathophysiology

Infectious mononucleosis is caused by the Epstein-Barr virus, a member of the herpesvirus group. Most cases are spread by the oropharyngeal route, but transmission by blood transfusion and during cardiac surgery is also possible. The disease is contagious from before symptoms develop until the patient's fever subsides and oropharyngeal lesions disappear.

The disorder develops this way:
• The virus invades the B cells of the oropharyngeal lymphoid tissues and then replicates.
• As the B cells die, the virus is released into the blood, causing fever and other symptoms.

Infectious mononucleosis lodges in the salivary glands and is spread by the oropharyngeal route. That's why people call it "kissing disease."

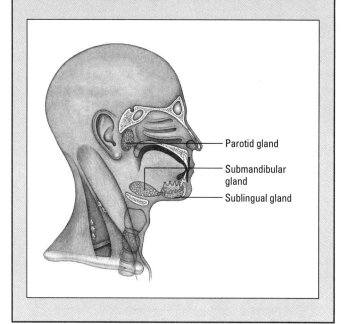

Glands affected by Epstein-Barr virus

The illustration shows the different glands that are affected by mononucleosis or the Epstein-Barr virus.

Parotid gland

Submandibular gland

Sublingual gland

Treating infectious mononucleosis

Because mononucleosis is hard to prevent and resists standard antimicrobial treatment, therapy is mainly supportive. It includes relief of symptoms, bed rest during the acute febrile period, and aspirin or another salicylate for headache and sore throat.

In cases of severe inflammation and airway obstruction, steroids can relieve swelling and prevent the need for a tracheotomy. About 20% of patients also have a streptococcal infection and need antibiotic therapy for at least 10 days.

- During this period, antiviral antibodies appear and the virus disappears from the blood, lodging mainly in the parotid gland, one of the salivary glands.
- Reproduction of the virus occurs in the parotid gland, accounting for its presence in saliva. (See *Glands affected by Epstein-Barr*.)

What to look for

After an incubation period of about 10 days in children and 30 to 50 days in adults, the patient experiences headache, malaise, profound fatigue, anorexia, myalgia (muscle tenderness), and abdominal discomfort. Three to 5 days later, he develops an extremely sore throat and dysphagia (difficulty swallowing). Fever usually peaks in the late afternoon or evening, reaching 101° to 102° F (38.3° to 38.9° C).

Inspection reveals exudative tonsillitis and pharyngitis. Petechiae on the palate, swollen eyes, a raised and red rash that resembles rubella, and jaundice also may occur. Cervical lymph nodes swell and are mildly tender when palpated. Inguinal and axillary nodes as well as the spleen and liver also may be swollen.

The prognosis for patients with infectious mononucleosis is excellent. (See *Treating infectious mononucleosis.*)

What tests tell you

The following diagnostic tests confirm infectious mononucleosis:
• WBC count is abnormally high (10,000 to 20,000/mm^3) during the second and third weeks of illness. From 50% to 70% of the total count consists of lymphocytes and monocytes, and 10% of the lymphocytes are atypical.
• Heterophil antibodies in serum drawn during the acute phase and at 3- to 4-week intervals increase to four times normal.
• Indirect immunofluorescence shows antibodies to Epstein-Barr virus and cellular antigens. This test is usually more definitive than that for heterophil antibodies, but it may not be necessary because most patients are heterophil-positive.

Lyme disease

Lyme disease occurs chiefly in the United States and is named for the Connecticut town in which it was first recognized in 1975. It affects multiple body systems and usually appears in summer or early fall with a skin lesion called erythema chronicum migrans. Weeks or months later, cardiac, neurologic, or joint abnormalities may develop, sometimes followed by arthritis. The incidence has increased in most states over the past 8 years.

Pathophysiology

Lyme disease is caused by the spirochete *Borrelia burgdorferi*, which is carried by deer ticks. Here is how the disease develops:
• The tick injects spirochete-laden saliva into the bloodstream or deposits fecal matter on the skin.
• After incubating for 3 to 32 days, the spirochetes migrate outward, causing a rash.

Ticks don't cause Lyme disease, they just carry and spread the bacteria that cause this infection.

• Spirochetes disseminate to other skin sites or organs through the bloodstream or lymphatic system. They may survive for years in the joints, or they may die after triggering an inflammatory response in the host.

What to look for

Lyme disease occurs in three stages. Signs and symptoms may take years to fully develop.

The bull's-eye-shaped rash is the beginning of the Lyme disease syndrome.

Stage 1

In stage one, a red lesion forms at the site of the tick bite — usually the axilla, thigh, or groin — and may expand to more than 20″ (50 cm) in diameter. The lesion has a white center and a bright red outer rim, and it may itch, sting, or burn. It usually disappears within 1 month.

After a few days, more lesions may erupt along with a migratory, ringlike rash and conjunctivitis. In 3 to 4 weeks, the lesions fade to small red blotches, which persist for several more weeks.

The rash is often accompanied by fatigue, headache, chills, fever, sore throat, stiff neck, nausea, and muscle and joint pain. In children, body temperature may rise to 104° F (40° C) and be accompanied by chills. At this stage, 10% of patients report symptoms such as palpitations and mild dyspnea. Severe headache and stiff neck, which suggest meningeal irritation, may also occur.

The rash may be accompanied by fatigue, headache, chills, fever, and other symptoms.

Stage 2

Weeks to months later, patients who aren't treated may enter the second stage of the disease. Meningitis, cranial nerve palsies, and peripheral neuropathy may occur. Fewer than 10% of patients have cardiac signs and symptoms. With neurologic involvement, neck stiffness usually occurs only with extreme flexion.

Stage 3

In the final stage, arthritis occurs 6 weeks to several years after the tick bite. Usually, only one or a few joints are affected, especially large ones such as the knees. Recurrent attacks may lead to chronic arthritis with severe cartilage and bone erosion. Half of all patients who aren't treated progress to this stage. (See *Treating Lyme disease.*)

In stage 3, the patient may develop chronic arthritis.

Battling illness

Treating Lyme disease

A 14- to 28-day course of oral tetracycline or doxycycline is the treatment of choice for adults infected with Lyme disease. Penicillin and erythromycin are alternatives. Children usually take oral penicillin. These drugs can minimize complications if taken early in the disease. In later stages, I.V. or I.M. ceftriaxone may be effective. Analgesics and antipyretics reduce inflammation and fever.

Memory jogger

To help you remember the progression of signs and symptoms, use the acronym LIME:

Lesions, lymph node swelling, like the flu (stage 1)

Innervation problems, such as meningitis and peripheral neuropathy (stage 2)

Movement problems such as arthritis (stage 3)

Everything else, such as myocarditis and arrhythmias (stage 3).

What tests tell you

The following tests are used to diagnose Lyme disease:
• Blood tests, including antibody titers, enzyme-linked immunosorbent assay, and Western blot assay, may be used to identify *Borrelia burgdorferi*. Unfortunately, blood tests don't always confirm Lyme disease, especially in the early stages before the body produces antibodies or seropositivity for *B. burgdorferi*. Mild anemia and elevated erythrocyte sedimentation rate, WBC count, serum immunoglobulin M (IgM) level, and aspartate aminotransferase level support the diagnosis.
• Cerebrospinal fluid (CSF) analysis may be used to detect antibodies to *B. burgdorferi* if the disease has affected the central nervous system (CNS).

Rabies

An acute CNS infection, rabies is transmitted by an animal bite and is almost always fatal once symptoms occur. Fortunately, immunization soon after infection may prevent death. (See *Treating rabies,* page 58.)

Pathophysiology

Rabies is caused by the rabies virus, a rhabdovirus. It's transmitted to a human from the bite of an infected animal through the skin or mucous membranes. Airborne droplets and infected tissue occasionally transmit the virus. Increased domestic animal control and vaccination in the United States has reduced cases of rabies in humans. Consequently, most human rabies can be traced to dog bites that occurred in other countries or bites from wild animals such as raccoons.

Here's how the disease progresses:
• The rabies virus begins replicating in the striated muscle cells at the bite site.
• The virus spreads along the nerve pathways to the spinal cord and brain, where it replicates again.
• The virus moves through the nerves into other tissues, including the salivary glands.

The incubation period for the rabies virus is hours to weeks.

What to look for

Signs and symptoms of rabies begin appearing in 1 to 3 months in patients who don't receive immunization in time.

Prodromal symptoms

At first, the patient complains of local or radiating pain or burning and a sensation of cold, pruritus, and tingling at the bite site. He may also report malaise, headache, anorexia, nausea, sore throat, a persistent loose cough, nervousness, anxiety, irritability, hyperesthesia, sensitivity to light and loud noises, and excessive salivation, tearing, and perspiration. He may have a fever of 100° to 102° F (37.8° to 38.9° C).

Excitation phase

About 2 to 10 days after prodromal signs and symptoms begin, an excitation phase occurs. This is marked by intermittent hyperactivity, anxiety, apprehension, pupillary dilation, shallow respirations, and altered level of consciousness. Cranial nerve dysfunction may cause ocular palsies, strabismus (deviation of the eye), asymmetrical pupillary dilation or constriction, absence of corneal reflexes, facial muscle weakness, and hoarseness. Temperature also rises to about 103° F (39.4° C).

Hydrophobia

About 50% of patients have hydrophobia, which causes forceful, painful pharyngeal muscle spasms that expel fluids from the mouth, resulting in dehydration and, possibly, apnea, cyanosis, and death. Swallowing problems cause

Battling illness

Treating rabies

Meticulous wound care and immunization with rabies immune globulin and human diploid cell vaccine as soon as possible after exposure are the only successful treatments for rabies.

Otherwise, treatment is mainly supportive, with special attention given to cardiovascular and respiratory systems.

Isolate the patient and wear a gown, gloves, mask, and protective eyewear when handling saliva and articles contaminated with saliva. Take precautions to avoid being bitten or scratched by the patient during the excitation phase.

The rabies virus travels along nerve pathways.

frothy drooling, and soon the sight, sound, or thought of water triggers uncontrollable pharyngeal muscle spasms and excessive salivation. Nuchal rigidity (rigidity of the neck muscles) and seizures accompanied by cardiac arrhythmias or arrest may occur.

Between excitatory and hydrophobic episodes, the patient usually remains cooperative and lucid. After about 3 days, these phases subside, and progressive paralysis leads to coma and death.

What tests tell you

There are no diagnostic tests for rabies before its onset. Histologic examination of brain tissue from human rabies victims shows perivascular inflammation of the gray matter, degeneration of neurons, and characteristic minute bodies, called Negri bodies, in the nerve cells.

Respiratory syncytial virus infection

Respiratory syncytial virus (RSV) occurs almost exclusively in infants and young children. In this population, it's the leading cause of lower respiratory tract infections, pneumonia, tracheobronchitis, bronchiolitis, otitis media, and fatal respiratory diseases.

Antibody titers suggest that most children under age 4 have contracted some form of RSV infection, although it may be mild. In fact, bronchiolitis associated with this disease peaks at age 2 months, making it the only viral disease that has its maximum impact during the first few months of life.

Pathophysiology

RSV infection results from an organism that belongs to a subgroup of a large group of viruses called the myxoviruses. Here are the key facts about RSV pathophysiology:
• The organism is transmitted from person to person by respiratory secretions.
• It has an incubation period of 4 to 5 days.
• Bronchiolitis or pneumonia occurs and, in severe cases, may damage the bronchiolar epithelium.
• Interalveolar thickening and filling of alveolar spaces with fluid may occur.

Respiratory syncytial virus is the only viral disease that has its maximum impact during the first few months of life.

What to look for

Signs and symptoms vary in severity. They include nasal congestion, coughing, wheezing, malaise, sore throat, earache, dyspnea, and fever. Although uncommon, signs of CNS infection, such as weakness, irritability, and nuchal rigidity, may also be observed.

Along with severe respiratory distress caused by increased mucus production, you may note nasal flaring, retraction, cyanosis, and tachypnea. With a lower respiratory tract infection, you may auscultate wheezes, rhonchi, and crackles caused by interstitial lung edema and bronchial spasm.

Reinfection is common and produces milder symptoms than the primary infection. School-age children, adolescents, and young adults with mild reinfections are probably the source of infection for infants and young children. (See *Treating respiratory syncytial virus infection.*)

What tests tell you

The following tests are used to diagnose RSV infection:
• Cultures of nasal and pharyngeal secretions may reveal the virus; however, this infection is so labile that cultures aren't always reliable.
• Serum antibody titers may be elevated, but in infants under age 6 months, the presence of maternal antibodies may nullify test results.

Battling illness

Treating respiratory syncytial virus infection

Treatment measures for respiratory syncytial virus support respiratory function, maintain fluid balance, and relieve symptoms. Ribavirin, a broad-spectrum antiviral agent, is used in infants with severe lower respiratory tract infection. The drug is given in aerosol form by way of tent, oxygen hood, mask, or ventilator for 2 to 5 days, 12 to 18 hours a day. This therapy reduces the severity of symptoms and improves oxygen saturation.

Rubella

Commonly called German measles, rubella is an acute, mildly contagious viral disease that produces a distinctive 3-day rash and enlarged lymph nodes. Worldwide in distribution, rubella flourishes during the spring, particularly in cities, with epidemics occurring sporadically.

Rubella occurs most commonly in children ages 5 to 9, adolescents, and young adults. The disease is self-limiting, and the prognosis is excellent except for congenital rubella, which can cause serious birth defects.

Pathophysiology

The rubella virus is transmitted through contact with the blood, urine, stool, or nasopharyngeal secretions of an infected person. It's communicable from about 10 days before the rash appears until 5 days after. It can also be transmitted transplacentally.

The virus replicates first in the respiratory tract and then spreads through the bloodstream. It has been detected in blood as early as 8 days before and up to 2 days after the rash appears. Shedding of virus from the oropharynx persists for up to 8 days after the onset of symptoms.

What to look for

The patient's history may reveal inadequate immunization, exposure to someone with a recent rubella infection, or recent travel to an endemic area without reimmunization.

Examination reveals a maculopapular, mildly itchy rash that usually begins on the face and then spreads rapidly, often covering the trunk and extremities within hours. Small, red macules on the soft palate (Forschheimer spots) may precede or accompany the rash. By the end of the second day, the rash begins to fade in the opposite order in which it appeared. It usually disappears on the third day but may persist for 4 or 5 days. The rapid appearance and disappearance of the rubella rash distinguishes it from rubeola (measles).

A low-grade fever (99° to 101° F [37.2° to 38.3° C]) may occur, but it usually disappears after the first day of the rash. Rarely, a patient's temperature may reach 104° F (40° C).

Children usually don't have prodromal symptoms, but adolescents and adults may have a headache, malaise, anorexia, sore throat, and cough before the rash appears. Palpation reveals suboccipital, postauricular, and postcervical lymph node enlargement, a hallmark sign.

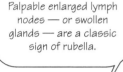

Palpable enlarged lymph nodes — or swollen glands — are a classic sign of rubella.

Congenital risk

In congenital rubella, the virus is transmitted through the placenta to the fetus by an infected mother. The fetus may have the virus in utero and for 6 to 31 months after birth.

Serious and potentially fatal complications include growth retardation, infiltration of the liver and spleen by hematopoietic tissue (tissue that forms blood cells), interstitial pneumonia, a decreased number of megakaryocytes (giant cells in the bone marrow that produce mature blood platelets), and structural malformations of the car-

Battling illness

Treating rubella

Because the rubella rash is self-limiting and only mildly pruritic, it doesn't require topical or systemic medication. Treatment consists of antipyretics and analgesics for fever and joint pain. Bed rest isn't necessary, but the patient should be isolated until the rash disappears.

Immunization with the live rubella virus prevents the disease. The vaccine should be given along with measles and mumps vaccines at age 15 months. Repeat immunizations should be given to anyone immunized before 1960 and then every 10 years thereafter.

Take precautions

Only hospital workers not at risk for rubella should provide patient care, and they should use isolation precautions until 5 days after the rash disappears. Infants with congenital rubella need to be isolated for 3 months, until three throat cultures are negative. Be sure to report confirmed cases of rubella to local public health officials.

Immune globulin may be administered I.M. or I.V. to staff and visitors who haven't been immunized. Before giving the vaccine to anyone, precautionary steps should be taken.

Before immunizing anyone with the rubella vaccine:
• make sure that they're not allergic to neomycin
• make sure that they're not pregnant
• advise women of childbearing age to avoid pregnancy by using an effective birth control method for at least 3 months after receiving the vaccine.

diovascular and central nervous systems. (See *Treating rubella*.)

What tests tell you

Clinical signs and symptoms are usually sufficient to make a diagnosis, so laboratory tests are seldom done. However, the following tests may be used:
• Cell cultures of the throat, blood, urine, and CSF, along with convalescent serum that shows a fourfold rise in antibody titers, confirms the diagnosis.
• Blood tests confirm rubella-specific IgM antibody. In congenital rubella, rubella-specific IgM antibody appears in umbilical cord blood.

Salmonellosis

More than 2 million people contract salmonellosis annually, making it one of the most common infections in the United States. It occurs as enterocolitis, bacteremia (bacteria in the blood), localized infection, typhoid fever and, rarely, paratyphoid fever.

Enterocolitis and bacteremia are especially common and virulent in infants, the elderly, and those already weakened by other infections, especially AIDS.

Got the fever

Typhoid fever, the most severe form of salmonellosis, usually lasts from 1 to 4 weeks. Most patients are under age 30, but most carriers are women over age 50. Typhoid fever is on the rise in this country due to travel to endemic areas.

Salmonella bacteria can survive for weeks in water, ice, sewage, and food.

Pathophysiology

Salmonellosis is caused by gram-negative bacilli of the genus *Salmonella,* a member of the Enterobacteriaceae family. The most common species of *Salmonella* include *S. typhi,* which causes typhoid fever; *S. enteritidis,* which causes enterocolitis; and *S. choleraesuis,* which causes bacteremia. Many *Salmonella* bacteria can survive for weeks in water, ice, sewage, and food.

Watch what you eat

Nontyphoidal salmonellosis usually follows ingestion of contaminated dry milk, chocolate bars, or pharmaceuticals of animal origin or of contaminated or inadequately processed foods, especially eggs, chicken, turkey, and duck. Proper cooking reduces the risk but doesn't eliminate it.

The disease may also spread through contact with infected people or animals and, in young children, through fecal-oral spread. Typhoid fever usually results from drinking water contaminated by the excretions of a carrier.

Typhoid fever progresses this way: After contaminated food is ingested, the bacteria pass the gastric barrier and invade the upper small bowel, producing a transient, asymptomatic bacteremia. The bacteria are ingested by mononuclear phagocytes and must survive and multiply within them to cause illness.

What to look for

All infections caused by bacteria other than *S. typhi* can cause acute diarrhea, septicemic syndrome, focal abscesses, meningitis, osteomyelitis, endocarditis, or mycotic aneurysm (an aneurysm infected by a fungus).

Persistent bacteremia kicks off the clinical phase of infection. The bacteria continue to multiply in the cells, and when the number passes a critical threshold, secondary bacteremia occurs, resulting in invasion of the gallbladder and intestine. Sustained bacteremia causes a persistent fever. Inflammatory responses to tissue invasion result in cholecystitis, intestinal hemorrhage, or perforation.

Nontyphoidal forms of salmonellosis usually produce mild to moderate illness, with low mortality. Mortality due to typhoid fever is about 3% in people who are treated and 10% in those who aren't. An attack of typhoid fever confers lifelong immunity. (See *Treating salmonellosis.*)

What tests tell you

The following tests are used to diagnose salmonellosis:
• Blood cultures isolate the organism in typhoid fever, paratyphoid fever, and bacteremia.
• Stool cultures isolate the organism in typhoid fever, paratyphoid fever, and enterocolitis.
• Cultures of urine, bone marrow, pus, and vomitus may show the presence of *Salmonella.*

In endemic areas, symptoms of enterocolitis allow a working diagnosis before the cultures are positive. *S. typhi* in stools 1 or more years after treatment indicates that the patient is a carrier (about 3% of patients).

Use extreme caution when handling materials that could be contaminated with *Salmonella* microbes.

Battling illness

Treating salmonellosis

Drugs used to treat typhoid fever, paratyphoid fever, or bacteremia include ampicillin, amoxicillin, chloramphenicol, ciprofloxacin, ceftriaxone, cefotaxime, and co-trimoxazole for extremely toxemic patients. Localized abscesses may need surgical drainage.

Camphorated opium tincture, kaolin and pectin mixtures, diphenoxylate, codeine, or small doses of morphine can relieve diarrhea and control cramps for patients who remain active.

When caring for a patient with salmonellosis, wear gloves and a gown when disposing of feces-contaminated materials, and wash your hands thoroughly after contact with the patient. Continue enteric precautions until three consecutive stool cultures are negative after antibiotic treatment. *Report all salmonellosis cases to public health officials.*

Toxoplasmosis

Seventy percent of people in the United States are infected with *Toxoplasma gondii*. Although it usually causes a localized infection, it may produce a generalized infection in neonates, patients with AIDS or lymphoma, patients who have undergone recent organ transplants, and those on immunosuppressant therapy.

Congenital and deadly

Congenital toxoplasmosis, characterized by CNS lesions, may cause stillbirth or serious birth defects. It's transmitted transplacentally from a mother who acquires primary toxoplasmosis shortly before or during pregnancy. The infection is more severe when it's acquired early in pregnancy.

Pathophysiology

Toxoplasmosis is caused by the intracellular parasite *Toxoplasma gondii,* which affects both birds and mammals. It's transmitted to humans by ingestion of tissue cysts in raw or undercooked meat or by fecal-oral contamination from infected cats. An infected cat may excrete as many as 100 million parasites per day, but a single cyst can cause infection.

Direct transmission can also occur during blood transfusions or organ transplants. The disease also occurs in people who don't eat meat and aren't exposed to cats, so an unknown means of transmission exists.

Here's how the infection works:
• When tissue cysts are ingested, parasites are released, which quickly invade and multiply within the GI tract.
• The parasitic cells rupture the invaded host cell and then disseminate to the CNS, lymphatic tissue, skeletal muscle, myocardium, retina, and placenta.
• As the parasites replicate and invade adjoining cells, cell death and focal necrosis occur, surrounded by an acute inflammatory response — the hallmarks of this infection.
• Once the cyst reaches maturity, the inflammatory process becomes undetectable, and the cysts remain latent within the brain until they rupture.

In the normal host, the immune response checks the infection, but this is not so with immunocompromised or fetal hosts. In these patients, focal destruction results in

Warn pregnant and hoping-to-be-pregnant women to have someone else change the cat's litter box.

Battling illness

Treating toxoplasmosis

Most effective during the acute stage, treatment for toxoplasmosis consists of drug therapy with a sulfonamide and pyrimethamine for 4 to 6 weeks. The patient may also receive folic acid to counteract the drugs' adverse effects, such as anemia and marrow toxicity. Unfortunately, these drugs don't eliminate already developed tissue cysts. For this reason, patients with acquired immunodeficiency syndrome (AIDS) need toxoplasmosis treatment for life.

An AIDS patient who can't tolerate sulfonamides may receive clindamycin instead. This drug is also the primary treatment in ocular toxoplasmosis.

Report all cases of toxoplasmosis to the local public health department.

necrotizing encephalitis, pneumonia, myocarditis, and organ failure.

What to look for

A patient with mild, localized toxoplasmosis may complain of malaise, myalgia, headache, fatigue, and sore throat. He'll also have a fever.

A patient with fulminating, generalized infection may complain of headache, vomiting, cough, and dyspnea. His temperature may run as high as 106° F (41.1° C). He may also have delirium and seizures (signs of encephalitis), a maculopapular rash (except on the palms, soles, and scalp), and cyanosis.

An infant with congenital toxoplasmosis may have hydrocephalus or microcephalus, seizures, jaundice, purpura, and rash. Other defects, which may not be apparent until months or years later, include strabismus, blindness, epilepsy, and mental retardation.

Once infected with toxoplasmosis, the patient may carry the organism for life. Reactivation of the acute infection can also occur. (See *Treating toxoplasmosis*.)

What tests tell you

Blood tests for the detection of a specific toxoplasma antibody is the primary diagnostic tool for detection of active infection.

Another test, which isolates the *T. gondii* in mice after their inoculation with human body fluids, reveals antibodies for the disease and confirms toxoplasmosis.

Quick quiz

1. The four types of microorganisms that cause infection are:

    A. bacteria, flora, microbes, and viruses.
    B. bacteria, viruses, fungi, and parasites.
    C. fungoids, spirochetes, mycoplasms, and parasites.

Answer: B. Bacteria, viruses, fungi, and parasites are pathogens that enter the body and cause infections.

2. An opportunistic infection occurs because:

    A. the host has an altered, weak immune system.
    B. the pathogen is especially persistent.
    C. a large number of pathogens attack the host cells.

Answer: A. Opportunistic infections occur in AIDS patients and others whose immune systems aren't functioning effectively.

3. The type of herpesvirus that causes cold sores is:

    A. herpes simplex Type 2.
    B. herpes simplex Type 1
    C. herpes zoster.

Answer: B. The herpes simplex Type 1 virus commonly causes cold sores, but it can also be spread to the genital area, eyes, and other organs.

4. The hallmarks of infectious mononucleosis are:

    A. fever, headache, and rash.
    B. exhaustion, sore throat, and nausea.
    C. fever, sore throat, and swollen cervical glands.

Answer: C. The disease also causes hepatic dysfunction, increased lymphocytes and monocytes, and the development of heterophil antibodies.

5. Lyme disease is named for:
  A. the kind of tick that causes it.
  B. the town where it originated.
  C. the characteristic greenish color of the rash.

Answer: B. The disease was first recognized in Lyme, Connecticut.

6. The most severe form of salmonellosis is:
  A. typhoid fever.
  B. bacteremia.
  C. paratyphoid fever.

Answer: A. Typhoid fever has a 3% mortality rate in people who are treated and a 10% mortality rate in those who aren't.

Scoring

☆☆☆   If you answered all six items correctly, give yourself a hand! Your enthusiasm for bacteria, viruses, fungi, and parasites is infectious!

☆☆   If you answered four or five correctly, give yourself a pat on the back! Your knowledge and understanding is thriving like parasites in the tropics!

☆   If you answered fewer than four correctly, practice, practice, and pump up on parasite power.

4

Immune system

Just the facts

In this chapter you'll learn:

♦ what structures are involved in the immune system

♦ how cell-mediated immunity, humoral immunity, and the complement system work

♦ the pathophysiology, signs and symptoms, diagnostic tests, and treatments for allergic rhinitis, anaphylaxis, lupus erythematosus, rheumatoid arthritis, and acquired immunodeficiency syndrome.

Understanding the immune system

The body protects itself from infectious organisms and other harmful invaders through an elaborate network of safeguards called the host defense system. This system has three lines of defense:

 physical and chemical barriers to infection

 the inflammatory response

 the immune response.

Physical barriers, such as the skin and mucous membranes, prevent most organisms from invading the body. Organisms that penetrate this first barrier simultaneously trigger the inflammatory and immune responses. Both responses involve stem cells in the bone marrow that form blood cells.

Four key structures

Four structures in the body make up the immune system:
- lymph nodes
- thymus
- spleen
- tonsils.

Lymph nodes

Lymph nodes are distributed along lymphatic vessels throughout the body. They filter lymphatic fluid that drains from body tissues and is later returned to the blood as plasma. The lymph nodes also remove bacteria and toxins from the circulatory system. This means that, on occasion, infectious agents can be spread by the lymphatic system.

Thymus

The thymus, located in the mediastinal area (between the lungs), produces B and T lymphocytes (B and T cells). These cells are responsible for humoral and cell-mediated immunity, described below.

Spleen

The largest lymphatic organ, the spleen, functions as a reservoir for blood. Cells in the splenic tissue called macrophages clear cellular debris and process hemoglobin.

Tonsils

The tonsils consist of lymphoid tissue and also produce lymphocytes. The location of the tonsils allows them to guard the body against airborne and ingested pathogens.

Types of immunity

Certain cells have the ability to distinguish between foreign matter and what belongs to the body — a skill known as immunocompetence. When foreign substances invade the body, two types of immune responses are possible: cell-mediated immunity and humoral immunity.

Cell-mediated immunity

In cell-mediated immunity, T cells respond directly to antigens (foreign substances such as bacteria or toxins that

Basically, my job is to destroy target cells through the release of lymphokines.

induce antibody formation). This response involves destruction of target cells — such as virus-infected cells and cancer cells — through the secretion of lymphokines (lymph proteins). Examples of cell-mediated immunity are rejection of transplanted organs and delayed immune responses that fight disease.

Eighty percent of blood cells are T cells. They probably originate from stem cells in the bone marrow; the thymus gland controls their maturity. In the process, a large number of antigen-specific cells are produced.

Kill, help, or suppress

My job is to produce antibodies that, in turn, will attack an antigen.

T cells can be killer, helper, or suppressor T cells.
- Killer cells bind to the surface of the invading cell, disrupt the membrane, and destroy it by altering its internal environment.
- Helper cells stimulate B cells to mature into plasma cells, which begin to synthesize and secrete immunoglobulin (proteins with known antibody activity).
- Suppressor cells reduce the humoral response.

Humoral immunity

B cells act in a different way than T cells to recognize and destroy antigens. B cells are responsible for humoral or immunoglobulin-mediated immunity. B cells originate in the bone marrow and mature into plasma cells that produce antibodies (immunoglobulin molecules that interact with a specific antigen). Antibodies destroy bacteria and viruses, thereby preventing them from entering host cells.

Get to know your immunoglobulin

Five major classes of immunoglobulin exist:

Immunoglobulin G (IgG) makes up about 80% of plasma antibodies. It appears in all body fluids and is the major antibacterial and antiviral antibody.

Immunoglobulin M (IgM) is the first immunoglobulin produced during an immune response. It's too large to

Keep in mind that I attack the antigen directly ...

... while I produce antibodies that incapacitate the antigen.

easily cross membrane barriers and is usually present only in the vascular system.

Different classes of antibodies perform different functions in the immune system.

Immunoglobulin A (IgA) is found mainly in body secretions, such as saliva, sweat, tears, mucus, bile, and colostrum. It defends against pathogens on body surfaces, especially those that enter the respiratory and GI tracts.

Immunoglobulin D (IgD) is present in plasma and is easily broken down. It's the predominant antibody on the surface of B cells and is mainly an antigen receptor.

Immunoglobulin E (IgE) is the antibody involved in immediate hypersensitivity reactions, or allergic reactions that develop within minutes of exposure to an antigen. IgE stimulates the release of mast cell granules, which contain histamine and heparin.

Thanks for the complement

The complement cascade plays a crucial role in the inflammatory response.

Part of humoral immunity, the complement system is a major mediator of the inflammatory response. It consists of 20 proteins circulating as functionally inactive molecules.

The system causes inflammation by increasing:
• vascular permeability
• chemostasis (chemical equilibrium)
• phagocytosis (engulfing of foreign particles by cells called phagocytes)
• lysis of the foreign cell.

In most cases, an antigen-antibody reaction is necessary for the complement system to activate, a process called the complement cascade.

Types of immune disorders

Disorders of the immune response are of three main types:

 immunodeficiency disorders

 hypersensitivity disorders

 autoimmune disorders.

Immunodeficiency disorders

Immunodeficiency disorders result from an absent or depressed immune system. Some examples include acquired immunodeficiency syndrome (AIDS), DiGeorge's syndrome, chronic fatigue syndrome, and immune dysfunction syndrome.

Hypersensitivity disorders

When an allergen (a substance the person is allergic to) enters the body, a hypersensitivity reaction occurs. It may be immediate or delayed. Hypersensitivity reactions are classified as:
- type I (IgE-mediated allergic reactions)
- type II (cytotoxic reactions)
- type III (immune complex reactions)
- type IV (cell-mediated reactions).

Autoimmune disorders

With autoimmune disorders, the body launches an immunologic response against itself. This leads to a sequence of tissue reactions and damage that may produce diffuse systemic signs and symptoms.

Examples of autoimmune disorders include rheumatoid arthritis, lupus erythematosus, dermatomyositis, and vasculitis.

Facts about AIDS

- Acquired immunodeficiency syndrome (AIDS) was first described by the Centers for Disease Control and Prevention (CDC) in 1981.

- Although 80% of those infected are men, women and children now contract the disease more than any other segment of the U.S. population.

- The prevalence of HIV infection isn't known, but the CDC estimates that 650,000 to 900,000 people in the United States are infected.

Immune disorders

The immune system disorders discussed below include:
- AIDS (an immunodeficiency disorder)
- allergic rhinitis and anaphylaxis (hypersensitivity disorders)
- lupus erythematosus and rheumatoid arthritis (autoimmune disorders).

AIDS

AIDS causes progressive impairment of the immune response and gradual destruction of cells — including T cells. Immunodeficiency makes the patient susceptible to opportunistic infections and unusual cancers. (See *Facts about AIDS.*)

The Centers for Disease Control and Prevention (CDC) has established criteria for making a diagnosis of AIDS. (See *Understanding an AIDS diagnosis*.) The course of AIDS can vary, but it usually results in death from opportunistic infections. Most experts believe that virtually everyone infected with human immunodeficiency virus (HIV) will eventually develop AIDS.

Pathophysiology

AIDS is caused by infection with the retrovirus HIV. The average time between HIV exposure and diagnosis of AIDS is 8 to 10 years.

Now I get it!

Understanding an AIDS diagnosis

According to the Centers for Disease Control and Prevention (CDC), the following three conditions must be met to establish a diagnosis of acquired immunodeficiency syndrome (AIDS):

1. presence of human immunodeficiency virus (HIV) infection
2. a CD4+ cell (T cell) count under 200
3. the presence of one or more specified conditions, listed under the categories A, B, or C. (These three categories classify conditions related to AIDS. Over time, more disorders have been included in these categories, increasing the number of patients who are diagnosed with AIDS.)

Category A conditions

Persistent, generalized lymph node enlargement or acute (primary) HIV infection with accompanying illness or history of acute HIV infection

Category B conditions

Bacillary angiomatosis, oropharyngeal or persistent vulvovaginal candidiasis, fever or diarrhea lasting over 1 month, idiopathic thrombocytopenic purpura, pelvic inflammatory disease (especially with a tubo-ovarian abscess), and peripheral neuropathy

Category C conditions

Candidiasis of the bronchi, trachea, lungs, or esophagus; invasive cervical cancer; disseminated or extrapulmonary coccidioidomycosis; extrapulmonary cryptococcosis; chronic intestinal cryptosporidiosis; cytomegalovirus (CMV) disease affecting organs other than the liver, spleen, or lymph nodes; CMV retinitis with vision loss; encephalopathy related to HIV; herpes simplex involving chronic ulcers or herpetic bronchitis, pneumonitis, or esophagitis; disseminated or extrapulmonary histoplasmosis; chronic, intestinal isosporiasis; Kaposi's sarcoma; Burkitt's lymphoma or its equivalent; immunoblastic lymphoma or its equivalent; primary brain lymphoma; disseminated or extrapulmonary *Mycobacterium avium* complex or *M. kansasii;* pulmonary or extrapulmonary *M. tuberculosis;* any other species of *Mycobacterium* (disseminated or extrapulmonary); *Pneumocystis carinii* pneumonia; recurrent pneumonia; progressive multifocal leukoencephalopathy; recurrent *Salmonella* septicemia; toxoplasmosis of the brain; wasting syndrome caused by HIV

HIV destroys CD4+ cells — also known as helper T cells — that regulate the normal immune response. The CD4+ antigen serves as a receptor for HIV and allows it to invade the cell. Afterward, the virus becomes latent or replicates, causing cell death.

HIV can infect almost any cell that has the CD4+ antigen on its surface, including monocytes, macrophages, bone marrow progenitors, and glial, gut, and epithelial cells. The infection can cause dementia, wasting syndrome, and blood abnormalities.

The CD4+ antigen on the cell surface allows HIV to affect the cell.

Modes of transmission

HIV is transmitted three ways:
- through contact with infected blood or blood products during transfusion or tissue transplantation (although routine testing of the blood supply since 1985 has cut the risk of contracting HIV this way) or by sharing a contaminated needle
- through contact with infected body fluids, such as semen and vaginal fluids, during unprotected sex (Anal intercourse is especially dangerous because it causes mucosal trauma.)
- across the placenta from an infected mother to a fetus or from an infected mother to an infant either through cervical or blood contact at delivery or through breast milk.

Although blood, semen, vaginal secretions, and breast milk are the body fluids that most readily transmit HIV, it has also been found in saliva, urine, tears, and feces. However, there's no evidence of transmission through these fluids.

What to look for

After initial exposure, the infected person may have no signs or symptoms, or he may have a flulike illness (seroconversion illness) and then remain asymptomatic for years. As the syndrome progresses, he may have neurologic symptoms from HIV encephalopathy or symptoms of an opportunistic infection, such as *Pneumocystis carinii* pneumonia, cytomegalovirus, or cancer. Eventually, repeated opportunistic infections overwhelm the patient's weakened immune defenses, invading every body system. (See *Opportunistic diseases associated with AIDS,* pages 76 to 80, and *Treating AIDS,* page 81.)

(Text continues on page 81.)

Opportunistic diseases associated with AIDS

The following chart describes some common infections associated with acquired immunodeficiency syndrome (AIDS), their characteristic signs and symptoms, and their treatments.

INFECTION	SIGNS AND SYMPTOMS	TREATMENT
Bacterial		
Tuberculosis A disease caused by *Mycobacterium tuberculosis*, an aerobic, acid-fast bacillus spread through inhalation of droplet nuclei that are aerosolized by coughing, sneezing, or talking	Fever, weight loss, night sweats, and fatigue, followed by dyspnea, chills, hemoptysis, and chest pain	Isoniazid, rifampin, pyrazinamide, and ethambutol or streptomycin are given during the first 2 months of therapy, followed by rifampin and isoniazid for a minimum of 9 months and for at least 6 months after culture is negative for bacteria.
***Mycobacterium avium* complex** A primary infection acquired by oral ingestion or inhalation; can infect the bone marrow, liver, spleen, GI tract, lymph nodes, lungs, skin, brain, adrenal glands, and kidneys; is chronic and may be both localized and disseminated in its course of infection	Multiple, nonspecific symptoms consistent with systemic illness: fever, fatigue, weight loss, anorexia, night sweats, abdominal pain, and chronic diarrhea. *Physical examination findings:* emaciation, generalized lymphadenopathy, diffuse tenderness, jaundice, and hepatosplenomegaly. *Laboratory findings:* anemia, leukopenia, and thrombocytopenia	Treatment regimens vary and can include two to six drugs. The Centers for Disease Control and Prevention currently recommends that every patient take either azithromycin or clarithromycin. Many experts prefer ethambutol as a second drug. Additional drugs include clofazimine, rifabutin, rifampin, ciprofloxacin, and sometimes amikacin.
Salmonellosis A disease acquired by ingestion of food or water contaminated by *Salmonella* but also linked to snake powders, pet turtles, and domestic turkeys; also can be spread by contaminated medications or diagnostic agents, direct fecal-oral transmission (especially during sexual activity), transfusion of contaminated blood products, and inadequately sterilized fiber-optic instruments used in upper GI endoscopic procedures	Nonspecific signs and symptoms, including fever, chills, sweats, weight loss, diarrhea, and anorexia	Although treatment of nontyphoid salmonellosis is usually unnecessary in immunocompetent individuals, it is required in persons with human immunodeficiency virus (HIV). Antibiotic selection depends on drug sensitivities. However, treatment may include co-trimoxazole, amoxicillin, fluoroquinolones, ampicillin, or third-generation cephalosporins.

This chart can help you identify various infections that accompany AIDS.

Opportunistic diseases associated with AIDS *(continued)*

INFECTION	SIGNS AND SYMPTOMS	TREATMENT
Fungal		
Coccidioidomycosis An infectious disease caused by the fungus *Coccidioides immitis*, which grows in soil in arid regions in the southwestern United States, Mexico, Central America, and South America	Flulike illness (malaise, fever, backache, headache, cough, arthralgia), periarticular swelling in knees and ankles, meningitis, bony lesions, skin findings, and genitourinary (GU) involvement	Fluconazole, itraconazole, amphotericin B, or ketoconazole is given.
Candidiasis A disease caused by the fungus *Candida albicans* that exists on teeth, gingivae, and skin and in the oropharynx, vagina, and large intestine; majority of infections endogenous and related to interruption of normal defense mechanisms; possible human-to-human transmission, including congenital transmission in neonates, in whom thrush develops after vaginal delivery	*Thrush, the most prevalent form in HIV-infected individuals:* creamy, curdlike, yellowish patches, surrounded by an erythematous base, found on buccal membrane and tongue surfaces. *Nail infection:* inflammation and tenderness of tissue surrounding the nails or the nail itself. *Vaginitis:* intense pruritus of the vulva and curdlike vaginal discharge	Nystatin suspension and clotrimazole troches are administered for thrush; nystatin suspension or pastilles, clotrimazole troches, or oral ketoconazole, fluconazole, or itraconazole for esophagitis; topical clotrimazole, miconazole, or ketoconazole for cutaneous candidiasis; topical imidazole or oral fluconazole, ketoconazole, or both for candidiasis of nails; and topical clotrimazole, miconazole, or oral ketoconazole for vaginitis.
Cryptococcosis An infectious disease caused by the fungus *Cryptococcus neoformans*; that can be found in nature; can be aerosolized and inhaled; settles in the lungs, where it can remain dormant or spread to other parts of the body, particularly the central nervous system (CNS); responsible for three forms of infection: pulmonary, CNS, and disseminated; most pulmonary cases found serendipitously	*Pulmonary cryptococcosis:* fever, cough, dyspnea, and pleuritic chest pain. *CNS cryptococcosis:* fever, malaise, headaches, stiff neck, nausea and vomiting, and altered mentation. *Disseminated cryptococcosis:* lymphadenopathy, multifocal cutaneous lesions. *Other symptoms:* macules, papules, skin lesions, oral lesions, placental infection, myocarditis, prostatic infection, optic neuropathy, rectal abscess, and lymph node infection	Primary therapy for initial infection is amphotericin B, administered I.V. for 6 to 8 weeks; sometimes amphotericin B and flucytosine are used in combination. Fluconazole and itraconazole are also used.
Histoplasmosis A disease caused by the fungus *Histoplasma capsulatum* that exists in nature, is readily airborne, and can reach the bronchioles and alveoli when inhaled	*Most common:* fever, weight loss, hepatomegaly, splenomegaly, and pancytopenia. *Less common:* diarrhea, cerebritis, chorioretinitis, meningitis, oral and cutaneous lesions, and GI mucosal lesions causing bleeding	Drug of choice is amphotericin B used as lifelong suppressive therapy, not a cure. Itraconazole is also used.

(continued)

Opportunistic diseases associated with AIDS *(continued)*

INFECTION	SIGNS AND SYMPTOMS	TREATMENT
Protozoan		
***Pneumocystis carinii* pneumonia** Pneumonia caused by *Pneumocystis carinii;* also has properties of fungal infection, exists in human lungs, and is transmitted by airborne exposure; the most common life-threatening opportunistic infection in individuals with AIDS	Fever, fatigue, and weight loss for several weeks to months before respiratory symptoms develop. *Respiratory symptoms:* dyspnea, usually noted initially on exertion and later at rest, and cough, usually starting out dry and nonproductive and later becoming productive	Co-trimoxazole may be given orally or I.V., although about 20% of AIDS patients are hypersensitive to sulfa drugs. I.V. pentamidine isethionate may be given but can cause many adverse effects, including permanent diabetes mellitus. Dapsone with trimethoprim, clindamycin, primaquine, atovaquone, or corticosteroids is also used. Prophylaxis following treatment includes co-trimoxazole, aerosolized pentamidine isethionate, or dapsone.
Cryptosporidiosis An intestinal infection by the protozoan *Cryptosporidium;* transmitted by person-to-person contact, water, food contaminants, and airborne exposure; most common site: small intestine	Abdominal cramping, flatulence, weight loss, anorexia, malaise, fever, nausea, vomiting, myalgia, and profuse, watery diarrhea	No effective therapy is known. Most medical therapy is palliative and directed toward symptom control, focusing on fluid replacement, occasionally total parenteral nutrition, correction of electrolyte imbalances, and analgesic, antidiarrheal, and antiperistaltic agents. Paromomycin, spiramycin, and octreotides are used.
Toxoplasmosis A disease caused by *Toxoplasma gondii;* major means of transmission through ingestion of undercooked meats and vegetables containing oocysts; causes focal or diffuse meningoencephalitis with cellular necrosis and progresses unchecked to the lungs, heart, and skeletal muscle	Localized neurologic deficits, fever, headache, altered mental status, and seizures	Sulfadiazine or clindamycin with pyrimethamine may be given; however, about 20% of AIDS patients are hypersensitive to sulfa drugs. Folinic acid may be given to prevent marrow toxicity from pyrimethamine.
Coccidiosis A disease caused by coccidian protozoan parasite *Isospora hominis* or *I. belli;* after ingestion, infects the small intestine and results in malabsorption and diarrhea	Watery, nonbloody diarrhea; crampy abdominal pain; nausea; anorexia; weight loss; weakness; occasional vomiting; and a low-grade fever	Fluconazole, itraconazole, or amphotericin B is used.

Opportunistic diseases associated with AIDS *(continued)*

INFECTION	SIGNS AND SYMPTOMS	TREATMENT
Viral		
Herpes simplex virus (HSV) Chronic infection caused by a herpesvirus; often a reactivation of an earlier herpes infection	Red, blisterlike lesions occurring in oral, anal, and genital areas; also found on the esophageal and tracheobronchial mucosa; pain, bleeding, and discharge	Acyclovir, famciclovir, or valacyclovir are given I.V. or P.O.
Cytomegalovirus (CMV) A viral infection of the herpesvirus that may result in serious, widespread infection; most common sites: lungs, adrenal glands, eyes, CNS, GI tract, male GU tract, and blood	Unexplained fever, malaise, GI ulcers, diarrhea, weight loss, swollen lymph nodes, hepatomegaly, splenomegaly, blurred vision, floaters, dyspnea (especially on exertion), dry nonproductive cough, and vision changes leading to blindness in patients with ocular infection	Ganciclovir or foscarnet is used to treat CMV. Ganciclovir has shown some anti-HIV properties. Foscarnet or intraocular ganciclovir implants may be used to treat CMV retinitis.
Progressive multifocal leukoencephalopathy (PML) Progressive demyelinating disorder caused by hyperactivation of a papovavirus that leads to gradual brain degeneration	Progressive dementia, memory loss, headache, confusion, weakness, and other possible neurologic complications such as seizures	No form of therapy for PML has been effective, but attempted therapies include prednisone, acyclovir, and adenine arabinoside administered both I.V. and intrathecally.
Herpes zoster A disease also known as acute posterior ganglionitis, shingles, zona, and zoster; acute infection caused by reactivation of the chickenpox virus	Small clusters of painful, reddened papules that follow the route of inflamed nerves; may be disseminated, involving two or more dermatomes	Herpes zoster is most often treated with oral acyclovir capsules until healed. Treatment may have to continue at lower doses indefinitely to prevent recurrence. I.V. acyclovir is effective in disseminated varicella zoster lesions in some patients. Medications may relieve pain associated with the infection and postherpetic neuropathies. Famciclovir and valacyclovir can also be used.

These infections can affect all body systems.

(continued)

Opportunistic diseases associated with AIDS *(continued)*

INFECTION	SIGNS AND SYMPTOMS	TREATMENT
Neoplasms		
Kaposi's sarcoma A generalized disease with characteristic lesions involving all skin surfaces, including the face (tip of the nose, eyelids), head, upper and lower limbs, soles of the feet, palms of the hands, conjunctivae, sclerae, pharynx, larynx, trachea, hard palate, stomach, liver, small and large intestines, and glans penis	Manifested cutaneously and subcutaneously; usually painless, nonpruritic tumor nodules that are pigmented and violaceous (red to blue), nonblanching and palpable; patchy lesions appearing early and possibly mistaken for bruises, purpura, or diffuse cutaneous hemorrhages	Treatment isn't indicated for all individuals. Indications include cosmetically offensive, painful, or obstructive lesions or rapidly progressing disease. Systemic chemotherapy using single or multiple drugs may be given to alleviate symptoms. Radiation therapy may be used to treat lesions. Intralesional therapy with vinblastine may be given for cosmetic purposes to treat small cutaneous lesions, and laser therapy and cryotherapy to treat small isolated lesions. Interferon alfa-2a and interferon alfa-2b are also used.
Malignant lymphomas Immune system cancer in which lymph tissue cells begin growing abnormally and spread to other organs; incidence in persons with AIDS: about 4% to 10%; diagnosed in HIV-infected individuals as widespread disease involving extranodal sites, most commonly in the GI tract, CNS, bone marrow, and liver	Unexplained fever, night sweats, or weight loss greater than 10% of patient's total body weight; signs and symptoms often confined to one body system: CNS (confusion, lethargy, and memory loss) or GI tract (pain, obstruction, changes in bowel habits, bleeding, and fever)	Individualized therapy may include a modified combination of methotrexate, bleomycin, doxorubicin, cyclophosphamide, vincristine, and dexamethasone. Radiation therapy, not chemotherapy, is used to treat primary CNS lymphoma.
Cervical neoplasm Emerging as a significant opportunistic complication of HIV infection as more women become infected with HIV and live longer with illness because of antiretroviral prophylaxis and treatment	*Possible indicators of early invasive disease:* abnormal vaginal bleeding, persistent vaginal discharge, or postcoital pain and bleeding. *Possible indicators of advanced disease:* pelvic pain, vaginal leakage of urine and feces from a fistula, anorexia, weight loss, and fatigue	Treatment is tailored to the disease stage. Preinvasive lesions may require total excisional biopsy, cryosurgery, laser destruction, conization (and frequent Papanicolaou test follow-up) and, rarely, hysterectomy. Invasive squamous cell carcinoma may require radical hysterectomy and radiation therapy.

What to look for in children

In children, the incubation period averages only 17 months. Signs and symptoms resemble those for adults, except that children are more likely to have a history of bacterial infections such as otitis media, lymphoid interstitial pneumonia as well as types of pneumonia not caused by *P. carinii*, sepsis, and chronic salivary gland enlargement.

An infected person can test negative for HIV antibodies for up to 35 months after exposure.

What tests tell you

The CDC recommends testing for HIV 3 months after a possible exposure — the approximate length of time before antibodies can be detected in the blood. However, an infected patient can test negative for as long as 35 months. Antibody tests in neonates may also be unreliable because transferred maternal antibodies persist for up to 10 months, causing a false-positive result.

The recommended protocol is initial screening with an enzyme-linked immunosorbent assay (ELISA). If results are positive, another ELISA is done on the same

Battling illness

Treating AIDS

There is no cure for acquired immunodeficiency syndrome (AIDS); however, several types of drugs are used to treat the disease:

• Antiretroviral drugs, such as zidovudine (AZT) and didanosine (ddI), inhibit or temporarily inactivate the human immunodeficiency virus (HIV).
• Immunomodulatory drugs, such as interferon beta, boost a weakened immune system.
• Anti-infective drugs (such as dapsone, rifabutin) and antineoplastic drugs (such as methotrexate) combat opportunistic infections and associated cancers. Some anti-infectives are used prophylactically against opportunistic infections.

Combination of the two (or more)

Many treatment protocols combine two or more drugs to produce the maximum benefit with the fewest adverse reactions. Combination therapy also helps inhibit the production of mutant HIV strains resistant to particular drugs. Many of these drugs are experimental. Examples of combination therapy include zidovudine plus lamivudine (Epivir), didanosine plus lamivudine, and zidovudine plus didanosine or zalcitabine.

Other treatments use protease inhibitors, such as indinavir (Crixivan), saquinavir (Invirase), ritonavir (Norvir), and nelfinavir (Viracept). These drugs can be used for children and adults. Protease inhibitors act on the virus to prevent packaging of viral proteins into infectious virions. They have been found to reduce the AIDS virus in the blood of adults by 98%.

Although many opportunistic infections respond to anti-infective drugs, infections tend to recur after treatment ends. Therefore, the patient usually requires continued prophylaxis until the drug loses its effectiveness or he can't tolerate it any longer.

blood. If still positive, findings are confirmed by the Western blot test or an immunofluorescence assay.

Other blood tests support the diagnosis and are used to evaluate the severity of immunosuppression. They include CD4+ and CD8+ cell (killer T cell) subset counts, erythrocyte sedimentation rate (ESR), complete blood count (CBC), serum beta (sub 2) microglobulin, p24 antigen, neopterin levels, and anergy testing.

Many opportunistic infections in AIDS patients are reactivations of previous infections. Therefore, patients may also be tested for syphilis, hepatitis B, tuberculosis, toxoplasmosis, and histoplasmosis.

Allergic rhinitis

Inhaled, airborne allergens may trigger an immune response in the upper airway. This causes two problems:
• rhinitis, or inflammation of the nasal mucous membrane
• conjunctivitis, or inflammation of the membrane lining the eyelids and covering the eyeball.

When allergic rhinitis occurs seasonally, it's called hay fever, even though hay doesn't cause it and no fever occurs. When it occurs year-round, it's called perennial allergic rhinitis.

More than 20 million Americans suffer from this disorder, making it the most common allergic reaction. Although it can affect anyone at any age, it's most common in young children and adolescents.

Pathophysiology

Allergic rhinitis is a type I, IgE-mediated hypersensitivity response to an environmental allergen in a genetically susceptible patient.

Hay fever occurs in the spring, summer, and fall and is usually induced by airborne pollens from trees, grass, and weeds. In the summer and fall, mold spores also cause it.

Major perennial allergens and irritants include house dust and dust mites, feathers, molds, fungi, tobacco smoke, processed materials or industrial chemicals, and animal dander.

Swelling of the nasal and mucous membranes may trigger secondary sinus infections and middle ear infections, especially in perennial allergic rhinitis. Nasal polyps

Recall that IgE is the antibody involved in immediate hypersensitivity reactions, or allergic reactions that develop within minutes of exposure to an antigen.

caused by edema and infection may increase nasal obstruction. Bronchitis and pneumonia are possible complications.

What to look for

In hay fever, the patient complains of sneezing attacks, rhinorrhea (profuse, watery nasal discharge), nasal obstruction or congestion, itching nose and eyes, and headache or sinus pain. An itchy throat, malaise, and fever also may occur.

Perennial allergic rhinitis causes chronic, extensive nasal obstruction or stuffiness, which can obstruct the eustachian tube, particularly in children. Conjunctivitis or other extranasal signs and symptoms can occur but are rare. (See *Types of rhinitis*.)

Both conditions can cause allergic shiners (dark circles under the eyes) from venous congestion in the maxillary sinuses. The severity of signs and symptoms may vary from year to year. (See *Treating allergic rhinitis,* page 84.)

What tests tell you

IgE levels may be normal or elevated. Microscopic examination of sputum and nasal secretions shows a high number of eosinophils. Eosinophils are granular leukocytes. The activity of eosinophils is not completely understood, but they are known to destroy parasitic organisms and play a role in allergic reactions.

Anaphylaxis

An acute type I allergic reaction, anaphylaxis causes sudden, rapidly progressive urticaria (hives) and respiratory distress. A severe reaction may lead to vascular collapse, systemic shock, and even death.

Pathophysiology

Anaphylactic reactions result from systemic exposure to sensitizing drugs or other antigens, such as the following:
- serums such as vaccines
- allergen extracts such as pollen
- enzymes such as L-asparaginase
- hormones

Types of rhinitis

Characteristics of three common disorders of the nasal mucosa are listed below.

Chronic vasomotor rhinitis

- Eyes aren't affected.
- Nasal discharge contains mucus.
- No seasonal variation.

Infectious rhinitis (common cold)

- Nasal mucosa is beet red.
- Nasal secretions contain exudate.
- Fever and sore throat occur.

Rhinitis medicamentosa

- Caused by excessive use of nasal sprays or drops.
- Nasal drainage and mucosal redness and swelling subside when the medication is discontinued.

Anaphylaxis can result from contact with many common substances.

Battling illness

Treating allergic rhinitis

Treatment of allergic rhinitis involves controlling signs and symptoms and preventing infection. Treatment may include removing the environmental allergen. Drug therapy and immunotherapy may be employed in treating perennial allergic rhinitis.

Annoying after-effects

Antihistamines are effective in stopping runny nose and watery eyes, but they usually produce sedation, dry mouth, nausea, dizziness, blurred vision, and nervousness. Changing antihistamines every season may keep patients from building up a tolerance to one drug. Nonsedating antihistamines, such as terfenadine and astemizole, produce fewer annoying effects.

Try these treatments, too

Topical intranasal corticosteroids may reduce local inflammation with minimal systemic adverse effects. Cromolyn sodium may help prevent allergic rhinitis, but it takes 4 weeks to produce a satisfactory effect and must be taken regularly during allergy season.

Long-term management includes immunotherapy or desensitization with injections of allergen extracts administered preseasonally, seasonally, or every year.

> Next time I'll use a nonsedating antihistamine.

- penicillin and other antibiotics
- local anesthetics
- salicylates
- polysaccharides such as iron dextran
- diagnostic chemicals such as radiographic contrast media
- foods, such as nuts and seafood
- sulfite-containing food additives
- insect venom, such as that from honeybees, wasps, and certain spiders
- a ruptured hydatid cyst (rare).

Penicillin is the most common anaphylaxis-causing antigen. It induces a reaction in 4 out of every 10,000 patients.

Here is how an anaphylactic reaction occurs:
- After initial exposure to an antigen, the immune system responds by producing IgE antibodies in the lymph nodes. Helper T cells enhance the process.
- Antibodies bind to membrane receptors located on mast cells in connective tissues and on basophils, which are a type of leukocyte.

• On reexposure, the antigen binds to adjacent IgE antibodies or cross-linked IgE receptors, activating inflammatory reactions such as the release of histamine.

Untreated anaphylaxis causes respiratory obstruction, systemic vascular collapse, and death — minutes to hours after the first symptoms occur. However, a delayed or persistent reaction may last up to 24 hours. (See *Treating anaphylaxis,* and *Understanding anaphylaxis,* page 86.)

In anaphylaxis, the degree of reaction is based on the amount of allergen exposure.

What to look for

Immediately after exposure, the patient may report a feeling of impending doom or fright, progressing to a fear of impending death, weakness, sweating, sneezing, dyspnea, nasal pruritus, and urticaria.

Skin effects

The skin may look cyanotic and pallid. Well-circumscribed, discrete, cutaneous wheals with red, raised wavy or indented borders and blanched centers usually appear and may merge to form giant hives.

Battling illness

Treating anaphylaxis

Anaphylaxis is always an emergency. It requires an immediate injection of epinephrine 1:1,000 aqueous solution, 0.1 to 0.5 ml, for mild signs and symptoms. If signs and symptoms are severe, repeat the dose every 5 to 20 minutes, as directed. Fifty to 100 mg of I.M. or I.V. diphenhydramine may be given for allergic signs and symptoms. Aminophylline helps relieve bronchospasm.

Emphasis on epinephrine

In the early stages of anaphylaxis, when the patient remains conscious and normotensive, give epinephrine I.M. or subcutaneously. Massage the injection site to speed the drug into the circulation. In severe reactions, when the patient is unconscious and hypotensive, give the drug I.V., as ordered. I.V. therapy is necessary to prevent vascular collapse after severe reaction is treated.

Also establish and maintain a patent airway. Watch for early signs of laryngeal edema, such as stridor, hoarseness, and dyspnea. If these occur, oxygen therapy along with endotracheal tube insertion or a tracheotomy is required.

Act fast with anaphylaxis! An epinephrine injection is the first line of defense.

Now I get it!

Understanding anaphylaxis

The chart below outlines the sequence of events that occurs during anaphylaxis.

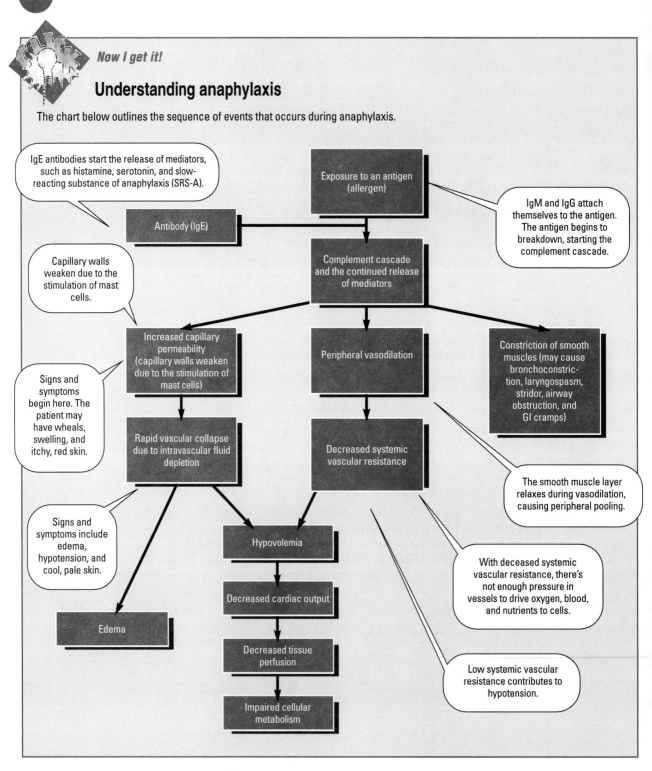

IgE antibodies start the release of mediators, such as histamine, serotonin, and slow-reacting substance of anaphylaxis (SRS-A).

Exposure to an antigen (allergen)

IgM and IgG attach themselves to the antigen. The antigen begins to breakdown, starting the complement cascade.

Antibody (IgE)

Capillary walls weaken due to the stimulation of mast cells.

Complement cascade and the continued release of mediators

Increased capillary permeability (capillary walls weaken due to the stimulation of mast cells)

Peripheral vasodilation

Constriction of smooth muscles (may cause bronchoconstriction, laryngospasm, stridor, airway obstruction, and GI cramps)

Signs and symptoms begin here. The patient may have wheals, swelling, and itchy, red skin.

Rapid vascular collapse due to intravascular fluid depletion

Decreased systemic vascular resistance

The smooth muscle layer relaxes during vasodilation, causing peripheral pooling.

Signs and symptoms include edema, hypotension, and cool, pale skin.

Hypovolemia

With deceased systemic vascular resistance, there's not enough pressure in vessels to drive oxygen, blood, and nutrients to cells.

Decreased cardiac output

Edema

Decreased tissue perfusion

Low systemic vascular resistance contributes to hypotension.

Impaired cellular metabolism

A lump in the throat

Angioedema may cause the patient to complain of a lump in his throat; swelling of the tongue and larynx also occur. You may hear hoarseness, stridor, or wheezing. Chest tightness signals bronchial obstruction. These are all early signs of potentially fatal respiratory failure.

Gastrointestinal effects

The patient may report severe stomach cramps, nausea, diarrhea, urinary urgency, and urinary incontinence.

Neurologic and cardiovascular effects

Neurologic signs and symptoms may include dizziness, drowsiness, headache, restlessness, and seizures. Cardiovascular effects include hypotension, shock, and cardiac arrhythmias, which may precipitate vascular collapse if untreated.

What tests tell you

No tests are needed. The patient's history and signs and symptoms establish a diagnosis of anaphylaxis.

Lupus erythematosus

A chronic, inflammatory, autoimmune disorder affecting the connective tissues, lupus erythematosus takes two forms: discoid and systemic. The first type affects only the skin; the second type affects multiple organs and can be fatal.

Discoid lupus erythematosus (DLE) causes superficial lesions — typically over the cheeks and bridge of the nose — that leave scars after healing. Systemic lupus erythematosus (SLE) is characterized by recurrent, seasonal remissions and exacerbations, especially during spring and summer.

The annual incidence in urban populations varies from 15 to 50 per 100,000 people. Lupus erythematosus strikes women 8 times more often than men and occurs 15 times more often during the childbearing years. It occurs worldwide but is most prevalent among Asians and Blacks.

Although there is no cure for lupus, the prognosis improves with early detection and treatment. Prognosis remains poor for patients who develop cardiovascular, renal,

In lupus, the body produces antibodies against its own cells.

or neurologic complications or severe bacterial infections. (See *Treating SLE.*)

Pathophysiology

The exact cause of lupus erythematosus remains a mystery, but autoimmunity is probably the primary cause, along with environmental, hormonal, genetic and, possibly, viral factors.

In autoimmunity, the body produces antibodies against its own cells. The formed antigen-antibody complexes can suppress the body's normal immunity and damage tissues. A significant feature of patients with SLE is their ability to produce antibodies against many different tissue components, such as red blood cells, neutrophils, platelets, lymphocytes, or almost any organ or tissue in the body.

Battling illness

Treating SLE

Drugs are the mainstay of treatment for systemic lupus erythematosus (SLE). In patients with mild disease, nonsteroidal anti-inflammatory drugs such as aspirin usually control arthritis and arthralgia. Skin lesions require sun protection and topical corticosteroid creams, such as triamcinolone and hydrocortisone.

Fluorinated steroids may control acute or discoid lesions. Stubborn lesions may respond to intralesional or systemic corticosteroids or antimalarials, such as hydroxychloroquine, chloroquine, and dapsone. Because hydroxychloroquine and chloroquine can cause retinal damage, patients receiving them should have an ophthalmologic examination every 6 months.

More medication information

Corticosteroids are the treatment of choice for systemic symptoms, acute generalized exacerbations, and injury to vital organs from pleuritis, pericarditis, nephritis, vasculitis, and CNS involvement. With initial prednisone doses of 60 mg or more, the patient's condition usually improves noticeably within 48 hours. Once symptoms are under control, the drug dosage is gradually reduced and then discontinued.

In some patients, cytotoxic drugs, such as azathioprine, cyclophosphamide, and methotrexate, may delay or prevent renal deterioration. Warfarin can prevent clotting in vascular structures. Antihypertensive drugs and dietary therapy may also be tried.

SLE susceptibility

Most people with SLE have a genetic predisposition. Other predisposing factors include stress, streptococcal or viral infections, exposure to sunlight or ultraviolet light, immunization, pregnancy, and abnormal estrogen metabolism. Drugs may also trigger or aggravate the disease. (See *Drugs that spark SLE.*)

What to look for

The onset of SLE may be acute or insidious, but the disease has no characteristic clinical pattern. Patients may complain of fever, anorexia, weight loss, malaise, fatigue, abdominal pain, nausea, vomiting, diarrhea, constipation, rashes, and polyarthralgia (multiple joint pain).

Blood disorders, such as anemia, leukopenia, lymphopenia, thrombocytopenia, and an elevated ESR, occur because of circulating antibodies. Women may report irregular menstruation or amenorrhea, particularly during exacerbations.

About 90% of patients have joint involvement that resembles rheumatoid arthritis. About 20% have Raynaud's phenomenon — intermittent, severe pallor of the fingers, toes, ears, or nose.

Warning!

Drugs that spark SLE

Many drugs can cause symptoms that resemble those of systemic lupus erythematosus (SLE). Others actually can trigger SLE or activate it. Drugs in this last group include the following:
- procainamide
- hydralazine
- isoniazid
- methyldopa
- anticonvulsants
- penicillins, sulfa drugs, and oral contraceptives (less commonly).

Cardiopulmonary effects

Cardiopulmonary signs and symptoms occur in about 50% of patients. They include chest pain, indicating pleuritis; dyspnea, suggesting parenchymal infiltrates and pneumonitis; tachycardia; central cyanosis; and hypotension. These signs and symptoms may signal pulmonary embolism.

Neurologic effects

Seizure disorders and mental dysfunction may indicate neurologic damage. Other central nervous system signs and symptoms include emotional lability, psychosis, headaches, irritability, and depression.

Urinary system effects

Be especially alert for infrequent urination, which may signal renal failure, and urinary frequency, painful urination, and bladder spasms, which are signs and symptoms of urinary tract infection. Urinary tract infections and renal failure are the leading causes of death for SLE patients.

Effects on the skin

The skin eruption characteristic of lupus erythematosus is a rash on areas exposed to light. The rash varies in severity from red areas to disc-shaped plaque. Patchy alopecia is also common in SLE.

The classic "butterfly rash" appears in about 50% of patients.

What tests tell you

The following tests are used to diagnose lupus erythematosus:
- CBC with differential may show anemia and a reduced white blood cell (WBC) count.
- Serum electrophoresis may show hypergammaglobulinemia.
- Other blood tests may show a decreased platelet count and an elevated ESR. Active disease is diagnosed by decreased serum complement levels, leukopenia, mild thrombocytopenia, and anemia.
- Chest X-rays may reveal pleurisy or lupus pneumonitis.

Rheumatoid arthritis

A chronic, systemic, inflammatory disease, rheumatoid arthritis usually attacks peripheral joints and surrounding muscles, tendons, ligaments, and blood vessels. Spontaneous remissions and unpredictable exacerbations mark the course of this potentially crippling disease.

Rheumatoid arthritis strikes women three times more often than men, and although it occurs in all age-groups, the peak onset for women is between ages 35 and 50.

Genetics? Infections? Hormones? The cause of rheumatoid arthritis is unknown.

Lifelong treatment

Rheumatoid arthritis usually requires lifelong treatment and sometimes surgery. In most patients, the disease is intermittent, allowing normal activity. However, 10% of patients suffer total disability from severe joint deformity, associated symptoms, or both. The prognosis worsens with the development of nodules, vasculitis (inflammation of a blood or lymph vessel), and high titers of rheumatoid factor.

Pathophysiology

The cause of rheumatoid arthritis isn't known, but infections, genetics, and endocrine factors may play a part.

When exposed to an antigen, a person susceptible to rheumatoid arthritis may develop abnormal or altered IgG antibodies. The body doesn't recognize these antibodies as "self," so it forms an antibody known as rheumatoid factor against them. By aggregating into complexes, rheumatoid factor causes inflammation.

Eventually, inflammation causes cartilage damage, Immune responses continue, including complement system activation. Complement system activation attracts leukocytes and stimulates the release of inflammatory mediators, which then exacerbate joint destruction.

IgG appears in all body fluids and is the major antibacterial and antiviral antibody.

Four inflammatory stages

If unarrested, joint inflammation occurs in four stages:

First, synovitis develops from congestion and edema of the synovial membrane and joint capsule.

Formation of pannus (thickened layers of granulation tissue) marks the onset of the second stage. Pannus covers and invades cartilage and eventually destroys the joint capsule and bone.

The third stage is characterized by fibrous ankylosis (fibrous invasion of the pannus and scar formation that occludes the joint space). Bone atrophy and misalignment cause visible deformities and restrict movement, causing muscle atrophy, imbalance and, possibly, partial dislocations.

In the fourth stage, fibrous tissue calcifies, resulting in bony ankylosis (fixation of a joint) and total immobility. Pain associated with movement may restrict active joint use and cause fibrous or bony ankylosis, soft-tissue contractures, and joint deformities. (See *Treating rheumatoid arthritis,* page 92.)

A lot more is known about the pathophysiology of rheumatoid arthritis than about its actual causes.

What to look for

At first, the patient may complain of nonspecific symptoms, including fatigue, malaise, anorexia, persistent low-grade fever, weight loss, and vague articular symptoms. As inflammation progresses through the four stages, specific symptoms develop, frequently in the fingers. They usually occur bilaterally and symmetrically and may extend to the wrists, elbows, knees, and ankles.

What tests tell you

Although no test can be used to definitively diagnose rheumatoid arthritis, the following are useful:

• X-rays may show bone demineralization and soft-tissue swelling and help determine the extent of cartilage and bone destruction, erosion, subluxations, and deformities.

• Rheumatoid factor test is positive in 75% to 80% of patients, as indicated by a titer of 1:160 or higher. Although the presence of rheumatoid factor doesn't confirm the disease, it helps in determining the prognosis. A patient with a high titer usually has more severe and progressive disease with extra-articular signs and symptoms.

• Synovial fluid analysis shows increased volume and turbidity but decreased viscosity and complement levels. The WBC count often exceeds 10,000/mm^3.

• Serum protein electrophoresis may show elevated serum globulin levels.

• ESR is elevated in 85% to 90% of patients. Because an elevated rate often parallels disease activity, this test helps monitor the patient's response to therapy.

Battling illness

Treating rheumatoid arthritis

Treatment measures are used to reduce pain and inflammation and preserve the patient's functional capacity and quality of life. Salicylates, especially aspirin, are the mainstay of therapy because they decrease inflammation and relieve joint pain. Nonsteroidal anti-inflammatory drugs are also used, including indomethacin, fenoprofen, and ibuprofen.

Other drugs that can be used are hydroxychloroquine, gold salts, penicillamine, and corticosteroids such as prednisone. Immunosuppressants — cyclophosphamide, methotrexate, and azathioprine — are used in the early stages of the disease.

Joint function can be preserved through range-of-motion exercises and a carefully individualized physical therapy program.

Surgery

Surgery may be used to reduce pain, realign joint surfaces, and redistribute stresses. Typical procedures include metatarsal head and distal ulnar resectional arthroplasty (joint reformation), arthrodesis (joint fusion), synovectomy (removal of destructive, proliferating synovium, usually in the wrists, fingers, and knees), and osteotomy (cutting of bone or excision of a wedge of bone). Tendon transfers may prevent deformities or relieve contractures. In advanced disease, the patient may need joint reconstruction or total joint arthroplasty.

Quick quiz

1. The CDC recommends that HIV antibody testing take place:
    A. immediately after a possible exposure.
    B. 3 months after a possible exposure.
    C. 6 months after a possible exposure.

Answer: B. HIV antibodies usually can't be detected in the blood until 3 months after exposure.

2. The immunoglobulin responsible for the hypersensitivity reaction is:
    A. IgA.
    B. IgG.
    C. IgE.

Answer: C. IgE is the immunoglobulin involved in the immediate hypersensitivity reactions of humoral immunity.

3. The type of rhinitis caused by excessive use of nasal sprays or drops is called:
    A. allergic rhinitis.
    B. rhinitis medicamentosa.
    C. infectious rhinitis.

Answer: B. Rhinitis medicamentosa can be resolved by discontinuing these medications.

4. The most common anaphylaxis-causing antigen is:
    A. bee sting.
    B. penicillin.
    C. shellfish.

Answer: B. Penicillin is the most common anaphylaxis-causing antigen because of its systemic effects on the body.

5. Lupus erythematosus is characterized by:
    A. ulnar rotation of the hand.
    B. butterfly rash.
    C. flulike syndrome.

Answer: B. About half of all patients get the characteristic butterfly rash over their noses and cheeks.

6. Common complaints associated with rheumatoid arthritis include:
- A. painful joints, weak muscles, and peripheral neuropathy.
- B. dizziness, dyspnea, and tachycardia.
- C. anemia, Raynaud's phenomenon, and amenorrhea.

Answer: A. These symptoms often develop in the fingers and extend bilaterally to the wrists, elbows, knees, and ankles.

Scoring

☆☆☆ If you answered all six items correctly, bravo! Your performance is immune to criticism.

☆☆ If you answered four or five correctly, congratulations! Clearly, the quick quiz provided a stimulus that evoked an effective and appropriate response.

☆ If you answered fewer than four correctly, that's okay. We still have plenty of body systems to cover, so now is no time to become hypersensitive.

⑤

Genetics

Just the facts

In this chapter you'll learn:

♦ the role of genes and chromosomes and how cells divide

♦ types of genetic abnormalities and how they're inherited

♦ the causes, pathophysiology, diagnostic tests, and treatments for several common genetic disorders.

Understanding genetics

Genetics is the study of heredity, the passing of traits from parents to their children. Physical traits such as eye color are inherited, as are biochemical and physiologic traits, including the tendency to develop certain diseases.

Transmitting an inheritance

Inherited traits are transmitted from parents to offspring in germ cells, or gametes. Human gametes are eggs, or ova, and sperm. A person's inheritance is determined at fertilization, when ovum and sperm are united.

In the nucleus of each germ cell are structures called chromosomes. Each chromosome contains a strand of genetic material called deoxyribonucleic acid (DNA). DNA is a long molecule that's made up of thousands of segments called genes. Each of the traits that a person inherits — from blood type to toe shape and a myriad of others in between — is carried by their genes.

Count on chromosomes

A human ovum contains 23 chromosomes. A sperm also contains 23 chromosomes; each similar in size and shape to a chromosome in the ovum. When ovum and sperm unite, the corresponding chromosomes pair up. The result is a fertilized cell with 46 chromosomes (23 pairs) in the nucleus.

The fertilized cell soon undergoes cell division (mitosis). In mitosis, each of the 46 chromosomes produces an exact duplicate of itself. The cell then divides and each new cell receives one set of 46 chromosomes. Each of the two cells that result likewise divide, and so on, eventually forming a many-celled human body. Therefore, each cell in a person's body (except the ova or sperm) contains 46 identical chromosomes.

In mitosis, one divided by itself equals two.

Gametes do it differently

The ova and sperm are formed by a different cell-division process called meiosis. In meiosis, each of the 23 pairs of chromosomes in a cell are split. The cell then divides and each new cell that results (an ovum or sperm) receives one set of 23 chromosomes.

Location, location, location

The location of a gene on a chromosome is called a locus. The locus of each gene is specific and doesn't vary from person to person. This allows each of the thousands of genes in an ovum to join the corresponding genes from a sperm when the chromosomes pair up at fertilization.

Pass it on

A person receives one set of chromosomes and genes from each parent. This means there are two genes for each trait that a person inherits. One gene may be more influential than the other in developing a specific trait. The more powerful gene is said to be dominant and the less influential gene is recessive.

For example, a child may receive a gene for brown eyes from one parent and a gene for blue eyes from the other parent. The gene for brown eyes is dominant; the gene for blue eyes is recessive. The dominant gene is most likely to be expressed. Therefore, the child is most likely to have brown eyes.

All about alleles

A variation of a gene and the trait it controls — such as brown, green, or blue eye color — is called an allele. When two different alleles are inherited, they're said to be heterozygous. When the alleles are identical, they're termed homozygous.

A dominant allele may be expressed when it's carried by only one of the chromosomes in a pair. A recessive allele is incapable of expression unless identical alleles are carried by both chromosomes in a pair.

Let's talk about sex

Of the 23 pairs of chromosomes in each living human cell, 22 are *not* involved in controlling a person's sex; they're called autosomes.

The two sex chromosomes of the 23rd pair determine a person's sex. In a female, both chromosomes are relatively large and each is designated by the letter X; females have two X chromosomes. In a male, one sex chromosome is an X chromosome and one is a smaller chromosome, designated by the letter Y.

Each gamete produced by a male contains either an X or a Y chromosome. When a sperm with an X chromosome fertilizes an ovum, the offspring is female. When a sperm with a Y chromosome fertilizes an ovum, the offspring is male.

An explanation of mutation

A mutation is a permanent change in genetic material. When a gene mutates, it produces a trait that's different from its original trait. The mutant gene is transmitted during reproduction. Some mutations cause serious or deadly defects that occur in three different forms:
- single-gene disorders
- chromosomal disorders
- multifactorial disorders.

Single-gene disorders

Single-gene disorders are inherited in clearly identifiable patterns. Two important inheritance patterns are called autosomal dominant and autosomal recessive. Because there are 22 pairs of autosomes and only 1 pair of sex

Most genetic disorders are caused by defects of the autosomes.

chromosomes, most hereditary disorders are caused by autosomal defects.

In a third inheritance pattern, sex-linked inheritance, single-gene disorders are passed through the X chromosome.

Also, keep the definitions of the terms dominant and recessive in mind. Dominant genes produce abnormal traits in offspring even if only one parent has the gene; recessive genes don't produce abnormal traits unless both parents have the gene and pass them to their offspring.

Autosomal dominant inheritance

The autosomal dominant inheritance pattern has the following characteristics:
- Male and female offspring are affected equally.
- One of the parents is also usually affected.
- If one parent is affected, half of the children will be affected.
- If both parents are affected, all of their children will be affected.

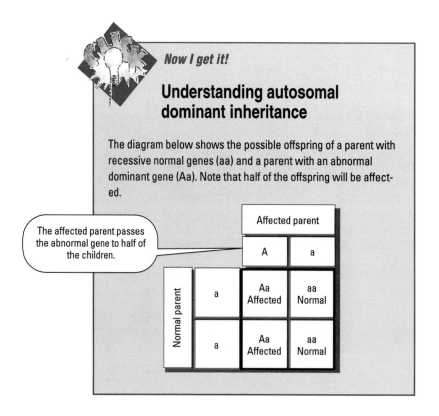

Now I get it!

Understanding autosomal dominant inheritance

The diagram below shows the possible offspring of a parent with recessive normal genes (aa) and a parent with an abnormal dominant gene (Aa). Note that half of the offspring will be affected.

The affected parent passes the abnormal gene to half of the children.

		Affected parent	
		A	a
Normal parent	a	Aa Affected	aa Normal
	a	Aa Affected	aa Normal

Marfan's syndrome is an example of an autosomal dominant disorder. (See *Understanding autosomal dominant inheritance.*)

Autosomal recessive inheritance

The autosomal recessive inheritance pattern has the following characteristics:
• Male and female offspring are affected equally.
• If both parents are unaffected but heterozygous for the trait (carriers), each of their offspring has a one in four chance of being affected.
• If both parents are affected, all of their offspring will be affected.
• If one parent is affected and the other is unaffected (non-carrier), all of their offspring will be unaffected but will carry the defective gene.
• If one parent is affected and the other is a carrier, half of their offspring will be affected (See *Understanding autosomal recessive inheritance.*)

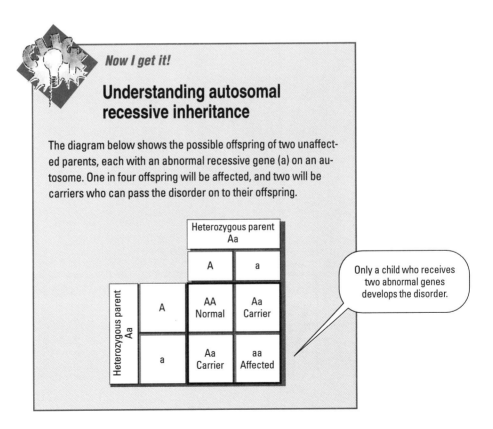

Now I get it!

Understanding autosomal recessive inheritance

The diagram below shows the possible offspring of two unaffected parents, each with an abnormal recessive gene (a) on an autosome. One in four offspring will be affected, and two will be carriers who can pass the disorder on to their offspring.

		Heterozygous parent Aa	
		A	a
Heterozygous parent Aa	A	AA Normal	Aa Carrier
	a	Aa Carrier	aa Affected

Only a child who receives two abnormal genes develops the disorder.

In many cases, no evidence of the trait appears in past generations. If you enjoy clinical jargon, you can say the patient has a "negative family history." Cystic fibrosis is an example of this type of inheritance disorder.

Think X-linked

Some genetic disorders are caused by genes located on the sex chromosomes and are termed sex-linked. Since the Y chromosome is not known to carry disease-causing genes, the terms X-linked and sex-linked are interchangeable.

Because females receive two X chromosomes (one from the father and one from the mother), they can be homozygous for a disease allele, homozygous for a normal allele, or heterozygous.

Because males have only one X chromosome, a single X-linked recessive gene can cause disease in a male. In comparison, a female needs two copies of the diseased gene. Therefore, males are more commonly affected by X-linked recessive diseases than females. (See *Understanding X-linked dominant inheritance* and *Understanding X-linked recessive inheritance.*)

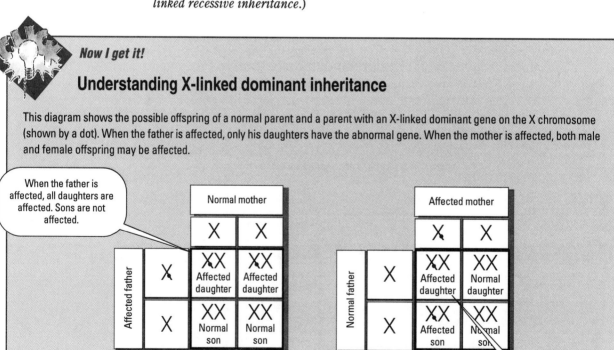

Now I get it!

Understanding X-linked dominant inheritance

This diagram shows the possible offspring of a normal parent and a parent with an X-linked dominant gene on the X chromosome (shown by a dot). When the father is affected, only his daughters have the abnormal gene. When the mother is affected, both male and female offspring may be affected.

When the father is affected, all daughters are affected. Sons are not affected.

When the mother is affected, both daughters and sons may also be affected.

X-linked dominant inheritance

Characteristics of X-linked dominant inheritance include the following:
- A person with the abnormal trait must have one affected parent.
- If a father has an X-linked dominant disorder, all of his daughters and none of his sons will be affected.
- If a mother has an X-linked dominant disorder, there is a 50% chance that each of her children will be affected.
- Evidence of the inherited trait should appear in the family history.

X-linked recessive inheritance

Here are the basic facts about the X-linked recessive inheritance pattern:
- In most cases, affected people are males with unaffected parents. In rare cases, the father is affected and the mother is a carrier.
- All of the daughters of an affected male will be carriers.

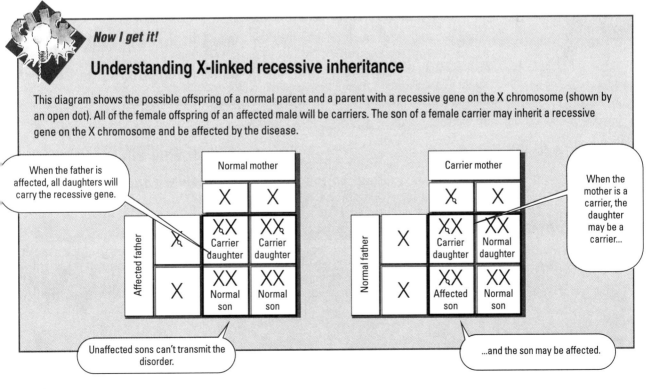

Now I get it!

Understanding X-linked recessive inheritance

This diagram shows the possible offspring of a normal parent and a parent with a recessive gene on the X chromosome (shown by an open dot). All of the female offspring of an affected male will be carriers. The son of a female carrier may inherit a recessive gene on the X chromosome and be affected by the disease.

When the father is affected, all daughters will carry the recessive gene.

When the mother is a carrier, the daughter may be a carrier...

Unaffected sons can't transmit the disorder.

...and the son may be affected.

- Sons of an affected male are unaffected. Unaffected sons can't transmit the disorder.
- The unaffected male children of a female carrier don't transmit the disorder.

Hemophilia is an example of an X-linked recessive inheritance disorder.

Chromosomal disorders

Disorders may also be caused by chromosomal aberrations — deviations in either the structure or the number of chromosomes. Deviations involve the loss, addition, rearrangement, or exchange of genes. If the remaining genetic material is sufficient to maintain life, an endless variety of clinical manifestations may occur.

A difficult separation

During cell division, chromosomes normally separate in a process called disjunction. Failure to do so — called nondisjunction — causes an unequal distribution of chromosomes between the two resulting cells. Gain or loss of chromosomes is usually due to nondisjunction of autosomes or sex chromosomes during meiosis.

One less or one extra

When chromosomes are gained or lost, the name of the affected cell contains the suffix "-somy." A cell that contains one less than the normal number of chromosomes is called a monosomy; this cell is nonviable. A cell that contains one extra chromosome is called a trisomy. (See *Understanding nondisjunction of chromosomes.*)

Mixed results

Nondisjunction may occur during very early cell divisions after fertilization and may or may not involve all the resulting cells. A mixture of both trisomic and normal cells results in *mosaicism* — the presence of two or more cell lines with different chromosomes. The effect on the offspring depends on the percentage of normal cells.

The incidence of nondisjunction increases with parental age. Miscarriages can also result from chromosomal defects. Fertilization of a defective ova by a defective sperm usually doesn't occur.

Now I get it!

Understanding nondisjunction of chromosomes

This illustration shows normal disjunction and nondisjunction of an ovum. The result is one trisomic cell and one monosomic (nonviable) cell.

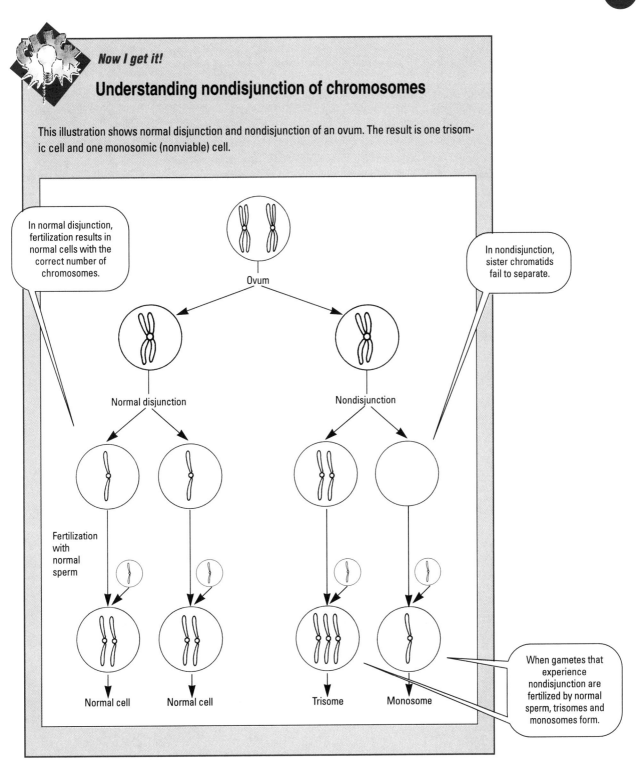

Shifty chromosomes

Another defect, translocation, is the shifting or moving of a chromosome. It occurs when chromosomes break and rejoin in an abnormal arrangement. The cells still have a normal amount of genetic material, so there are no visible abnormalities. However, the children of parents whose chromosomes are translocated may have serious genetic defects, such as monosomies or trisomies. Parental age doesn't seem to be a factor.

Multifactorial disorders

Disorders caused by both genetic and environmental factors are classified as multifactorial. Examples are cleft lip, cleft palate, and myelomeningocele (spina bifida with a portion of the spinal cord and membranes protruding). Environmental factors that contribute include the following:
• maternal age
• use of chemicals (such as drugs, alcohol, or hormones) by mother or father
• maternal infections during pregnancy or existing diseases in the mother
• maternal or paternal exposure to radiation
• maternal nutritional factors
• general maternal or paternal health
• other factors, including high altitude, maternal smoking, maternal-fetal blood incompatibility, and poor-quality prenatal care.

Genetic disorders

This section describes six single-gene disorders, including Marfan's syndrome, cystic fibrosis, phenylketonuria, sickle cell anemia, Tay-Sachs disease, and hemophilia. You'll also find information on Down syndrome, a chromosomal disorder, and cleft lip and cleft palate, multifactorial disorders.

Cleft lip and cleft palate

Cleft lip and cleft palate deformities occur in about 1 in 800 births. Cleft lip with or without cleft palate is more common in males, and cleft palate alone is more common in females. Nutritional intake is affected by an abnormal lip and palate. Furthermore, children with cleft palates often have hearing problems caused by middle ear damage or infection. (See *Treating cleft lip and cleft palate.*)

Pathophysiology

A multifactorial genetic disorder, cleft lip and cleft palate originate in the second month of gestation when the front and sides of the face and the shelves of the palate fuse imperfectly. These deformities fall into four categories:
• clefts of the lip (unilateral or bilateral)
• clefts of the palate (along the midline)
• unilateral clefts of the lip, alveolus (gum pad), and palate
• bilateral clefts of the lip, alveolus, and palate.

What to look for

The deformity may range from a simple notch to a complete cleft that extends from the lip through the floor of the nostril on either side of the midline. A complete cleft palate may involve the soft palate, the bones of the maxilla (upper jawbone), and the cavity on one or both sides of the premaxilla (front of the upper jawbone).

In a double cleft, the most severe of all cleft deformities, the cleft runs from the soft palate forward to either side of the nose, separating the maxilla and the premaxilla into free-moving segments. The tongue and other muscles can displace these segments, enlarging the cleft. (See *Types of cleft lip and cleft palate*, page 106.)

Another cleft disorder, Pierre Robin syndrome, occurs when abnormal smallness of the jaw (micrognathia) and downward dropping of the tongue (glossoptosis) accompany cleft palate.

Battling illness

Treating cleft lip and cleft palate

Cleft deformities must be treated with a combination of speech therapy and surgery. The timing of surgery varies. When a wide horseshoe defect makes surgery impossible, a contoured speech bulb (attached to the posterior of a denture to occlude the nasopharynx) helps the child develop intelligible speech.

Infants with Pierre Robin syndrome should never be placed on their backs because the tongue can fall back and obstruct the airway. Special consideration must be made when feeding infants with cleft lip or cleft palate. They usually eat better with a large, soft nipple with large holes, known as a lamb's nipple.

Because the palate is essential to speech, structural changes can permanently affect speech, even after surgical repair.

What tests tell you

Not much. Unfortunately, there are no diagnostic tests for cleft lip and cleft palate.

Types of cleft lip and cleft palate

The illustrations below show four variations of cleft lip and cleft palate.

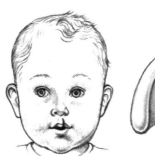 

Notch in vermilion border (junction of the lip and surrounding skin)

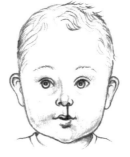 

Unilateral cleft lip and palate

 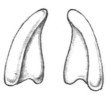

Bilateral cleft lip and palate

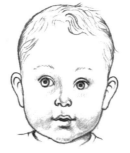 

Cleft palate

Cystic fibrosis

A chronic, progressive, inherited disease, cystic fibrosis is the most common fatal genetic disease in white children. When both parents carry the recessive gene, they have a 25% chance of transmitting the disease with each pregnancy.

The incidence of cystic fibrosis is highest in people of northern European ancestry (about 1 in 1,200 births). It's less common in Blacks (1 in 17,000 births), Native Americans, and Asians. It occurs with equal frequency in both sexes.

Pathophysiology

Cystic fibrosis is inherited as an automsomal recessive trait. One implicated gene is located on chromosome 7. Research now suggests that there may be as many as 300 genes that code for cystic fibrosis.

Most cases arise from a mutation that affects the genetic coding for a single amino acid, resulting in a protein that doesn't function properly. The abnormal protein — the cystic fibrosis transmembrane conductor regulator — resembles other transmembrane transport proteins. However, it lacks a phenylalanine (an essential amino acid) that is usually produced by normal genes. This abnormal protein may interfere with chloride transport by preventing adenosine triphosphate from binding to the protein or by interfering with activation by protein kinase. The lack of an essential amino acid leads to dehydration and mucosal thickening in the respiratory and intestinal tracts.

What to look for

Cystic fibrosis increases the viscosity of bronchial, pancreatic, and other mucus gland secretions, obstructing glandular ducts.

Respiratory effects

The accumulation of thick, tenacious secretions in the bronchioles and alveoli causes the following respiratory changes:
- frequent upper respiratory tract infections
- dyspnea
- paroxysmal (sudden) cough
- frequent bouts of pneumonia.

Respiratory effects may eventually lead to collapsed lungs (atelectasis) or emphysema.

Gastrointestinal effects

Cystic fibrosis also affects the intestines, pancreas, and liver in the following ways:
• diabetes and pancreatitis may result from insult to the pancreas.
• hepatic failure and cholecystitis may result from blockage of pancreatic ducts
• deficiency of the enzymes trypsin, amylase, and lipase (also a result of obstruction of pancreatic ducts) may prevent the conversion and absorption of fat and protein in the intestinal tract, which interferes with food digestion and absorption of the fat-soluble vitamins A, D, E, and K.

Reproductive effects

Males may experience lack of sperm in the semen (azoospermia); females may experience secondary amenorrhea and increased mucus in the reproductive tracts that blocks passage of the ovum.

Effects in children

A child with cystic fibrosis may have a barrel chest, cyanosis, clubbing of the fingers and toes, and a distended abdomen. He may cough and expectorate tenacious, yellow-green sputum and have wheezy respirations and crackles on auscultation.

Although scientists have not yet found a cure for this disease, medical research has greatly increased life expectancy. (See *Treating cystic fibrosis*.)

What tests tell you

According to the Cystic Fibrosis Foundation, a definitive diagnosis of cystic fibrosis requires:
• two clearly positive sweat tests, using pilocarpine solution (a sweat inducer)
• the presence of an obstructive pulmonary disease
• confirmed pancreatic insufficiency or failure to thrive
• a family history of cystic fibrosis.
    Tests that support the diagnosis include:
• chest X-rays that show early signs of lung obstruction

Battling illness

Treating cystic fibrosis

Cystic fibrosis has no cure, so treatment aims to help the patient lead as normal a life as possible. Specific treatments depend on the organ systems involved.

Salt supplements are used to combat electrolyte loss through sweat, and oral pancreatic enzymes taken with meals and snacks offset deficiencies. Broad-spectrum antibiotics and oxygen therapy are used as needed. To manage pulmonary dysfunction, chest physiotherapy, including postural drainage and chest percussion over all lobes, is usually performed several times a day.

- stool specimen analysis that shows the absence of trypsin, suggesting pancreatic insufficiency
- DNA testing to detect the abnormal protein that causes cystic fibrosis. This test can also be used to detect carriers and for prenatal diagnosis in families with an affected child.

Down syndrome

The first disorder attributed to a chromosomal aberration, Down syndrome produces mental retardation, characteristic facial features, and distinctive physical abnormalities. It's also associated with heart defects and other congenital disorders.

Life expectancy and quality of life for patients with Down syndrome have increased significantly because of improved treatment of related complications and better developmental education programs. Nevertheless, up to 44% of patients with congenital heart disease die before they're 1 year old.

A statement of stats

Overall, Down syndrome occurs in 1 per 800 to 1,000 births, but the incidence increases with maternal age. For instance, at age 20, a mother has about 1 chance in 2,000 of having a child with Down syndrome; by age 49, she has 1 chance in 12. Although women over age 35 account for fewer than 8% of all births, they bear 20% of all children with Down syndrome. The incidence of Down syndrome has decreased with the widespread use of prenatal testing such as amniocentesis.

Pathophysiology

Also called trisomy 21, Down syndrome is caused by an aberration in which chromosome 21 has three copies instead of two because of nondisjunction. (See *Understanding Down syndrome,* page 110.) Although the incidence of nondisjunction increases with maternal age, the extra chromosome originates from the mother only 80% of the time. About 4% of the time, Down syndrome results from translocation and infusion of the long arm of chromosome 21 and 14. This phenomenon is known as Robertsonian translocation.

Studies suggest that in some cases, the abnormality results from deterioration of the oocyte (primitive egg), which can be caused by age or the cumulative effects of environmental factors, such as radiation and viruses.

Patients with Down syndrome exhibit a karyotype (chromosomal arrangement) of 47 chromosomes instead of the normal 46. Mortality rates are high in both the fetus and neonate, and early death usually results from complications of associated heart defects. If the patient survives to adulthood, premature senile dementia, similar to Alzheimer's disease, usually occurs in the fourth decade. Patients are also prone to leukemia, acute and chronic infections, diabetes mellitus, and thyroid disorders.

Now I get it!

Understanding Down syndrome

Human chromosomes are arranged in seven groups, designated by the letters A through G. The illustration below shows the arrangement of chromosomes (karyotype) in a normal person and one with Down syndrome. In Down syndrome, an extra chromosome occurs at locus 21.

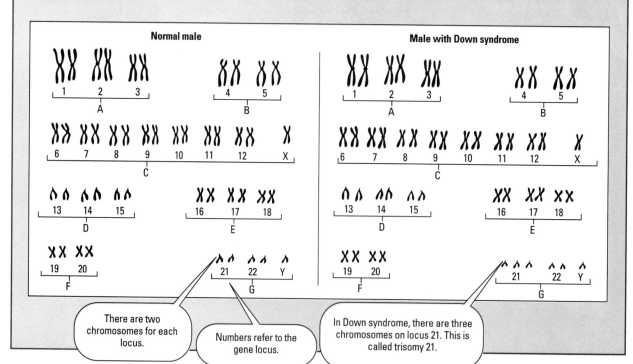

What to look for

The physical signs of Down syndrome are apparent at birth. The infant is lethargic, with the following craniofacial anomalies:

- slanting, almond-shaped eyes
- protruding tongue
- small, open mouth
- single, transverse crease on the palm (called a simian crease)
- small white spots on the iris (Brushfield's spots)
- small skull
- flat nose bridge
- flattened face
- small ears
- short neck with excess skin.

Ongoing inspection

Characteristic findings include dry, sensitive skin with decreased elasticity, umbilical hernia, short stature, and short extremities. The hands and feet are broad, flat, and squarish, and the feet have a wide space between the first and second toes. Decreased muscle tone in limbs impairs reflex development.

Many infants have congenital heart disease, duodenal obstruction, clubfoot, imperforate anus, cleft lip and cleft palate, Hirschsprung's disease (congenital colon enlargement), myelomeningocele, and pelvic bone abnormalities.

Effects in later years

As the child grows, dental development is slow, with abnormal or absent teeth. Strabismus and, occasionally, cataracts occur. The genitalia develop poorly, and puberty is delayed; however, the female with Down syndrome may menstruate and be fertile. The male is infertile with low serum testosterone levels and, in many cases, undescended testes.

Patients may have an IQ as low as 30, although some are educable with IQs approaching 80. Social performance is usually beyond that expected for their mental age. However, intellectual development slows with age. (See *Treating Down syndrome*.)

Battling illness

Treating Down syndrome

Treatment for Down syndrome includes surgery to correct cardiac defects and other congenital abnormalities, antibiotic therapy for recurrent infections, and thyroid hormone replacement for hypothyroidism. Plastic surgery is performed to correct protruding tongue, cleft lip, and cleft palate. These measures improve the patient's appearance and speech and reduce susceptibility to cavities and orthodontic problems.

Special education programs, mandated in most communities, promote self-esteem and help children achieve their potential.

What tests tell you

A karyotype showing the chromosomal abnormality confirms the diagnosis of Down syndrome. The following tests reveal Down syndrome before birth:

• Prenatal ultrasonography suggests Down syndrome if a duodenal obstruction or an atrioventricular canal defect is seen.

• Blood tests show reduced alpha-fetoprotein levels indicative of Down syndrome.

• Amniocentesis or chorionic villi sampling are recommended for pregnant women past age 35 and for pregnant women of any age when they or the father are known carriers of a translocated chromosome.

Hemophilia

A hereditary bleeding disorder, hemophilia affects only males. It's the most common X-linked genetic disease, occurring in about 125 in 1 million male births.

Two types of hemophilia may occur:

• hemophilia A, or classic hemophilia, which affects more than 80% of all hemophiliacs

• hemophilia B, or Christmas disease, which affects 15% of all hemophiliacs.

The severity and prognosis of the disease vary with the degree of deficiency or nonfunction and the site of bleeding. Advances in treatment have greatly improved the prognosis, and many patients live normal life spans.

Pathophysiology

Hemophilia A is caused by deficiency or nonfunction of factor VIII; hemophilia B is caused by deficiency or nonfunction of factor IX. Both are inherited as X-linked recessive traits. This means that female carriers have a 50% chance of transmitting the gene to each daughter, making her a carrier, and a 50% chance of transmitting the gene to each son, who would be born with hemophilia.

Hemophilia produces mild to severe abnormal bleeding. After a platelet plug develops at a bleeding site, the lack of clotting factor prevents a stable fibrin clot from forming. Although hemorrhaging doesn't usually happen immediately, delayed bleeding is common.

Effects of increased bleeding

Bleeding into joints and muscles causes pain, swelling, extreme tenderness, limited range of motion and, sometimes, permanent deformity.

Bleeding near peripheral nerves may cause peripheral neuropathies, pain, paresthesia, and muscle atrophy.

If bleeding impairs blood flow through a major vessel, it can cause ischemia and gangrene.

What to look for

Signs and symptoms vary, depending on the severity of the patient's condition:

• Severe hemophilia causes spontaneous bleeding. Often, the first sign is excessive bleeding after circumcision. Later, spontaneous or severe bleeding after minor trauma may produce large subcutaneous and deep intramuscular hematomas.

• Moderate hemophilia causes symptoms similar to severe hemophilia, but spontaneous bleeding occurs only occasionally.

• Mild hemophilia often goes undiagnosed until adulthood because the patient doesn't bleed spontaneously or after minor trauma. However, major trauma or surgery can cause prolonged bleeding. Blood may ooze slowly or stop and start for up to 8 days after surgery. Patients with mild hemophilia have the best prognosis.

Warning! Warning! Warning!

Patients with undiagnosed hemophilia usually complain of pain and swelling in a weight-bearing joint, such as the hip, knee, or ankle. The patient history may reveal prolonged bleeding after surgery, dental extractions, or trauma. The patient may also have signs and symptoms of internal bleeding, such as abdominal, chest, or flank pain; hematuria (bloody urine) or hematemesis (bloody vomit); and tarry stools. Inspection may reveal hematomas on the extremities, torso, or both and joint swelling if bleeding has occurred there.

The perils of poor tissue perfusion

Signs and symptoms of decreased tissue perfusion include restlessness, anxiety, confusion, pallor, cool and clammy skin, chest pain, decreased urine output, hypotension, and tachycardia. (See *Treating hemophilia*, page 114.)

Battling illness

Treating hemophilia

Hemophilia isn't curable, but treatment can prevent crippling deformities and prolong life. Increasing plasma levels of deficient clotting factors helps prevent disabling deformities caused by repeated bleeding into muscles, joints, and organs.

Clotting factor

In hemophilia A, administration of cryoprecipitate antihemophilic factor (AHF) and lyophilized AHF (AHF that's been frozen and dehydrated under high vacuum) can result in normal hemostasis (arrest of bleeding). In hemophilia B, administration of recombinant factor VIII and purified factor IX promote hemostasis. Doses should be large enough to raise clotting factor levels. Clotting factor levels must be kept within the desired range until the wound heals. Fresh-frozen plasma can also be given, but it has drawbacks, such as the potential for volume overload and a transfusion reaction.

A hemophiliac who undergoes surgery needs factor replacement before and after the procedure. This may be necessary even for minor surgery such as a dental extraction.

Other treatments

Joint pain may be controlled with an analgesic. However, don't give the drug by I.M. injection because hematomas may form at the site. Also, never give aspirin or aspirin-containing drugs because they decrease platelet adherence and may increase bleeding.

To help avoid injury, young children should wear clothing with padded patches on the knees and elbows. Older children should avoid contact sports such as football. Warn parents to notify the doctor immediately after even a minor injury, especially to the head, neck, or abdomen. Early recognition is the key to stopping bleeding.

What tests tell you

Specific coagulation factor assays can diagnose the type and severity of hemophilia. A positive family history can also aid in diagnosis.

The following tests help diagnose hemophilia A:
- Factor VIII assay reveals 0% to 25% of normal factor VIII.
- Activated partial thromboplastin time is prolonged.
- Platelet count and function, bleeding time, and prothrombin time are normal.

The following tests help diagnose hemophilia B:
- Factor IX assay shows deficiency.
- Baseline coagulation result is similar to that of hemophilia A, with normal factor VIII.

Marfan's syndrome

Patients with Marfan's syndrome exhibit three characteristic abnormalities: skeletal malformation, displaced ocular lenses (ectopia lentis), and aortic aneurysm, usually at the base of the aorta. Signs and symptoms range from mild to severe. The disorder affects men and women equally, with an incidence of about 1 in 10,000 people.

Pathophysiology

This genetic disorder has been mapped to a specific chromosome location. It's inherited as an autosomal dominant trait. Patients are heterozygous for the mutation. This means that patients have one gene that codes for the mutation and one that doesn't. Patients with Marfan's syndrome may have other chromosomal abnormalities.

Marfan's syndrome is caused by a mutation in a single allele of a gene located on chromosome 15 (the 15th of the 22 autosomes in the human cell). Mutations occur in a gene that codes for fibrillin, a component of connective tissue. These small fibers are abundant in the large blood vessels and the suspensory ligaments of the ocular lenses. The exact function of fibrillin, however, is unknown.

What to look for

Genetic defects may result in the functional and structural changes described below.

Skeletal effects

Skeletal malformations include:
• increased height (patients are usually taller than family members)
• unusually long extremities
• arachnodactyly — a spiderlike appearance of the hands and fingers
• chest depression (pectus excavatum)
• chest protrusion (pectus carinatum)
• chest asymmetry
• scoliosis and kyphosis
• arched palate
• joint hypermobility.

Ocular effects

Lens displacement usually isn't progressive, but it may contribute to cataract formation. An elongated ocular globe causes nearsightedness in most patients. Retinal detachments may also develop, as well as retinal tears. However, most patients have adequate vision with corrective lenses.

Cardiac effects

Cardiac abnormalities may occur in the valves and in the aorta.

Valvular abnormalities result from anatomic defects, such as redundancy to the leaflets, stretching of the chordae tendineae, and dilation of the valvulae annulus. Mitral valve prolapse (dropping down of the cusps of the mitral valve into the left atrium during systole) develops early in life. It can progress to severe mitral valve regurgitation (backflow of blood from the left ventricle into the left atrium).

Dilation of the aortic root and ascending aorta may cause aortic regurgitation (backflow of blood into the left ventricle), dissection (separation of the layers of the aortic wall), and rupture. In adults, dilation may be accelerated by physical and emotional stress as well as by pregnancy.

Miscellaneous effects

Marfan's syndrome may also cause the following:
• thin body build with little subcutaneous fat
• striae over the shoulders and buttocks
• spontaneous pneumothorax
• inguinal and incisional hernias
• dilation of the dural sac.

No established treatment exists for Marfan's syndrome. (See *Treating Marfan's syndrome.*)

What tests tell you

To date, no tests exist to diagnose Marfan's syndrome. Diagnostic tests based on detection of fibrillin defects in cultured skin fibroblasts or DNA analysis of the gene may be available in the near future.

Currently, a diagnosis may be made if the patient and other family members have dislocated lenses, aortic dilation, kyphoscoliosis or other chest deformities, and long, thin extremities.

Diagnosis may also be made if ectopia lentis and an aneurysm of the ascending aorta occur without other symptoms or familial tendency. Diagnosis of ectopia lentis is made by pupillary dilation and slit lamp examination. Cardiac problems may be discovered by an echocardiogram.

Battling illness

Treating Marfan's syndrome

No established treatment exists for Marfan's syndrome. Propranolol or other beta-adrenergic blocking agents may delay or prevent aortic dilation. Surgical replacement of the aorta, aortic valve, and mitral valve have also been successful in some patients.

Scoliosis is progressive and should be treated by mechanical bracing and physical therapy if the curvature is greater than 20 degrees. Surgery may be required if the curvature progresses beyond 45 degrees.

Phenylketonuria

Phenylketonuria (PKU) is an inborn error in metabolism of the amino acid phenylalanine. It causes high serum levels of phenylalanine and increased urine concentrations of phenylalanine and its by-products, and results in cerebral damage and mental retardation.

The disorder occurs once in about 14,000 births in the United States, and about 1 person in 60 is an asymptomatic carrier. PKU has a low incidence among Blacks and Ashkenazic Jews and a high incidence among people of Irish and Scottish descent.

Pathophysiology

PKU is transmitted through an autosomal recessive gene. Patients with classic PKU have almost no activity of phenylalanine hydroxylase, an enzyme that helps convert phenylalanine to tyrosine. As a result, phenylalanine accumulates in the blood and urine, and tyrosine levels are low.

Usually, no abnormalities are apparent at birth, when blood phenylalanine levels are essentially normal. But levels begin to rise within a few days, and by the time they reach about 30 mg/dl, cerebral damage has begun. Irreversible damage is probably complete by age 2 or 3, although early detection and treatment can minimize it. Patients may have a family history of PKU. (See *Treating PKU,* page 118.)

What to look for

By age 4 months, the untreated child begins to show signs of arrested brain development, including mental retardation. Later, personality disturbances occur, such as schizoid and antisocial behavior and uncontrollable temper. About one-third of patients have seizures, which usually begin between ages 6 and 12 months. Many patients also show a precipitous decrease in IQ.

Other signs include macrocephaly; eczematous skin lesions or dry, rough skin; hyperactivity; irritability; purposeless, repetitive motions; and an awkward gait. You may also note a musty odor from skin and urine excretion of phenylacetic acid.

Battling illness

Treating PKU

Treating this genetic disorder requires a strict diet!

To prevent or minimize brain damage from phenylketonuria (PKU), phenylalanine blood levels are kept between 3 and 15 mg/dl by restricting dietary intake of phenylalanine. During the first month of life, a special, low-phenylalanine amino acid mixture is substituted for most dietary protein, supplemented with a small amount of natural foods.

Dietary restrictions

Patients must avoid bread, cheese, eggs, flour, meat, poultry, fish, nuts, milk, legumes, and aspartame (Nutrasweet). Even with this diet, slight central nervous system dysfunction may still occur. Frequent tests for urine phenylpyruvic acid and blood phenylalanine levels evaluate the diet's effectiveness. *Patients need careful monitoring because overzealous dietary restrictions can induce phenylalanine deficiency, causing lethargy, anorexia, anemia, skin rashes, diarrhea, and even death.*

What tests tell you

Several tests are used to diagnose PKU:
• The Guthrie screening test on a capillary blood sample reliably detects PKU and is required by most states at birth. However, because phenylalanine levels may be normal at birth, infants should be reevaluated 24 to 48 hours after they begin protein feedings.
• Fluorometric or chromatographic assays provide additional diagnostic information.
• Electroencephalography is abnormal in about 80% of affected children.
• DNA-based tests are used in prenatal diagnosis of PKU.

Sickle cell anemia

Sickle cell anemia is a congenital hematologic disease that causes impaired circulation, chronic ill health, and premature death. It's most common in tropical Africans and in people of African descent, but it also occurs in Puerto Rico, Turkey, India, the Middle East, and the Mediterranean. If two carriers have offspring, each child has a one in four chance of developing the disease. About 1 in 10 Blacks carries the abnormal gene, and 1 in every 400 to 600 Black children has sickle cell anemia.

Pathophysiology

Sickle cell anemia results from homozygous inheritance of an autosomal recessive gene that produces a defective hemoglobin molecule (hemoglobin S). Hemoglobin S causes red blood cells (RBCs) to become sickle-shaped. Sickle cell trait, which results from heterozygous inheritance of this gene, causes few or no symptoms. However, people with this trait are carriers who can pass the gene to their offspring. This defective gene may have persisted because the heterozygous sickle cell trait provides resistance to malaria.

Sickle cell anemia occurs as a result of a change in the gene that encodes the beta chain of hemoglobin. This change, in turn, causes a structural change in hemoglobin. A single amino acid change from glutamic acid to valine occurs in the sixth position of the beta-hemoglobin chain. (See *Distinguishing between sickled cells and normal cells.*)

Now I get it!

Distinguishing between sickled cells and normal cells

Normal red blood cells and sickled cells vary in more ways than shape. They also differ in life span, oxygen-carrying ability, and the rate at which they're destroyed.

Normal red blood cells

Sickled cells

- 120-day life span
- Hemoglobin has normal oxygen-carrying capacity
- 12 to 14 g of hemoglobin per milliliter
- Red blood cells destroyed at normal rate

- 30- to 40-day life span
- Hemoglobin has decreased oxygen-carrying capacity
- 6 to 9 g of hemoglobin per milliliter
- Red blood cells destroyed at accelerated rate

Rigid and rough

When hypoxia (oxygen deficiency) occurs, the hemoglobin S in the RBCs becomes insoluble. As a result, the blood cells get rigid and rough, forming an elongated crescent or sickle shape. Sickling can cause hemolysis (cell destruction).

Sickle cells also accumulate in capillaries and smaller blood vessels, causing occlusions and increasing blood viscosity. This impairs normal circulation, causing pain, tissue infarctions (tissue death), swelling, and anoxic changes that lead to further sickling and obstruction.

Triggering a crisis

Each patient with sickle cell anemia has a different hypoxic threshold and different factors that trigger a sickle cell crisis. Illness, cold exposure, stress, anything that induces an acidotic state, or any pathophysiologic process that pulls water out of sickle cells will precipitate a crisis in most patients. (See *Understanding sickle cell crisis.*)

What to look for

Signs and symptoms of sickle cell anemia usually don't develop until after age 6 months because large amounts of fetal hemoglobin protect infants until then. Fetal hemoglobin has a higher oxygen concentration and inhibits sickling.

The patient's history includes chronic fatigue, unexplained dyspnea or dyspnea on exertion, joint swelling, aching bones, severe localized and generalized pain, leg ulcers (especially on the ankles), and frequent infections. Men often develop priapism, or unexplained, painful erections.

In sickle cell crisis, symptoms include severe pain, hematuria, lethargy, irritability, and pale lips, tongue, palms, and nail beds.

Categories of crisis

Various types of sickle cell crises can occur. Here's how to determine what type the patient is having:
• Painful crisis is the hallmark of sickle cell anemia, appearing periodically after age 5. It results from blood vessel obstruction by rigid, tangled sickle cells, leading to tissue anoxia and, possibly, necrosis. It's characterized by severe abdominal, thoracic, muscle, or bone pain and,

possibly, increased jaundice, dark urine, or a low-grade fever.

• Aplastic crisis results from bone marrow depression and is associated with infection (usually viral). It's charac-

Now I get it!

Understanding sickle cell crisis

Sickle cell crisis is triggered by infection, cold exposure, high altitudes, overexertion, and other conditions that cause cellular oxygen deprivation. Here's what happens:

• Deoxygenated, sickle-shaped erythrocytes adhere to the capillary wall and to each other, blocking blood flow and causing cellular hypoxia.

• The crisis worsens as tissue hypoxia and acidic waste products cause more sickling and cell damage.

• With each new crisis, organs and tissues are destroyed and areas of tissue die slowly — especially in the spleen and kidneys.

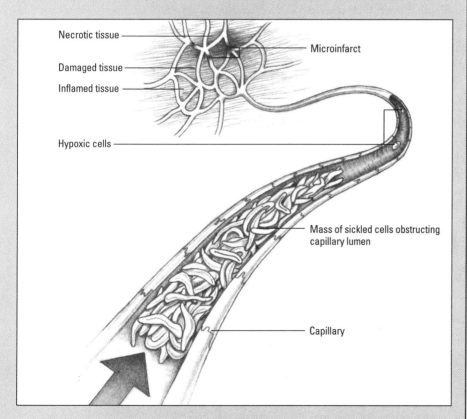

terized by pallor, lethargy, sleepiness, dyspnea, possible coma, markedly decreased bone marrow activity, and RBC hemolysis (destruction).

• Acute sequestration crisis (rare) occurs in infants between ages 8 months and 2 years and may cause sudden, massive entrapment of RBCs in the spleen and liver. Lethargy and pallor progress to hypovolemic shock and death if untreated.

• *Hemolytic crisis* (rare) usually affects patients who have glucose-6-phosphate dehydrogenase deficiency along with sickle cell anemia. It probably results from complications of sickle cell anemia, such as infection, rather than from the disease itself. In this crisis, degenerative changes cause liver congestion and enlargement, and chronic jaundice worsens.

In the past, many patients with this disease died in their early 20s, but now 50% to 70% live into their 40s and 50s. (See *Treating sickle cell anemia*.)

What tests tell you

A positive family history and typical clinical features suggest sickle cell anemia. The following tests are also done:

• Stained blood smear shows sickle cells, and hemoglobin electrophoresis shows hemoglobin S. These tests confirm the disease.

• Additional blood tests show low RBC counts, elevated white blood cell and platelet counts, decreased erythrocyte sedimentation rate, increased serum iron levels, decreased RBC survival, and reticulocytosis. Hemoglobin levels may be low or normal.

• Lateral chest X-ray detects the "Lincoln log" deformity in the vertebrae of many adults and some adolescents.

Tay-Sachs disease

Tay-Sachs disease results from a congenital enzyme deficiency. It causes progressive mental and motor deterioration and is always fatal, usually before age 5. It occurs in fewer than 100 infants born each year in the United States and strikes people of Ashkenazic Jewish ancestry about 100 times more often than the general population.

In this ethnic group, it occurs in about 1 in 3,600 live births. About 1 in 30 are heterozygous carriers of the de-

Battling illness

Treating sickle cell anemia

Treatment for sickle cell anemia aims to alleviate symptoms and prevent painful crisis. Analgesics, such as meperidine or morphine, and warm compresses may help relieve the pain from vasoocclusive crises. I.V. therapy may be needed to prevent dehydration and vessel occlusion. Patients should be encouraged to drink plenty of fluids. Iron and folic acid supplements may help prevent anemia. If the patient's hemoglobin level drops, blood transfusions may be necessary.

Complications
Complications resulting from the disease and from transfusion therapy may be reduced using certain vaccines, anti-infectives such as low dose penicillin, or chelating agents such as deferoxamine.

New developments
Researchers are seeking to increase the solubility of hemoglobin S. Studies indicate that certain drugs, such as hydroxyurea, may reduce frequency of crisis.

fective gene. If two carriers have children, each of their offspring has a 25% chance of having Tay-Sachs disease.

Pathophysiology

Tay-Sachs disease is an autosomal recessive disorder in which the enzyme hexosaminidase A is deficient. Hexosaminidase A is necessary for metabolism of gangliosides, water-soluble glycolipids found primarily in tissues of the central nervous system (CNS). Without hexosaminidase A, accumulating lipid pigments distend and progressively demyelinate (remove the protective myelin sheath) and destroy CNS cells.

What to look for

The child usually looks normal at birth; abnormal clinical signs appear between age 5 and 6 months. Progressive weakness of the neck, trunk, arm, and leg muscles prevents him from sitting up or lifting his head. He has trouble turning over, can't grasp objects, and has vision loss progressing to blindness. He also is easily startled by loud sounds.

By age 18 months, the patient may have seizures, generalized paralysis, and spasticity. His pupils are always dilated and, although blind, he may hold his eyes wide open and roll his eyeballs. Decerebrate rigidity and a complete vegetative state follow. Around age 2, the patient contracts recurrent bronchopneumonia, which is usually fatal before age 5.

Other findings include an enlarged head circumference, optic nerve atrophy, and a distinctive cherry-red spot on the retina.

Most children with Tay-Sachs disease are treated in long-term care facilities because they need round-the-clock care. (See *Treating Tay-Sachs disease*.)

What tests tell you

Serum analysis shows deficient hexosaminidase A in affected infants.

Diagnostic screening is essential for all couples of Ashkenazic Jewish ancestry and for others with a family history of the disease. A simple blood test can identify carriers.

Battling illness

Treating Tay-Sachs disease

Tay-Sachs disease has no known cure, so all treatment is supportive. Treatment includes tube feedings with nutritional supplements, suctioning and postural drainage to remove secretions, skin care to prevent pressure ulcers once the child becomes bedridden, and mild laxatives to relieve neurogenic constipation. Anticonvulsants usually don't prevent seizures.

Quick quiz

1. A patient with an aortic aneurysm at the base of the aorta, spiderlike extremities, and displaced ocular lenses probably has:

    A. an autosomal recessive inheritance disorder.

    B. Marfan's syndrome.

    C. Down syndrome.

Answer: B. Marfan's syndrome is an autosomal dominant disorder caused by a mutation in a single allele of the gene that codes for fibrillin.

2. Cystic fibrosis affects the exocrine glands, causing:

    A. increased viscosity of bronchial, pancreatic, and other mucus gland secretions, leading to atelectasis and emphysema.

    B. increased sperm in males.

    C. overhydration.

Answer: A. Cystic fibrosis also causes dehydration and absence of sperm in the semen.

3. If PKU isn't diagnosed and treated early, it can cause:

    A. cerebral damage and mental retardation.

    B. spontaneous hemorrhaging and hypovolemic shock.

    C. thick, tenacious secretions of bronchial ducts, atelectasis, and respiratory distress.

Answer: A. Untreated infants show signs of arrested brain development by age 4 months, but early detection and treatment can minimize cerebral damage.

Scoring

☆☆☆   If you answered all three items correctly, outasight! You display such a brilliant grasp of genetic concepts, we're thinking of having you cloned!

☆☆   If you answered fewer than three correctly, don't worry! You're actually a genetic genius and your performance on the quick quiz is just a mutation.

Endocrine system

Just the facts

In this chapter you'll learn:

♦ the role of the hypothalamus in regulating hormones

♦ how the adrenal glands, pancreas, pituitary gland, and thyroid gland function

♦ the pathophysiology, signs and symptoms, diagnostic tests, and treatments for several endocrine disorders.

Understanding the endocrine system

The endocrine system consists of:
• glands — specialized cell clusters
• hormones — chemical transmitters secreted by the glands in response to stimulation.

Together with the central nervous system (CNS), the endocrine system regulates and integrates the body's metabolic activities and maintains homeostasis.

The functions of the endocrine and nervous systems are interrelated.

At its heart, the hypothalamus

The hypothalamus is the integrative center for the endocrine and autonomic (involuntary) nervous systems. It helps control some endocrine glands by neural and hormonal pathways.

Neural pathways

Neural pathways connect the hypothalamus to the posterior pituitary gland. Neural stimulation of the posterior pituitary gland in turn causes the secretion of two effector hormones — antidiuretic hormone (ADH) and oxytocin.

When ADH is secreted, the body retains water. Oxytocin stimulates uterine contractions during labor and milk secretion in lactating women.

Please release me

The hypothalamus also exerts hormonal control at the anterior pituitary gland by releasing or inhibiting hormones. Hypothalamic hormones stimulate the posterior pituitary gland to release trophic (gland-stimulating) hormones:

☝ adrenocorticotropic hormone (ACTH), also called corticotropin

✌ thyroid-stimulating hormone (TSH)

🤟 luteinizing hormone (LH)

🖐 follicle-stimulating hormone (FSH).

Why are antidiuretic hormone and oxytocin called effector hormones?

The secretion of trophic hormones stimulates their respective target glands, such as the adrenal cortex, the thyroid gland, and the gonads.

Hypothalamic hormones also control the release of effector hormones from the pituitary gland. Examples are growth hormone (GH) and prolactin.

Getting feedback

A negative feedback system regulates the endocrine system. This system may be simple or complex. (See *The feedback loop*.)

Assessing dysfunction

A patient with a possible endocrine disorder needs careful assessment to identify the cause of the dysfunction. Dysfunction may result from defects:
• in the gland
• in the release of trophic or effector hormones
• in hormone transport
• of the target tissue.

Because they produce an effect when secreted.

What causes endocrine disorders?

Endocrine disorders may be caused by:
• hyposecretion or hypofunction (hormone deficiency)
• hyporesponsiveness (lack of hormone receptors in target cells), which has the same effect as hyposecretion
• hypersecretion or hyperfunction (hormone overproduction).

Other causes may include:
- inflammation
- tumor.

Now I get it!

The feedback loop

This diagram shows the negative feedback mechanism that helps regulate the endocrine system.

From simple...

Simple feedback occurs when the level of one substance regulates secretion of hormones (simple loop). For example, a low serum calcium level stimulates the parathyroid gland to release parathyroid hormone (PTH). PTH, in turn, promotes resorption of calcium. A high serum calcium level inhibits PTH secretion.

...to complex

When the hypothalamus receives feedback from target glands, the mechanism is more complicated (complex loop). Complex feedback occurs through an axis established between the hypothalamus, pituitary gland, and target organ. For example, secretion of the hypothalamic corticotropin-releasing hormone stimulates release of pituitary corticotropin, which in turn stimulates cortisol secretion by the adrenal gland (the target organ). A rise in serum cortisol levels inhibits corticotropin secretion by decreasing corticotropin-releasing hormone.

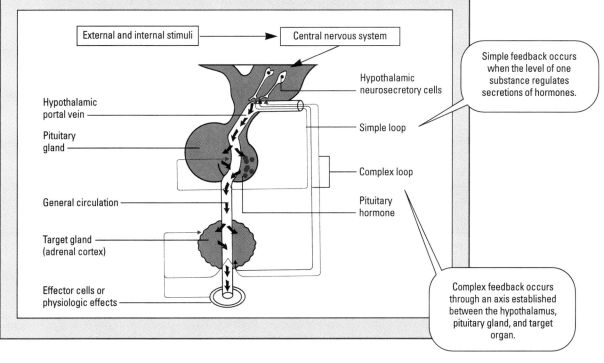

Hypofunction or hyperfunction

Hypofunction or hyperfunction may originate in the hypothalamus, the pituitary effector glands, or the target gland.

Hyporesponsiveness

In hyporesponsiveness, the cells of the target organ don't have appropriate receptors for a hormone. This means the effects of the hormone are not detected. Because the receptors don't detect the hormone, there is no feedback mechanism to turn the hormone off. Blood levels of the hormone are normal or high. Hyporesponsiveness causes the same clinical symptoms as hyposecretion.

Inflammation

Inflammation is usually chronic, often resulting in glandular hypofunction. However, it may be acute or subacute, as in thyroiditis.

Tumors

Tumors can occur within a gland, as in thyroid carcinoma. In addition, tumors occurring in other areas of the body can cause abnormal hormone production (ectopic hormone production). For example, certain lung tumors secrete ADH or parathyroid hormone.

Glands

Glands discussed below include the adrenal glands, the pancreas, the pituitary gland, and the thyroid gland.

> Once hormones are released into the circulatory system by the endocrine glands, they are distributed throughout the body.

Adrenal glands

The adrenal glands produce steroids, amines, epinephrine, and norepinephrine. Hyposecretion or hypersecretion of these substances causes a variety of disorders and complications that range from psychiatric and sexual problems to coma and death.

The adrenal cortex is the outer layer of the adrenal gland. It secretes three types of steroidal hormones:

 mineralocorticoids

 glucocorticoids

 androgens.

Aldosterone in action

Aldosterone is a mineralocorticoid. It regulates the reabsorption of sodium and the excretion of potassium by the kidneys.

Aldosterone may be involved with other hormones in the development of hypertension.

Cortisol in action

Cortisol, a glucocorticoid, carries out the following four important functions:
• stimulation of gluconeogenesis (formation of glycogen from noncarbohydrate sources), which occurs in the liver in response to low carbohydrate intake or starvation
• increased protein breakdown and free fatty acid mobilization
• suppression of immune response
• assistance with stress response.

Androgens in action

Androgens (sex steroid hormones) promote male traits, especially secondary sex characteristics. Examples of such characteristics are facial hair and a low-pitched voice.

A gland with a lot of nerve

The hormone epinephrine affects our response to stress.

The adrenal medulla is the inner portion of the adrenal gland. It's an aggregate of nerve tissue that produces the catecholamine hormones epinephrine and norepinephrine that cause vasoconstriction.

Epinephrine causes the response to physical or emotional stress called the fight-or-flight response. This response produces marked dilation of bronchioles and increased blood pressure, blood glucose level, and heart rate.

Pancreas

The pancreas produces glucagon and insulin. Glucagon is a hormone released when a person is in a fasting state. It stimulates the release of stored glucose from the liver to raise blood glucose levels.

Insulin is a hormone released during the postprandial (nourished) state. It aids glucose transport into the cells and promotes glucose storage. It also stimulates protein synthesis and enhances free fatty acid uptake and storage. Insulin deficiency or resistance causes diabetes mellitus.

The pancreas produces the hormone glucagon, which helps cells raise blood glucose levels.

Pituitary gland

The posterior pituitary gland secretes two effector hormones:

oxytocin, which stimulates uterine contractions during labor and causes the milk let-down reflex in lactating women

ADH, which controls the concentration of body fluids by altering the permeability of the distal renal tubules and collecting ducts in the kidneys, thereby conserving water.

ADH secretion depends on plasma osmolality (concentration), which is monitored by hypothalamic neurons. Hypovolemia and hypotension are the most powerful stimulators of ADH release. Other stimulators include pain, stress, trauma, nausea, and the use of morphine, tranquilizers, certain anesthetics, and positive-pressure breathing apparatus.

In addition to the trophic hormones, ACTH, TSH, LH, and FSH, the anterior pituitary secretes prolactin and GH. Prolactin stimulates milk secretion in lactating females. GH affects most body tissues. It triggers growth by increasing protein production and fat mobilization and decreasing carbohydrate use.

> Low blood pressure and low blood volume stimulate the release of ADH.

Thyroid gland

The thyroid gland secretes the iodine-containing hormones thyroxine (T_4) and triiodothyronine (T_3). Thyroid hormones are necessary for normal growth and development. They also act on many tissues by increasing metabolic activity and protein synthesis.

Diseases of the thyroid are caused by thyroid hormone overproduction or deficiency and gland inflammation and enlargement. Most patients have a good prognosis with treatment. Untreated, thyroid disease may progress to an emergency, as in thyroid crisis or storm. It can also cause irreversible disabilities such as vision loss.

> The thyroid secretes hormones that a person needs to grow and develop normally.

Endocrine disorders

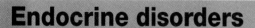

Two adrenal disorders are discussed below: Addison's disease and Cushing's syndrome.

Diabetes insipidus, a pituitary disorder of water metabolism, and diabetes mellitus, a pancreatic disorder, are also discussed.

The pathophysiology of three thyroid gland disorders — hyperthyroidism, hypothyroidism, and simple goiter — is also described.

> In this disorder, the adrenal glands don't secrete enough steroid hormones.

Addison's disease

Addison's disease is also called adrenal hypofunction or adrenal insufficiency. It's a relatively uncommon disorder that occurs in people of all ages and in both sexes.

Primary form

This form of Addison's disease originates within the adrenal glands. It's characterized by decreased mineralocorticoid, glucocorticoid, and androgen secretion.

Secondary form

The secondary form of Addison's disease is caused by a disorder outside the gland, such as a pituitary tumor with corticotropin deficiency. In secondary forms of the disorder, aldosterone secretion may be unaffected.

Adrenal crisis

Also called addisonian crisis, adrenal crisis is a critical deficiency of mineralocorticoids and glucocorticoids. It's a medical emergency that requires immediate, vigorous treatment. (See *Understanding adrenal crisis*, page 132.)

Pathophysiology

In primary Addison's disease, more than 90% of both adrenal glands are destroyed. Massive destruction usually results from an autoimmune process whereby circulating antibodies attack adrenal tissue. Destruction of the gland may also be idiopathic (with no known cause.)

Other causes of primary Addison's disease include:
- tuberculosis
- removal of both adrenal glands
- hemorrhage into the adrenal gland
- neoplasms
- infections, such as human immunodeficiency virus, histoplasmosis, meningococcal pneumonia, and cytomegalovirus.

Now I get it!

Understanding adrenal crisis

Adrenal crisis (acute adrenal insufficiency) is the most serious complication of Addison's disease. It may occur gradually or suddenly.

Who's at risk
This potentially lethal condition usually develops in patients who:
• don't respond to hormone replacement therapy
• undergo extreme stress without adequate glucocorticoid replacement
• abruptly stop hormone therapy
• undergo trauma

• undergo bilateral adrenalectomy
• develop adrenal gland thrombosis after a severe infection (Waterhouse-Friderichsen syndrome).

What happens
In adrenal crisis, destruction of the adrenal cortex leads to a rapid decline in the steroid hormones cortisol and aldosterone. This directly affects the liver, stomach, and kidneys. The flowchart below illustrates what happens in adrenal crisis.

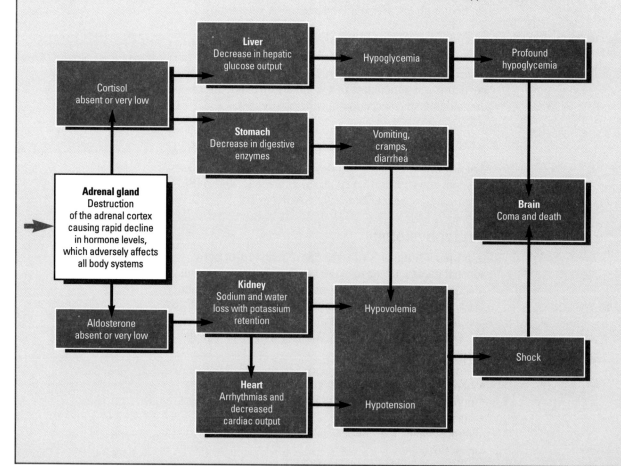

In rare cases, a familial tendency toward autoimmune disease predisposes a patient to Addison's disease and other endocrine disorders.

The signs and symptoms of primary adrenal insufficiency result from decreased glucocorticoid production. Signs and symptoms may develop slowly and stay unrecognized if production is adequate for the normal demands of life. It may progress to adrenal crisis when trauma, surgery, or other severe physical stress completely exhausts the body's store of glucocorticoids.

In primary disease, destruction of the adrenal glands usually results from an autoimmune process.

Secondary disease

Secondary Addison's disease may result from:
• hypopituitarism, which may lead to decreased corticotropin secretion (usually occurs when long-term corticosteroid therapy is abruptly stopped, because such therapy suppresses pituitary corticotropin secretion and causes adrenal gland atrophy)
• removal of a nonendocrine corticotropin-secreting tumor
• disorders in hypothalamic-pituitary function that diminish the production of corticotropin.

Secondary disease usually results from a disorder outside the glands.

 ## What to look for

The following signs and symptoms may indicate Addison's disease:
• muscle weakness
• hypotension
• hypoglycemia
• hyperpigmentation
• GI disturbances and weight loss
• hyperkalemia
• confusion.

Patient history may reveal synthetic steroid use, adrenal surgery, or recent infection. The patient may complain of fatigue, light-headedness when rising from a chair or bed, cravings for salty food, decreased tolerance for even minor stress, anxiety, irritability, and various GI disturbances, such as nausea, vomiting, anorexia, and chronic diarrhea. He may also have reduced urine output and other symptoms of dehydration. Women may have decreased libido from reduced androgen production and amenorrhea.

Physical exam findings

During your examination, you may detect poor coordination, dry skin and mucous membranes, and sparse axillary and pubic hair in women. The patient's skin is typically a deep bronze, especially in the creases of the hands and on the knuckles, elbows, and knees. He may also have a darkening of scars, areas of vitiligo (an absence of pigmentation), and increased pigmentation of the mucous membranes, especially in the mouth.

Abnormal coloration results from decreased secretion of cortisol, a glucocorticoid, which causes the pituitary gland to secrete excessive amounts of melanocyte-stimulating hormone (MSH) and corticotropin. Secondary adrenal hypofunction doesn't cause hyperpigmentation because corticotropin and MSH levels are low.

With early diagnosis and treatment, the prognosis for a patient with either primary or secondary Addison's disease is good. (See *Treating Addison's disease*.)

What tests tell you

The following laboratory tests are used in diagnosing adrenal hypofunction:

• Plasma and urine tests are used to detect decreased corticosteroid concentrations.

• Measurement of corticotropin levels is used to classify the disease as primary or secondary. A high level indicates the primary disorder, and a low level points to the secondary disorder.

• Rapid corticotropin test (ACTH stimulation test) demonstrates plasma cortisol response to corticotropin. After obtaining plasma cortisol samples, an I.V. infusion of cosyntropin is administered. Plasma samples are taken 30 and 60 minutes later. If the plasma cortisol level doesn't increase, adrenal insufficiency is suspected.

Is it a crisis?

In a patient with typical symptoms of Addison's disease, the following laboratory findings strongly suggest adrenal crisis:

• reduced serum sodium levels

• increased serum potassium, serum calcium, and blood urea nitrogen levels

• elevated hematocrit and elevated lymphocyte and eosinophil counts

If your patient has a change in skin color, you may think it's jaundice. But if he has Addison's disease, discoloration is caused by a lack of cortisol.

Battling illness

Treating Addison's disease

Lifelong corticosteroid replacement is the main treatment for patients with primary or secondary Addison's disease. In general, cortisone or hydrocortisone are given because they have a mineralocorticoid effect. Fludrocortisone, a synthetic drug that acts as a mineralocorticoid, may also be given to prevent dehydration and hypotension. Women with muscle weakness and decreased libido may benefit from testosterone injections but risk masculinizing effects.

In stressful situations, the patient may need to double or triple his usual corticosteroid dose. These situations include acute illness (because a fever increases the basal metabolic rate), injury, or psychologically stressful episodes.

Crisis control

Treatment for adrenal crisis is prompt I.V. bolus administration of 100 mg of hydrocortisone, followed by hydrocortisone diluted with dextrose in normal saline solution and given I.V. until the patient's condition stabilizes. Up to 300 mg/day of hydrocortisone and 3 to 5 L of I.V. normal saline solution may be required during the acute stage.

With proper treatment, the crisis usually subsides quickly, with blood pressure stabilizing and water and sodium levels returning to normal. Afterward, maintenance doses of hydrocortisone keep the patient's condition stable.

• X-rays showing a small heart and adrenal calcification
• decreased plasma cortisol levels (less than 10 mcg/dl in the morning, with lower levels at night). Because this test is time-consuming, crisis therapy shouldn't be delayed while waiting for results.

Cushing's syndrome

Cushing's syndrome is a cluster of physical abnormalities caused when the adrenal glands secrete excess glucocorticoids. It may also be caused by excessive androgen secretion. When the glucocorticoid excess is due to pituitary-dependent conditions, it's called Cushing's disease.

Pathophysiology

Cushing's syndrome appears in three forms:

☝ primary, caused by a disease of the adrenal cortex

✌ secondary, caused by hyperfunction of corticotropin-secreting cells of the anterior pituitary

🖐 tertiary, caused by hypothalamic dysfunction or injury.

In most patients, Cushing's syndrome results from overproduction of the hormone corticotropin.

Excess corticotropin

In about 70% of patients, Cushing's syndrome results from an excess of corticotropin. This leads to hyperplasia (excessive cell proliferation) of the adrenal cortex. Corticotropin overproduction may stem from:
• pituitary hypersecretion (Cushing's disease)
• a corticotropin-producing tumor in another organ, especially a malignant tumor of the pancreas or bronchus
• administration of synthetic glucocorticoids.

In the remaining 30% of patients, Cushing's syndrome results from a cortisol-secreting adrenal tumor, which is usually benign. In infants, the usual cause is adrenal carcinoma.

Administration of steroids during treatment can also lead to Cushing's syndrome.

Cushing's complications

Complications associated with Cushing's syndrome are due to the effects of cortisol, the principal glucocorticoid. These complications may include the following:
• Increased calcium resorption from bone may lead to osteoporosis and pathologic fractures.
• Peptic ulcer may result from increased gastric secretions, pepsin production, and decreased gastric mucus.
• Lipidosis (a disorder of fat metabolism) usually occurs.
• Increased hepatic gluconeogenesis and insulin resistance can impair glucose tolerance.

Complications of Cushing's syndrome may include osteoporosis, peptic ulcer, high blood pressure, impaired fat metabolism, and impaired glucose tolerance.

Immune complications

Frequent infections or slow wound healing due to decreased lymphocyte production and suppressed antibody formation may occur. Suppressed inflammatory response may mask infection.

Cardiovascular complications

Hypertension due to sodium and water retention is common in Cushing's syndrome. It may lead to ischemic heart disease and heart failure.

Sexual and psychological complications

Menstrual disturbances and sexual dysfunction also occur because of increased adrenal androgen secretion. Decreased ability to handle stress may result in psychiatric problems, ranging from mood swings to psychosis.

What to look for

If your patient has some or all of the following signs, he might have Cushing's syndrome:
• buffalo hump
• thinning extremities with muscle wasting and fat mobilization
• thin, fragile skin
• thinning scalp hair
• moon face and ruddy complexion
• hirsutism
• truncal obesity
• broad purple striae
• bruising
• impaired wound healing.

The prognosis depends on early diagnosis, identification of the underlying cause, and effective treatment. (See *Treating Cushing's syndrome,* page 138.)

What tests tell you

Diagnosis of Cushing's syndrome depends on a demonstrated increase in cortisol production and the failure to suppress endogenous cortisol secretion after dexamethasone is given. The following tests are performed:
• Low-dose dexamethasone suppression test or 24-hour urine test is used to determine the free cortisol excretion rate. Failure to suppress plasma and urine cortisol levels confirms the diagnosis.
• High-dose dexamethasone suppression test is used to determine if Cushing's syndrome results from pituitary dysfunction. If dexamethasone suppresses plasma cortisol levels, the test result is positive. Failure to suppress plasma cortisol levels indicates an adrenal tumor or a nonen-

Battling illness

Treating Cushing's syndrome

Restoring hormone balance and reversing Cushing's syndrome may require radiation, drug therapy, or surgery. Treatment depends on the cause of the disease.

A patient with pituitary-dependent Cushing's disease and adrenal hyperplasia may require hypophysectomy (removal of the pituitary gland) or pituitary irradiation. If these treatments are unsuccessful or impractical, bilateral adrenalectomy may be performed.

Diverse drugs

A patient with a nonendocrine corticotropin-producing tumor requires excision of the tumor, followed by drug therapy with mitotane, metyrapone, or aminoglutethimide. A combination of a aminoglutethimide, cyproheptadine, and ketoconazole decreases cortisol levels and helps some patients. Aminoglutethimide, alone or with metyrapone, may also be useful in metastatic adrenal carcinoma.

An ounce of prevention

Before surgery, the patient needs to control edema, diabetes, hypertension, and other cardiovascular manifestations and prevent infection. Glucocorticoids given before surgery can help prevent acute adrenal insufficiency during surgery. Cortisol therapy is essential during and after surgery to combat the physiologic stress imposed by removal of the pituitary or adrenal glands.

If normal cortisol production resumes, steroid therapy may gradually be tapered and eventually discontinued. However, the patient who has a bilateral adrenalectomy or a total hypophysectomy requires lifelong steroid replacement.

docrine, corticotropin-secreting tumor. This test can produce false-positive results.
• Radiologic evaluation locates a causative tumor in the pituitary gland or adrenals. Tests include ultrasonography, computed tomography (CT) scan, and magnetic resonance imaging (MRI) enhanced with gadolinium.

Diabetes insipidus

Diabetes insipidus is a disorder of water metabolism caused by a deficiency of antidiuretic hormone (ADH), also called vasopressin. The absence of ADH allows filtered water to be excreted in the urine instead of reabsorbed. The disease causes excessive urination and excessive thirst and fluid intake. It may first appear in childhood or early adulthood and is more common in men than women.

Without ADH, the kidneys are unable to reabsorb fluid...

Without ADH

Pathophysiology

Some drugs as well as injury to the posterior pituitary gland can cause abnormalities in ADH secretion. A less common cause is failure of the kidneys to respond to ADH. Lesions of the hypothalamus, infundibular stem, and posterior pituitary can also interfere with ADH synthesis, transport, or release. Lesions may be caused by brain tumor, removal of the pituitary gland (hypophysectomy), aneurysm, thrombus, or infection.

...which causes excretion of large quantities of dilute urine and excessive thirst.

Normally, ADH is synthesized in the hypothalamus and then stored by the posterior pituitary gland. Once it's released into the general circulation, ADH increases the water permeability of the distal and collecting tubules of the kidneys, causing water reabsorption. If ADH is absent, the filtered water is excreted in the urine instead of being reabsorbed, and the patient excretes large quantities of dilute urine. (See *Understanding ADH,* page 140.)

What to look for

The patient's history shows the following:
• abrupt onset of extreme polyuria (usually 4 to 16 L/day of dilute urine, but sometimes as much as 30 L/day)
• polydipsia (extreme thirst) and consumption of extraordinarily large volumes of fluid.

In severe cases, fatigue occurs because sleep is interrupted by the need to void and drink fluids. Children often have enuresis (involuntary urination), sleep disturbances, irritability, anorexia, and decreased weight gain and linear growth.

Additional signs and symptoms may include:
• weight loss

- dizziness
- weakness
- constipation
- increased serum sodium and osmolality.

The prognosis is good for uncomplicated diabetes insipidus with adequate water replacement, and patients usually lead normal lives. However, when the disease is

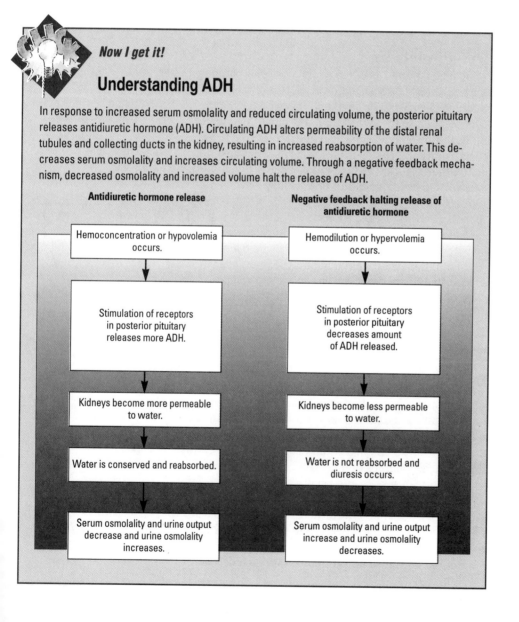

Now I get it!

Understanding ADH

In response to increased serum osmolality and reduced circulating volume, the posterior pituitary releases antidiuretic hormone (ADH). Circulating ADH alters permeability of the distal renal tubules and collecting ducts in the kidney, resulting in increased reabsorption of water. This decreases serum osmolality and increases circulating volume. Through a negative feedback mechanism, decreased osmolality and increased volume halt the release of ADH.

Antidiuretic hormone release

Hemoconcentration or hypovolemia occurs.

↓

Stimulation of receptors in posterior pituitary releases more ADH.

↓

Kidneys become more permeable to water.

↓

Water is conserved and reabsorbed.

↓

Serum osmolality and urine output decrease and urine osmolality increases.

Negative feedback halting release of antidiuretic hormone

Hemodilution or hypervolemia occurs.

↓

Stimulation of receptors in posterior pituitary decreases amount of ADH released.

↓

Kidneys become less permeable to water.

↓

Water is not reabsorbed and diuresis occurs.

↓

Serum osmolality and urine output increase and urine osmolality decreases.

complicated by an underlying disorder such as cancer, the prognosis varies. (See *Treating diabetes insipidus.*)

Yikes! Hypovolemia, circulatory collapse, CNS damage

Untreated diabetes insipidus can produce hypovolemia, hyperosmolality, circulatory collapse, loss of consciousness, and CNS damage. These complications are most likely to occur if the patient has an impaired or absent thirst mechanism.

A prolonged urine flow may produce chronic complications, such as bladder distention, enlarged calyces, hydroureter (distention of the ureter with fluid), and hydronephrosis (collection of urine in the kidney). Complications may also result from underlying conditions, such as metastatic brain lesions, head trauma, and infections.

What tests tell you

The following tests distinguish diabetes insipidus from other disorders causing polyuria:
• Urinalysis reveals almost colorless urine of low osmolality (50 to 200 mOsm/kg of water, less than that of plasma) and low specific gravity (less than 1.005).
• Dehydration test differentiates ADH deficiency from other forms of polyuria by comparing urine osmolality after dehydration and after ADH administration. Diabetes insipidus is diagnosed if the increase in urine osmolality

Battling illness

Treating diabetes insipidus

Until the cause of diabetes insipidus is identified and eliminated, patients are given various forms of vasopressin to control fluid balance and prevent dehydration. The following may be administered:

• Aqueous vasopressin is a replacement agent administered by subcutaneous injection. It's used in the initial management of diabetes insipidus after head trauma or a neurosurgical procedure.
• Desmopressin acetate, a synthetic vasopressin analogue, affects prolonged antidiuretic activity and has no pressor effects. A long-acting drug, desmopressin acetate is administered intranasally.
• Lypressin is a synthetic vasopressin replacement given as a short-acting nasal spray. It has significant disadvantages, including a variable absorption rate, nasal congestion and irritation, nasal passage ulceration with repeated use, substernal chest tightness, coughing, and dyspnea after accidental inhalation of large doses.

after ADH administration exceeds 9%. Patients with pituitary diabetes insipidus have decreased urine output and increased urine specific gravity. Those with nephrogenic diabetes insipidus show no response to ADH.

• Plasma or urinary ADH evaluation may be performed after fluid restriction or hypertonic saline infusion.

If the patient is critically ill, diagnosis may be based on these laboratory values alone:

• urine osmolality of 200 mOsm/kg
• urine specific gravity of 1.005
• serum osmolality of 300 mOsm/kg
• serum sodium of 147 mEq/L.

Note that diabetes insipidus is a pituitary disorder...

Diabetes mellitus

Diabetes mellitus is a chronic disease of insulin deficiency or resistance. It's characterized by disturbances in carbohydrate, protein, and fat metabolism. This leads to hyperglycemia (increased blood glucose).

The disease occurs in two primary forms:

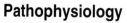 type 1 — insulin-dependent diabetes mellitus

type 2 — non-insulin-dependent diabetes mellitus — the more prevalent form.

Several secondary forms also exist, caused by conditions such as pancreatic disease, pregnancy (gestational diabetes mellitus), hormonal or genetic problems, and certain drugs or chemicals.

Diabetes mellitus affects about 6% of the U.S. population (16 million people), about half of whom are undiagnosed. The incidence increases with age.

...while diabetes mellitus is a pancreatic disorder.

Pathophysiology

Normally, insulin allows glucose to travel into cells. There, it's used for energy and stored as glycogen. It also stimulates protein synthesis and free fatty acid storage in adipose tissue. Insulin deficiency blocks tissues' access to essential nutrients for fuel and storage. Type 1, type 2, and secondary diabetes all result from insulin deficiency, although the pathophysiology of each is different.

When a person lacks sufficient insulin, body tissues have less access to essential nutrients.

Type 1 diabetes

In type 1 diabetes, the beta cells in the pancreas are destroyed or suppressed. Type 1 diabetes is subdivided into idiopathic and immune-mediated types.

A local or organ-specific deficit may induce an autoimmune attack on beta cells. This attack, in turn, causes an inflammatory response in the pancreas called insulitis.

Islet cell antibodies may be present long before symptoms become apparent. These immune markers also precede evidence of beta cell deficiency. Autoantibodies against insulin have also been noted.

By the time the disease becomes apparent, 80% of the beta cells are gone. Some experts believe that the beta cells aren't destroyed by the antibodies but that they're disabled and might later be reactivated.

In type 1 diabetes mellitus, cells in the pancreas that produce insulin are damaged.

Type 2 diabetes

Type 2 diabetes may be caused by the following:
• resistance to insulin action in target tissues
• abnormal insulin secretion
• inappropriate hepatic gluconeogenesis.

Type 2 diabetes may also develop as a consequence of obesity.

Secondary diabetes

Three common causes of secondary diabetes are:
• physical or emotional stress, which may cause prolonged elevation in levels of the stress hormones cortisol, epinephrine, glucagon, and GH. This, in turn, raises blood glucose and increases demands on the pancreas.
• pregnancy, which causes weight gain and high levels of estrogen and placental hormones.
• use of adrenal corticosteroids, oral contraceptives, and other drugs that antagonize the effects of insulin.

Acute complications

Two acute metabolic complications of diabetes are diabetic ketoacidosis (DKA) and hyperosmolar hyperglycemic nonketotic syndrome (HHNS). These life-threatening conditions require immediate medical intervention. (See *Understanding DKA and HHNS,* page 144.)

Chronic complications

Patients with diabetes mellitus also have higher risk for various chronic complications affecting virtually all body

Now I get it!

Understanding DKA and HHNS

Diabetic ketoacidosis (DKA) and hyperosmolar hyperglycemic nonketotic syndrome (HHNS) are acute complications of hyperglycemic crisis that may occur with diabetes. If not treated properly, either may result in coma or death.

DKA occurs most often with type1 diabetes and may be the first evidence of the disease. HHNS occurs most often in patients with type 2 diabetes, but it also occurs in anyone whose insulin tolerance is stressed and in patients who've undergone certain therapeutic procedures, such as peritoneal dialysis, hemodialysis, tube feedings, or total parenteral nutrition.

Acute insulin deficiency (absolute in DKA; relative in HHNS) precipitates both conditions. Causes include illness, stress, infection and, in patients with DKA, failure to take insulin.

Buildup of glucose

Inadequate insulin hinders glucose uptake by fat and muscle cells. Because the cells can't take in glucose to convert to energy, glucose accumulates in the blood. At the same time, the liver responds to the demands of the energy-starved cells by converting glycogen to glucose and releasing glucose into the blood, further increasing the blood glucose level. When this level exceeds the renal threshold, excess glucose is excreted in the urine.

Still, the insulin-deprived cells can't utilize glucose. Their response is rapid metabolism of protein, which results in loss of intracellular potassium and phosphorus and excessive liberation of amino acids. The liver converts these amino acids into urea and glucose.

As a result of these processes, blood glucose levels are grossly elevated. The aftermath is increased serum osmolarity and glucosuria (high amounts of glucose in the urine), leading to osmotic diuresis. Glucosuria is higher in HHNS than in DKA because blood glucose levels are higher in HHNS.

A deadly cycle

The massive fluid loss from osmotic diuresis causes fluid and electrolyte imbalances and dehydration. Water loss exceeds glucose and electrolyte loss, contributing to hyperosmolarity. This, in turn, perpetuates dehydration, decreasing the glomerular filtration rate and reducing the amount of glucose excreted in the urine. This leads to a deadly cycle: Diminished glucose excretion further raises blood glucose levels, producing hyperosmolarity and dehydration and finally causing shock, coma, and death.

DKA complication

All of these steps hold true for both DKA and HHNS, but DKA involves an additional, simultaneous process that leads to metabolic acidosis. The absolute insulin deficiency causes cells to convert fats into glycerol and fatty acids for energy. The fatty acids can't be metabolized as quickly as they're released, so they accumulate in the liver, where they're converted into ketones (ketoacids). These ketones accumulate in the blood and urine and cause acidosis. Acidosis leads to more tissue breakdown, more ketosis, more acidosis, and eventually shock, coma, and death.

systems. The most common chronic complications are cardiovascular disease, peripheral vascular disease, eye disease (retinopathy), kidney disease, skin disease (diabetic dermopathy), and peripheral and autonomic neuropathy.

Research shows that glucose readings don't have to be as high as previously thought for complications to develop. So meticulous blood glucose control is essential to help prevent acute and chronic complications. (See *Screening guidelines*.)

What to look for

Patients with type 1 diabetes usually report rapidly developing symptoms, including muscle wasting and loss of subcutaneous fat.

With type 2 diabetes, symptoms are generally vague and long-standing and develop gradually. In type 2 diabetes, patients generally report a family history of diabetes mellitus, gestational diabetes, delivery of a baby weighing more than 9 lb (4 kg), severe viral infection, another endocrine disease, recent stress or trauma, or use of drugs that increase blood glucose levels. Obesity, especially in the abdominal area, is also common.

Patients with type 1 or type 2 diabetes may report symptoms related to hyperglycemia, such as:
- excessive urination (polyuria)
- excessive thirst (polydipsia)
- excessive eating (polyphagia)
- weight loss
- fatigue
- weakness
- vision changes
- frequent skin infections
- dry, itchy skin
- vaginal discomfort.

Research shows that glucose readings don't have to be as high as previously thought for complications to develop.

Both types

Patients with either type of diabetes may have poor skin turgor, dry mucous membranes related to dehydration, decreased peripheral pulses, cool skin temperature, and decreased reflexes. Patients in crisis with DKA may have a fruity breath odor because of increased acetone production. (See *Treating diabetes mellitus*, page 146.)

Screening guidelines

The guidelines below, from the American Diabetes Association, are also endorsed by the National Institutes of Health.

- Adults should be tested for diabetes every 3 years starting at age 45. Those who get a high glucose reading should have the test repeated on another day.

- People at increased risk may need to be tested earlier or more often. Higher-risk groups include Native Americans, Blacks, Asians, Hispanics, and anyone who is overweight or has high blood pressure, high cholesterol, or a strong family history of diabetes.

- The cutoff used for declaring someone as diabetic should be lowered from 140 mg/dl glucose to 126 mg/dl glucose.

Battling illness

Treating diabetes mellitus

Effective treatment for diabetes optimizes blood glucose levels and decreases complications. In type 1 diabetes, treatment includes insulin replacement, meal planning, and exercise. Current forms of insulin replacement include single-dose, mixed-dose, split-mixed-dose, and multiple-dose regimens. The multiple-dose regimens may use an insulin pump.

Insulin action

Insulin may be rapid-acting (Humalog), fast-acting (Regular), intermediate-acting (NPH and Lente), long-acting (Ultralente), or a premixed combination of fast-acting and intermediate-acting. Insulin may be derived from beef, pork, or human sources. Purified human insulin is used commonly today.

Personalized meal plan

Treatment for both types of diabetes also requires a meal plan to meet nutritional needs, to control blood glucose levels, and to help the patient reach and maintain his ideal body weight. A dietitian estimates the total amount of energy a patient needs per day based on his ideal body weight. Then she plans meals with the appropriate carbohydrate, fat, and protein content. For the diet to work, the patient must follow it consistently and eat at regular times.

In type1 diabetes, the calorie allotment may be high, depending on the patient's growth stage and activity level. Weight reduction is a goal for the obese patient with type 2 diabetes.

Other treatments

Exercise is also useful in managing type 2 diabetes because it increases insulin sensitivity, improves glucose tolerance, and promotes weight loss. In addition, patients with type 2 diabetes may need oral antidiabetic drugs to stimulate endogenous insulin production and increase insulin sensitivity at the cellular level.

Treatment for long-term complications may include dialysis or kidney transplantation for renal failure, photocoagulation for retinopathy, and vascular surgery for large vessel disease. Pancreas transplantation is also an option.

Under investigation

Researchers are investigating ways to restore normal insulin secretion by transplanting pancreatic animal tissue. Ideally, a semipermeable, or one-way, membrane filtration system would allow insulin to flow out of a device grafted onto the vascular system without allowing the patient's immune system to destroy the transplanted cells.

What tests tell you

In nonpregnant adults a diagnosis of diabetes mellitus may be confirmed by:
- symptoms of diabetes and a random blood glucose level equal to or above 200 mg/dl
- a fasting plasma glucose level equal to or greater than 126 mg/dl on at least two occasions
- a blood glucose level above 200 mg/dl on the second hour of the glucose tolerance test and on at least one other occasion during a glucose tolerance test.

Three other tests may be done:
- An ophthalmologic examination may show diabetic retinopathy.
- Urinalysis shows the presence of acetone.

• Blood tests for glycosylated hemoglobin are used to monitor the long-term effectiveness of diabetes therapy. These tests show variants in hemoglobin levels that reflect average blood glucose levels during the preceding 2 to 3 months.

Goiter

A goiter is an enlargement of the thyroid gland. It's not caused by inflammation or neoplasm and not initially associated with hyperthyroidism or hypothyroidism. This condition is commonly referred to as simple goiter or nontoxic goiter. It's classified two ways:

☝ endemic, caused by lack of iodine in the diet

✌ sporadic, related to ingestion of certain drugs or food; this type of goiter occurs randomly.

Simple goiter is most common in females, especially during adolescence, pregnancy, and menopause. At these times, the demand for thyroid hormone increases.

Pathophysiology

Simple goiter occurs when the thyroid gland can't secrete enough thyroid hormone to meet metabolic needs. As a result, the thyroid mass increases to compensate. This usually overcomes mild to moderate hormonal impairment.

TSH levels in simple goiter are generally normal. Enlargement of the gland probably results from impaired hormone production in the thyroid and depleted iodine, which increases the thyroid gland's reaction to TSH.

Getting down to specifics

Endemic goiter usually results from inadequate dietary intake of iodine, which leads to inadequate synthesis of thyroid hormone. In Japan, goiter resulting from iodine excess has been found, from excessive ingestion of seaweed.

Sporadic goiter commonly results from ingestion of large amounts of goitrogenic foods or use of goitrogenic drugs. These foods and drugs contain agents that decrease T_4 production. Such foods include rutabagas, cabbage, soybeans, peanuts, peaches, peas, strawberries, spinach, and radishes.

Goitrogenic drugs include:
- propylthiouracil
- iodides
- phenylbutazone
- aminosalicylic acid
- cobalt
- lithium.

Some areas, called goiter belts, have a high incidence of endemic goiter. This is caused by iodine deficient soil and water. Goiter belts include areas in the Midwest, Northwest, and Great Lakes region.

A closer look

Here's a more detailed account of what happens in goiter:
- Depletion of glandular organic iodine along with impaired hormone synthesis increases the thyroid's responsiveness to normal TSH levels.
- Resulting increases in both thyroid mass and cellular activity overcome mild impairment of hormone synthesis. Although the patient has a goiter, his metabolic function is normal.
- When the underlying disorder is severe, compensatory responses may cause both a goiter and hypothyroidism. (See *Treating simple goiter*.)

What to look for

A simple goiter causes these signs and symptoms:
- single or multinodular, firm, irregular enlargement of the thyroid gland
- stridor
- respiratory distress and dysphagia from compression of the trachea and esophagus
- dizziness or syncope when arms are raised above the head, due to obstructed venous return.

Complications

The enlarged thyroid gland frequently undergoes exacerbations and remissions, with areas of hypervolution and involution. Fibrosis may alternate with hyperplasia, and nodules containing thyroid follicles may develop. Production of excessive amounts of thyroid hormone may lead to thyrotoxicosis.

Battling illness

Treating simple goiter

The goal of treatment for simple goiter is to reduce thyroid hyperplasia.

Hormone replacement
Thyroid hormone replacement with levothyroxine, dessicated thyroid, or liothyronine is the treatment of choice because it inhibits thyroid-stimulating hormone secretion and allows the gland to rest. Small doses of iodide (Lugol's solution or potassium iodide solution) often relieve goiters caused by iodine deficiency.

Diet
Patients with sporadic goiters must avoid known goitrogenic drugs or food.

Radiation
Radioiodine ablation therapy to the thyroid gland is used to destroy the cells that concentrate iodine to make thyroxine.

Surgery
In rare cases of a large goiter unresponsive to treatment, partial removal of the thyroid gland may relieve pressure on surrounding structures.

Complications from a large retrosternal goiter mainly result from the compression and displacement of the trachea or esophagus. Thyroid cysts and hemorrhage into the cysts may increase the pressure on and compression of the surrounding tissues and structures. Large goiters may obstruct venous return, produce venous engorgement and, rarely, cause collateral circulation of the chest.

With treatment, the prognosis is good for patients with either endemic or sporadic goiter.

> Pressure from a goiter can displace or obstruct surrounding tissues and structures, such as the trachea and veins.

What tests tell you

The following tests are used to diagnose simple goiter and rule out other diseases with similar clinical effects:

• Serum thyroid hormone levels are usually normal. Abnormalities in T_3, T_4, and TSH levels rule out this diagnosis. Transient increased levels of TSH, which occur infrequently, may be missed by diagnostic tests.
• Thyroid antibody titers are usually normal. Increases indicate chronic thyroiditis.
• ^{131}I uptake is usually normal but may increase in the presence of iodine deficiency or a biosynthetic defect.
• Urinalysis may show low urinary excretion of iodine.
• Radioisotope scanning identifies thyroid neoplasms.

Hyperthyroidism

When thyroid hormone is overproduced, it creates a metabolic imbalance called hyperthyroidism or thyrotoxicosis. Excess thyroid hormone can cause various thyroid disorders; Graves' disease is the most common. (See *Types of hyperthyroidism,* page 150.)

Graves' disease

Graves' disease is an autoimmune disorder that causes goiter and multiple systemic changes. It occurs mostly in people ages 30 to 60, especially when their family histories include thyroid abnormalities. Only 5% of patients are under age 15.

Taken by storm

Thyrotoxic crisis, also known as thyroid storm, is an acute exacerbation of hyperthyroidism. This is a medical emergency that may lead to life-threatening cardiac, hepatic, or renal failure. Inadequately treated hyperthyroidism and stressful conditions, such as surgery, infection, toxemia of pregnancy, and DKA, can lead to thyrotoxic crisis. (See *Understanding thyroid storm.*)

Pathophysiology

In Graves' disease, thyroid-stimulating antibodies bind to and stimulate the TSH receptors of the thyroid gland.

The trigger for this autoimmune response is unclear; it may have several causes. Genetic factors may play a part; the disease tends to occur in identical twins. Immunologic factors may also be the culprit; the disease occasionally coexists with other autoimmune endocrine abnormalities, such as type 1 diabetes mellitus, thyroiditis, and hyperparathyroidism.

Graves' disease is also associated with the production of several autoantibodies formed because of a defect in suppressor T-lymphocyte function.

Types of hyperthyroidism

Besides Graves' disease, other forms of hyperthyroidism include toxic adenoma, thyrotoxicosis factitia, functioning metastatic thyroid carcinoma, thyroid-stimulating hormone (TSH)–secreting pituitary tumor, and subacute thyroiditis.

Toxic adenoma
The second most common cause of hyperthyroidism, toxic adenoma is a small, benign nodule in the thyroid gland that secretes thyroid hormone. The cause of toxic adenoma is unknown; its incidence is highest in elderly people. Clinical effects are similar to those of Graves' disease except that toxic adenoma doesn't induce ophthalmopathy, pretibial myxedema, or acropachy (clubbing of fingers or toes). A radioactive iodine (^{131}I) uptake and a thyroid scan show a single hyperfunctioning nodule suppressing the rest of the gland. Treatment includes ^{131}I therapy or surgery to remove the adenoma after antithyroid drugs restore normal gland function.

Thyrotoxicosis factitia
Thyrotoxicosis factitia results from chronic ingestion of thyroid hormone for TSH suppression in patients with thyroid carcinoma. It may also result from thyroid hormone abuse by people trying to lose weight.

Functioning metastatic thyroid carcinoma
This rare disease causes excess production of thyroid hormone.

TSH-secreting pituitary tumor
This form of hyperthyroidism causes overproduction of thyroid hormone.

Subacute thyroiditis
A virus-induced granulomatous inflammation of the thyroid, subacute thyroiditis produces transient hyperthyroidism associated with fever, pain, pharyngitis, and tenderness of the thyroid gland.

Hyperthyroidism in hiding

In a person with latent hyperthyroidism, excessive iodine intake and, possibly, stress can cause active hyperthyroidism.

Graves' disease may result from both genetic and immunologic influences.

What to look for

The onset of signs and symptoms often follows a period of acute physical or emotional stress. The classic features of Graves' disease are:

- an enlarged thyroid gland
- exophthalmos (abnormal protrusion of the eye)
- nervousness
- heat intolerance
- weight loss despite increased appetite
- excessive sweating
- diarrhea
- tremors

Now I get it!

Understanding thyroid storm

Thyrotoxic crisis — also known as thyroid storm — usually occurs in patients with preexisting, though often unrecognized, thyrotoxicosis. Left untreated, it's invariably fatal.

Pathophysiology

The thyroid gland secretes the thyroid hormones triiodothyronine (T_3) and thyroxine (T_4). When T_3 and T_4 are overproduced, systemic adrenergic activity increases. The result is epinephrine overproduction and severe hypermetabolism, leading rapidly to cardiac, GI, and sympathetic nervous system decompensation.

Assessment findings

Initially, the patient may have marked tachycardia, vomiting, and stupor. If left untreated, he may experience vascular collapse, hypotension, coma, and death. Other findings may include a combination of irritability and restlessness, visual disturbance such as diplopia, tremor and weakness, angina, shortness of breath, cough, and swollen extremities. Palpation may disclose warm, moist, flushed skin and a high fever that begins insidiously and rises rapidly to a lethal level.

Precipitating factors

Onset is almost always abrupt and evoked by a stressful event, such as trauma, surgery, or infection. Other, less common precipitating factors include:

- insulin-induced ketoacidosis
- hypoglycemia or diabetic ketoacidosis
- cerebrovascular accident
- myocardial infarction
- pulmonary embolism
- sudden discontinuation of antithyroid drug therapy
- initiation of radioactive iodine therapy
- preeclampsia
- subtotal thyroidectomy with accompanying excessive intake of synthetic thyroid hormone.

- palpitations.

Because thyroid hormones have widespread effects on almost all body systems, many other signs and symptoms are common with hyperthyroidism:

- central nervous system—difficulty concentrating, anxiety, excitability or nervousness, fine tremor, shaky handwriting, clumsiness, emotional instability, and mood swings ranging from occasional outbursts to overt psychosis (These symptoms are most common in younger patients.)
- cardiovascular system—arrhythmias, especially atrial fibrillation; cardiac insufficiency; cardiac decompensation; and resistance to the usual therapeutic dose of a digitalis glycoside (These symptoms are most common in elderly patients.)
- skin, hair, and nails—vitiligo and skin hyperpigmentation; warm, moist, flushed skin with a velvety texture; fine, soft hair; premature graying; hair loss in both sexes; fragile nails; onycholysis (separation of distal nail from nail bed); and pretibial myxedema producing raised, thickened skin and plaquelike or nodular lesions
- respiratory system—dyspnea on exertion and, possibly, at rest and breathlessness when climbing stairs
- GI system—anorexia, nausea, and vomiting
- musculoskeletal system—muscle weakness, generalized or localized muscle atrophy, acropachy (soft-tissue swelling with underlying bone changes where new bone formation occurs), and osteoporosis
- reproductive system—menstrual abnormalities, impaired fertility, decreased libido, and gynecomastia (abnormal development of mammary glands in men)
- eyes—infrequent blinking, lid lag, reddened conjunctiva and cornea, corneal ulcers, impaired upward gaze, convergence, strabismus (eye deviation), and exophthalmos, which causes the characteristic staring gaze.

Touch tells you

On palpation, the thyroid gland may feel asymmetrical, lobular, and enlarged to three or four times its normal size. The liver may also be enlarged. Hyperthyroidism may cause tachycardia, often accompanied by a full, bounding, palpable pulse. Hyperreflexia is also present.

Learn from listening

Auscultation of the heart may detect paroxysmal supraventricular tachycardia and atrial fibrillation, especially in elderly patients. Occasionally, a systolic murmur occurs at the left sternal border. Wide pulse pressures may be audible when blood pressure readings are taken. In Graves' disease, an audible bruit over the thyroid gland indicates thyrotoxicity but, occasionally, it may also be present in other hyperthyroid disorders.

With treatment, most patients can lead normal lives. (See *Treating hyperthyroidism.*)

What tests tell you

The following laboratory tests confirm Graves' disease:
• Radioimmunoassay shows increased serum T_3 and T_4 concentrations.

Battling illness

Treating hyperthyroidism

In Graves' disease, the most common hyperthyroid disorder, treatment consists of drugs, radioactive iodine, and surgery.

Drug therapy
Antithyroid drugs, such as propylthiouracil and methimazole, are used for children, young adults, pregnant women, and patients who refuse other treatments.

Radiation therapy
A single oral dose of radioactive iodine (^{131}I) is the treatment of choice for women past reproductive age or men and women not planning to have children (small amounts concentrate in the gonads).

During treatment, the thyroid gland picks up the radioactive element as it would regular iodine. Subsequently, the radioactivity destroys some of the cells that normally concentrate iodine and produce thyroxine, thus decreasing thyroid hormone production and normalizing thyroid size and function.

In most patients, hypermetabolic symptoms diminish within 6 to 8 weeks. However, some patients may require a second dose.

Surgery
Partial thyroidectomy is indicated for the patient under age 40 who has a very large goiter and whose hyperthyroidism has re-peatedly relapsed after drug therapy. The surgery involves removal of part of the thyroid gland, decreasing its size and capacity for hormone production.

Preoperatively, the patient may receive iodide (Lugol's solution or potassium iodide solution), antithyroid drugs, or high doses of propranolol to help prevent thyroid storm. If normal thyroid function isn't achieved, surgery should be delayed and propranolol administered to decrease the risk of cardiac arrhythmias.

Other treatments
Therapy for hyperthyroid ophthalmopathy includes local applications of topical drugs but may require high doses of corticosteroids. A patient with severe exophthalmos that causes pressure on the optic nerve may require surgical decompression to reduce pressure on the orbital contents.

Treatment for thyrotoxic crisis includes giving an antithyroid drug, I.V. propranolol to block sympathetic effects, a corticosteroid to inhibit the conversion of triiodothyronine to thyroxine and replace depleted cortisol, and an iodide to block the release of thyroid hormones. Supportive measures include the administration of nutrients, vitamins, fluids, and sedatives.

• TSH level is low in primary hyperthyroidism and elevated when excessive TSH secretion is the cause.
• Thyroid scan reveals increased uptake of radioactive iodine (^{131}I).

Other tests show increased serum protein-bound iodine and decreased serum cholesterol and total lipid levels.

Hypothyroidism

In thyroid hormone deficiency (hypothyroidism) in adults, metabolic processes slow down. That's because of a deficit in the hormones T_3 or T_4, which regulate metabolism. The disorder is most prevalent in women, and its incidence in the United States is increasing in people ages 40 to 50.

Primary or secondary

Hypothyroidism is classified as primary or secondary. The primary form stems from a disorder of the thyroid gland itself. The secondary form stems from a failure to stimulate normal thyroid function. This form may progress to myxedema coma, a medical emergency. (See *Understanding myxedema coma.*)

Pathophysiology

Primary hypothyroidism has several possible causes:

 thyroidectomy

 inflammation from radiation therapy

other inflammatory conditions, such as amyloidosis and sarcoidosis

chronic autoimmune thyroiditis (Hashimoto's disease).

Secondary hypothyroidism is caused by a failure to stimulate normal thyroid function. For example, the pituitary may fail to produce TSH (thyrotropin) or the hypothalamus may fail to produce thyrotropin-releasing hormone.

Secondary hypothyroidism may also be caused by an inability to synthesize thyroid hormones because of iodine deficiency (usually dietary) or the use of antithyroid medications.

Low thyroid hormone levels in the blood may result from a problem in the pituitary gland, thyroid gland, or hypothalamus.

Primary hypothyroidism may result from gland dysfunction related to surgery, radiation therapy, inflammation, or chronic autoimmune thyroiditis.

Because insufficient synthesis of thyroid hormones affects almost every organ system in the body, signs and symptoms vary according to the organs involved as well as the duration and severity of the condition.

 ## What to look for

The signs and symptoms of hypothyroidism may be vague and varied. Here are some early ones:
- energy loss
- fatigue
- forgetfulness
- sensitivity to cold
- unexplained weight gain
- constipation.

As the disease progresses, the patient may have:
- anorexia
- decreased libido
- menorrhagia (painful menstruation)
- paresthesia (numbness, prickling, or tingling)
- joint stiffness
- muscle cramping.

Other signs and symptoms include:
- central nervous system — psychiatric disturbances, ataxia (loss of coordination), intention tremor (tremor during voluntary motion), carpal tunnel syndrome, benign intracranial hypertension, and behavior changes, ranging from slight mental slowing to severe impairment
- skin, hair, and nails — dry, flaky, inelastic skin; puffy face, hands, and feet; dry, sparse hair with patchy hair loss and loss of the outer third of the eyebrow; thick, brittle nails with transverse and longitudinal grooves; and a thick, dry tongue, causing hoarseness and slow, slurred speech
- cardiovascular system — hypercholesterolemia (high cholesterol) with associated arteriosclerosis and ischemic heart disease, poor peripheral circulation, heart enlargement, heart failure, and pleural and pericardial effusions
- GI system — achlorhydria (absence of free hydrochloric acid in the stomach), pernicious anemia, and adynamic (weak) colon, resulting in megacolon (extremely dilated colon) and intestinal obstruction
- reproductive system — impaired fertility

Now I get it!

Understanding myxedema coma

A medical emergency, myxedema coma often has a fatal outcome. Progression is usually gradual but when stress, such as infection, exposure to cold, or trauma, aggravates severe or prolonged hypothyroidism, coma may develop abruptly. Other precipitating factors are thyroid medication withdrawal and the use of sedatives, narcotics, or anesthetics.

What happens
Patients in myxedema coma have significantly depressed respirations, so their partial pressure of carbon dioxide in arterial blood may rise. Decreased cardiac output and worsening cerebral hypoxia may also occur. The patient becomes stuporous and hypothermic. Vital signs reflect bradycardia and hypotension. Lifesaving interventions are necessary.

• eyes and ears — conductive or sensorineural deafness and nystagmus
• circulatory system — anemia, which may result in bleeding tendencies and iron deficiency anemia.

Going to extremes

Severe hypothyroidism, or myxedema, is characterized by thickening of the facial features and induration of the skin. The skin may feel rough, doughy, and cool. Other signs and symptoms are weak pulse, bradycardia, muscle weakness, sacral or peripheral edema, and delayed reflex relaxation time (especially in the Achilles tendon). Unless a goiter is present, the thyroid tissue itself may not be easily palpable.

Hyponatremia (low blood sodium) may result from impaired water excretion and from poor regulation of ADH secretion. (See *Treating hypothyroidism*.)

What tests tell you

Primary hypothyroidism is confirmed by an elevated TSH level and low serum free T_4 index. Additional tests may include the following:
• Serum TSH levels are used to determine whether the disorder is primary or secondary. An increased serum TSH level is due to thyroid insufficiency; a decreased or normal level is due to hypothalamic or pituitary insufficiency.
• Serum antithyroid antibodies show elevated levels in autoimmune thyroiditis.
• Perchlorate discharge tests are used to identify enzyme deficiency within the thyroid gland. A deficiency will affect the uptake of iodine.
• Radioisotope scanning is used to identify ectopic thyroid tissue.
• Skull X-ray, CT scan, and MRI are used to locate pituitary or hypothalamic lesions that may cause secondary hypothyroidism.

Battling illness

Treating hypothyroidism

Treatment consists of gradual thyroid hormone replacement with a synthetic hormone. Treatment begins slowly, particularly in elderly patients, to avoid adverse cardiovascular effects. The dosage is increased every 2 to 3 weeks until the desired response is obtained.

In underdeveloped areas, prophylactic iodine supplements have successfully decreased the incidence of iodine-deficient goiter.

No time to waste
Rapid treatment is necessary for patients with myxedema coma and those having emergency surgery. These patients need both I.V. administration of levothyroxine and hydrocortisone therapy.

Quick quiz

1. A patient with weight loss, GI disturbances, dehydration, fatigue, and a craving for salty food probably has:
    A. Cushing's syndrome.
    B. Addison's disease.
    C. hypothyroidism.

Answer: B. Other classic symptoms of Addison's disease are muscle weakness, anxiety, light-headedness, and amenorrhea.

2. Cushing's syndrome may be caused by:
    A. a destruction of more than 90% of the adrenal gland.
    B. thyroid hormone overproduction.
    C. glucocorticoid excess.

Answer: C. Cushing's syndrome is also caused by excess androgen secretion.

3. The two primary forms of diabetes mellitus are:
    A. type A and type B.
    B. type 1 and type 2.
    C. insulitis and insipidus.

Answer: B. Type 1 diabetes is insulin-dependent and type 2 diabetes is non-insulin-dependent.

4. Exophthalmos causes the eyes to:
    A. protrude more than usual.
    B. focus differently.
    C. become red.

Answer: A. A classic symptom of Graves' disease, exophthalmos causes a characteristic stare.

5. Simple goiter is more common in:
    A. older adults.
    B. blacks.
    C. females.

Answer: C. The disorder occurs more often in females, especially during adolescence, pregnancy, and menopause, when the demand on the body for thyroid hormone increases.

Scoring

☆☆☆   If you answered all five items correctly, hooray! You're hyper-informed about hormones!

☆☆   If you answered three or four correctly, exceptional! Your endocrine expertise is indeed endearing.

☆   If you answered fewer than three correctly, just remember, never give up. Hopefully, your glands will release test-taking hormones that will stimulate you to stick with it.

Respiratory system

Just the facts

In this chapter you'll learn:

♦ the function of respiratory passages and lungs

♦ about the neurochemical ventilation control system

♦ the pathophysiology, diagnostic tests, and treatments for several respiratory diseases.

Understanding the respiratory system

The respiratory system consists of two lungs, conducting airways, and associated blood vessels.

The major function of the respiratory system is gas exchange. During gas exchange, air is taken into the body on inhalation (inspiration) and travels through respiratory passages to the lungs. Oxygen (O_2) in the lungs replaces carbon dioxide (CO_2) in the blood, and the CO_2 is expelled from the body on exhalation (expiration).

Disease or trauma may interfere with the respiratory system's vital work, affecting any of the following structures and functions:

 conducting airways

 lungs

 breathing mechanics

 neurochemical control of ventilation.

Gas exchange, that's the name of the game.

Conducting airways

The conducting airways allow air into and out of structures within the lung that perform gas exchange. The conducting airways include the upper airway and the lower airway.

Upper airway

The upper airway consists of the:
- nose
- mouth
- pharynx
- larynx.

The upper airway allows air flow into and out of the lungs. It warms, humidifies, and filters inspired air and protects the lower airway from foreign matter.

Oh, my! A blocked oxygen supply

Upper airway obstruction occurs when the nose, mouth, pharynx, or larynx becomes partially or totally blocked, cutting off the oxygen supply. Several conditions can cause upper airway obstruction, including trauma, tumors, and foreign objects.

If not treated promptly, upper airway obstruction can lead to hypoxemia (insufficient oxygen in the blood) and then progress quickly to severe hypoxia (lack of oxygen available to body tissues), loss of consciousness, and death.

Lower airway

The lower airway consists of the:
- trachea
- right and left mainstem bronchi
- five secondary bronchi
- bronchioles.

The lower airways facilitate gas exchange. Each bronchiole descends from a lobule and contains terminal bronchioles, alveolar ducts, and alveoli. Terminal bronchioles are "anatomic dead spaces" because they don't participate in gas exchange. The alveoli are the chief units of gas exchange. (See *A close look at a lobule.*)

Like the upper airway, the lower airway can become partially or totally blocked as a result of inflammation, tumors, foreign bodies, or trauma. This can lead to respiratory distress and failure.

A close look at a lobule

As illustrated below, each lobule contains terminal bronchioles and the acinus, consisting of respiratory bronchioles and alveolar sacs.

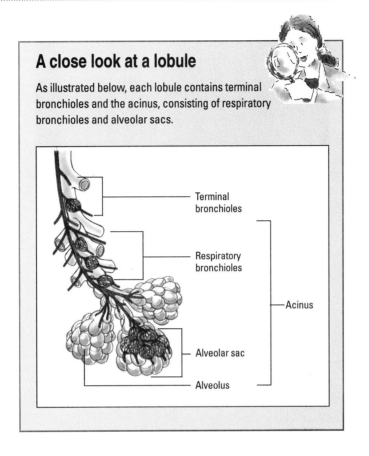

- Terminal bronchioles
- Respiratory bronchioles
- Acinus
- Alveolar sac
- Alveolus

Guarding the lungs

In addition to warming, humidifying, and filtering inspired air, the lower airway provides the lungs with the following defense mechanisms:

- clearance mechanisms (cough reflex, mucociliary system)
- immunologic responses
- pulmonary injury responses.

The mucociliary system produces mucus, which traps foreign particles. Foreign matter is then swept to the upper airway for expectoration. A breakdown in the epithelium of the lungs or the mucociliary system can cause the defense mechanisms to malfunction. This allows atmospheric pollutants and irritants to enter and inflame the lungs.

Lungs

The lungs are air-filled, spongelike organs. They're divided into lobes (three lobes on the right, two lobes on the left). Lobes are further divided into lobules and segments.

The lungs contain millions of pulmonary alveoli, which are grapelike clusters of air-filled sacs at the ends of the respiratory passages. Here, gas exchange takes place by diffusion (the passage of gas molecules through respiratory membranes). In diffusion, O_2 is passed to the blood for circulation through the body. At the same time, CO_2 — a cellular waste product that's gathered by the blood as it circulates — is collected from the blood for disposal out of the body through the lungs.

That's me. air-filled and spongelike...

All about alveoli

Alveoli consist of type I and type II epithelial cells:

☝ Type I cells form the alveolar walls, through which gas exchange occurs.

✌ Type II cells produce surfactant, a lipid-type substance that coats the alveoli. During inspiration, the alveolar surfactant allows the alveoli to expand uniformly. During expiration, the surfactant prevents alveolar collapse.

O_2 - CO_2 swap

How much O_2 and CO_2 trade places in the alveoli? That depends largely on the amount of air in the alveoli (ventilation) and the amount of blood in the pulmonary capillaries (perfusion). The ratio of ventilation to perfusion is called the $\dot{V}/\dot{Q}$ ratio. The $\dot{V}/\dot{Q}$ ratio expresses the effectiveness of gas exchange.

For effective gas exchange, ventilation and perfusion must match as closely as possible. In normal lung function, the alveoli receive air at a rate of about 4 L/minute while the capillaries supply blood to the alveoli at a rate of about 5 L/minute, creating a $\dot{V}/\dot{Q}$ ratio of 4:5 or 0.8. (See *Understanding ventilation and perfusion.*)

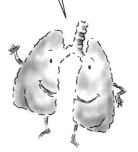

What's my secret? Millions of pulmonary alveoli that perform gas exchange.

Mismatch mayhem

A $\dot{V}/\dot{Q}$ mismatch, resulting from ventilation-perfusion dysfunction or altered lung mechanics, accounts for most of the impaired gas exchange in respiratory disorders. Ineffective gas exchange between the alveoli and the pul-

Now I get it!

Understanding ventilation and perfusion

Effective gas exchange depends on the relationship between ventilation and perfusion, or the $\dot{V}/\dot{Q}$ ratio. The diagrams below show what happens when the $\dot{V}/\dot{Q}$ ratio is normal and abnormal.

A high $\dot{V}/\dot{Q}$ ratio results from reduced or absent alveolar perfusion.

Normal ventilation and perfusion

When ventilation and perfusion are matched, unoxygenated blood from the venous system returns to the right ventricle through the pulmonary artery to the lungs, carrying carbon dioxide. The arteries branch into the alveolar capillaries. Gas exchange takes place in the alveolar capillaries.

Inadequate perfusion (dead-space ventilation)

When the $\dot{V}/\dot{Q}$ ratio is high, as shown here, ventilation is normal, but alveolar perfusion is reduced or absent. Note the narrowed capillary, indicating poor perfusion. This commonly results from a perfusion defect, such as pulmonary embolism or a disorder that decreases cardiac output.

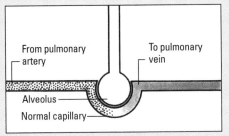

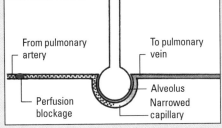

Inadequate ventilation (shunt)

When the $\dot{V}/\dot{Q}$ ratio is low, pulmonary circulation is adequate but not enough oxygen is available to the alveoli for normal diffusion. A portion of the blood flowing through the pulmonary vessels does not become oxygenated.

Inadequate ventilation and perfusion (silent unit)

The silent unit indicates an absence of ventilation and perfusion to the lung area. The silent unit may help compensate for a $\dot{V}/\dot{Q}$ imbalance by delivering blood flow to better-ventilated lung areas.

A silent unit can stem from several causes, including pulmonary embolism and chronic alveolar collapse.

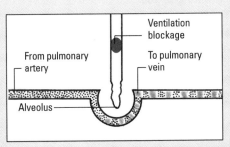

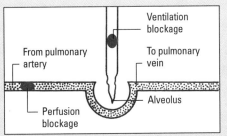

 Blood with CO_2 ▭ Blood with O_2 ▱ Blood with CO_2 and O_2

monary capillaries can affect all body systems by altering the amount of oxygen delivered to living cells.

Ineffective gas exchange from an abnormality causes three outcomes:

- shunting (reduced ventilation to a lung unit)
- dead-space ventilation (reduced perfusion to a lung unit)
- silent unit (combination of the above).

Don't shunt me out

Shunting causes the movement of unoxygenated blood to the left side of the heart. A shunt may occur from a physical defect that allows unoxygenated blood to bypass fully functioning alveoli. Or it may result when airway obstruction prevents oxygen from reaching an adequately perfused area of the lung.

Respiratory disorders commonly are classified as shunt-producing if the $\dot{V}/\dot{Q}$ ratio falls below 0.8 and dead space–producing if the $\dot{V}/\dot{Q}$ ratio exceeds 0.8.

Breathing mechanics

The amount of air that reaches the lungs carrying O_2 and then departs carrying CO_2 depends on three factors:

 lung volume and capacity

 compliance (the lungs' ability to expand)

 resistance to air flow.

No room for expansion

Changes in compliance can occur in either the lung or the chest wall. Destruction of the lung's elastic fibers, which occurs in adult respiratory distress syndrome, decreases lung compliance. The lungs become stiff, making breathing difficult. The alveolocapillary membrane may also be affected, causing hypoxia. Chest-wall compliance is affected by thoracic deformity, muscle spasm, and abdominal distention.

La pièce de résistance

Resistance refers to opposition to air flow. Changes in resistance may occur in the lung tissue, chest wall, or airways. Airway resistance accounts for about 80% of all respiratory system resistance. It's increased in obstructive diseases such as asthma, chronic bronchitis, and emphysema.

With increased resistance, a person has to work harder to breathe, especially during expiration, to compensate for narrowed airways and diminished gas exchange.

Neurochemical control

The central nervous system's respiratory center is located in the lateral medulla oblongata of the brain stem. Impulses travel down the phrenic nerves to the diaphragm, and then down the intercostal nerves to the intercostal muscles between the ribs. There, they change the rate and depth of respiration.

The central nervous system's respiratory center is located in the lateral medulla oblongata of the brain stem.

Getting to know your neurons

The respiratory center consists of different groups of neurons:
• The dorsal respiratory group of neurons determines the autonomic rhythm of respiration.
• The ventral respiratory group of neurons is inactive during normal respiration but becomes active when increased ventilatory effort is needed. It contains both inspiratory and expiratory neurons.
• The pneumotaxic center and apneustic center don't generate a rhythm but modulate an established rhythm. The pneumotaxic center affects the inspiratory effort by limiting the volume of air inspired. The apneustic center prevents excessive inflation of the lungs.

Factors of influence

Chemoreceptors respond to the hydrogen ion concentration of arterial blood (pH), the partial pressure of arterial carbon dioxide ($Paco_2$), and the partial pressure of arterial oxygen (Pao_2). Central chemoreceptors respond indirectly to arterial blood by sensing changes in the pH of cerebrospinal fluid (CSF).

$Paco_2$ also helps regulate ventilation (by impacting the pH of CSF). If $Paco_2$ is high, the respiratory rate increases. If $Paco_2$ is low, the respiratory rate decreases.

Peripheral information

The respiratory center also receives information from peripheral chemoreceptors in the carotid and aortic bodies (small neurovascular structures in the carotid arteries and on either side of the aorta). These chemoreceptors respond to decreased PaO_2 and decreased pH. Either change results in increased respiratory drive within minutes.

Respiratory disorders

The respiratory disorders discussed below include adult respiratory distress syndrome (ARDS), asbestosis, asthma, chronic bronchitis, emphysema, cor pulmonale, pneumonia, pneumothorax, pulmonary edema, and tuberculosis.

Adult respiratory distress syndrome

ARDS is a form of pulmonary edema that can quickly lead to acute respiratory failure. Also known as shock, stiff, white, wet, or Da Nang lung, ARDS may follow a direct or indirect lung injury. It's difficult to diagnose and can prove fatal within 48 hours of onset if not promptly diagnosed and treated.

In ARDS, fluid builds up in the lungs, causing them to stiffen.

Pathophysiology

Trauma is the most common cause of ARDS. Trauma-related factors, such as fat emboli, sepsis, shock, pulmonary contusions, and multiple transfusions, may increase the likelihood that microemboli will develop.

Other causes of ARDS include the following:
- anaphylaxis
- aspiration of gastric contents
- diffuse pneumonia, especially viral pneumonia
- drug overdose (for example, heroin, aspirin, and ethchlorvynol)
- idiosyncratic drug reaction to ampicillin or hydrochlorothiazide
- inhalation of noxious gases (such as nitrous oxide, ammonia, or chlorine)
- near-drowning
- oxygen toxicity
- coronary artery bypass grafting
- hemodialysis
- leukemia

- acute miliary tuberculosis
- pancreatitis
- thrombotic thrombocytopenic purpura (embolism and thrombosis of the small blood vessels of the brain)
- uremia
- venous air embolism.

Effects of fluid accumulation

In ARDS, fluid accumulates in the lung interstitium, alveolar spaces, and small airways, causing the lung to stiffen. This impairs ventilation and reduces oxygenation of pulmonary capillary blood. Here's what happens:

- Injury reduces normal blood flow to the lungs, allowing platelets to aggregate. These platelets release substances, such as serotonin, bradykinin, and histamine, that inflame and damage the alveolar membrane and later increase capillary permeability.
- Histamines and other inflammatory substances increase capillary permeability. Fluids shift into the interstitial space.
- As capillary permeability increases, proteins and more fluid leak out, causing pulmonary edema.
- Fluid in the alveoli and decreased blood flow damage surfactant in the alveoli. This reduces the alveolar cells' ability to produce more surfactant. Without surfactant, alveoli collapse, impairing gas exchange.
- The patient breathes faster, but sufficient O_2 can't cross the alveolocapillary membrane. CO_2, however, crosses more easily and is lost with every exhalation. Both O_2 and CO_2 levels in the blood decrease.
- Pulmonary edema worsens. Meanwhile, inflammation leads to fibrosis, which further impedes gas exchange. The resulting hypoxemia leads to metabolic acidosis.

Don't forget to see our full-color depiction of the pathophysiology of ARDS on pages 188 and 189.

Memory jogger

To remember the progression of ARDS, use this mnemonic:

Assault to the pulmonary system

Respiratory distress

Decreased lung compliance

Severe respiratory failure.

What to look for

ARDS initially produces rapid, shallow breathing and dyspnea within hours to days of the initial injury.

Hypoxemia develops, causing an increased drive for ventilation. Because of the effort required to expand the stiff lung, intercostal and suprasternal retractions result.

Fluid accumulation produces crackles and rhonchi. Worsening hypoxemia causes restlessness, apprehension, mental sluggishness, motor dysfunction, and tachycardia.

Severe ARDS causes overwhelming hypoxemia. If uncorrected, this results in hypotension, decreased urine

Battling illness

Treating ARDS

Therapy focuses on correcting the cause of acute respiratory distress syndrome (ARDS) and preventing progression of hypoxemia and respiratory acidosis. Supportive care consists of administering humidified oxygen by continuous positive pressure. However, this therapy alone seldom fulfills the patient's ventilatory requirements, so several other treatments are used.

Ventilation
The primary treatment for ARDS is mechanical ventilation and intubation to increase lung volume, open airways, and improve oxygenation. Positive end-expiratory pressure may be added to increase lung volume and open alveoli.

New techniques
Two new techniques can maximize the benefits and minimize the risks of mechanical ventilation:
• Pressure-controlled inverse ratio ventilation reverses the conventional inspiration to expiration ratio and minimizes the risk of barotrauma. The mechanical breaths are pressure-limited.
• Permissive hypercapnia limits peak inspiratory pressure. Although CO_2 removal is compromised, no treatment is given for subsequent changes in blood hydrogen and oxygen concentration.

Drugs
During mechanical ventilation, sedatives, narcotics, or neuromuscular blocking agents, such as vecuronium, may be ordered. These drugs minimize restlessness, oxygen consumption, and carbon dioxide production and facilitate ventilation.

When ARDS results from fatty emboli or a chemical injury, a short course of high-dose corticosteroids may be given. Sodium bicarbonate may reverse severe metabolic acidosis, and fluids and vasopressors help maintain blood pressure. Nonviral infections require treatment with antimicrobial drugs.

Additional support
Supportive measures include diuretic therapy, correction of electrolyte and acid-base imbalances, and fluid restriction (even small increases in capillary pressures from I.V. fluids can greatly increase interstitial and alveolar edema).

Patients who recover from ARDS may have little or no permanent lung damage.

output, and respiratory and metabolic acidosis. Eventually, ventricular fibrillation or standstill may occur. (See *Treating ARDS*.)

What tests tell you

Arterial blood gas (ABG) analysis with the patient breathing room air initially shows a reduced Pao_2 (less than 60 mm Hg) and a decreased $Paco_2$ (less than 35 mm Hg). Hypoxemia despite increased supplemental oxygen is the hallmark of ARDS. The resulting blood pH usually reflects respiratory alkalosis.

As ARDS worsens, ABG values show the following:
• respiratory acidosis: increasing $Paco_2$ (more than 45 mm Hg)
• metabolic acidosis: decreasing HCO_3 (less than 22 mEq/L)
• declining Pao_2 despite oxygen therapy.

Testing...one, two...testing

Other diagnostic tests include the following:
• Pulmonary artery catheterization is used to identify the cause of edema by measuring pulmonary capillary wedge pressure (12 mm Hg or less in ARDS). Pulmonary artery mixed venous blood shows hypoxia.
• Serial chest X-rays in early stages show bilateral infiltrates. In later stages, they show lung fields with a ground-glass appearance and, with irreversible hypoxemia, "whiteouts" of both lung fields.

Differential diagnosis

A differential diagnosis rules out cardiogenic pulmonary edema, pulmonary vasculitis, and diffuse pulmonary hemorrhage. The following tests are performed to aid in the diagnosis:
• sputum analysis, including Gram stain and culture and sensitivity tests
• blood cultures to identify infectious organisms
• toxicology tests to screen for drug ingestion
• serum amylase tests to rule out pancreatitis.

Asbestosis

Asbestosis is characterized by diffuse interstitial pulmonary fibrosis. Prolonged exposure to airborne asbestos particles causes pleural plaques and tumors of the pleura and peritoneum. Asbestosis may develop 15 to 20 years after the period of regular exposure to asbestos has ended. (See *A close look at asbestosis*.)

Ninety times the risk

Asbestosis occurs when lung spaces become filled with asbestos fibers.

Asbestos is a potent co-carcinogen, increasing a cigarette smoker's risk for lung cancer. An asbestos worker who smokes is 90 times more likely to develop lung cancer than a smoker who never worked with asbestos.

Pathophysiology

Asbestosis is caused by prolonged inhalation of asbestos fibers. People at high risk include workers in the mining, milling, construction, fireproofing, and textile industries. Asbestos is also used in paints, plastics, and brake and clutch linings.

Family members of asbestos workers may develop asbestosis from exposure to stray fibers shaken off the workers' clothing. The general public may be exposed to fibrous asbestos dust in deteriorating buildings or in waste piles from asbestos plants.

Down the airway

Here's what happens in asbestosis:
• Inhaled asbestos fibers travel down the airway and penetrate respiratory bronchioles and alveolar walls.
• Fibers become encased in a brown, iron-rich, protein-like sheath in sputum or lung tissue.
• Interstitial fibrosis may develop in lower lung zones, affecting lung parenchyma and the pleurae.
• Raised hyaline plaques may form in the parietal pleura, the diaphragm, and the pleura adjacent to the pericardium.

What to look for

Signs and symptoms of asbestosis include:
• dyspnea on exertion
• dyspnea at rest with extensive fibrosis
• severe, nonproductive cough in nonsmokers
• productive cough in smokers
• finger clubbing
• chest pain (often pleuritic)
• recurrent respiratory tract infections

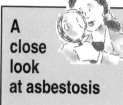

A close look at asbestosis

After years of exposure to asbestos, healthy lung tissue progresses to massive pulmonary fibrosis, as shown below.

Healthy lung tissue

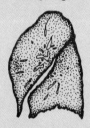

Simple asbestosis

Progressive massive fibrosis

- pleural friction rub and crackles on auscultation
- decreased lung inflation
- recurrent pleural effusions
- decreased forced expiratory volume and vital capacity.

Asbestosis may progress to pulmonary fibrosis with respiratory failure and cardiovascular complications, including pulmonary hypertension and cor pulmonale. (See *Treating asbestosis*.)

What tests tell you

The following tests are used to diagnose asbestosis:
- Chest X-rays may show fine, irregular, linear, diffuse infiltrates. With extensive fibrosis, the lungs have a honeycomb or ground-glass appearance. Other findings include pleural thickening and calcification, bilateral obliteration of costophrenic angles and, in later disease stages, an enlarged heart with a classic "shaggy" border.
- Pulmonary function tests may identify decreased vital capacity, forced vital capacity (FVC), and total lung capacity; decreased or normal forced expiratory volume in 1 second (FEV_1); a normal ratio of FEV_1 to FVC; and reduced diffusing capacity for carbon monoxide when fibrosis destroys alveolar walls and thickens the alveolocapillary membrane.
- ABG analysis may reveal decreased Pao_2 and $Paco_2$ from hyperventilation.

The moral of the story: Inhaled asbestos fibers are bad news.

Asthma

Asthma is a chronic reactive airway disorder. It causes episodic airway obstruction resulting from bronchospasms, increased mucus secretion, and mucosal edema.

Asthma is one type of chronic obstructive pulmonary disease (COPD), a long-term pulmonary disease characterized by air flow resistance. Chronic bronchitis and emphysema (discussed below) are other types of COPD. (See *Understanding COPD*, page 172.)

Although asthma can strike at any age, about half of all patients are under age 10. In this age-group, twice as many boys as girls are affected. About one-third of patients get asthma between ages 10 and 30. In this group, incidence is the same in both sexes. About one-third of

Battling illness

Treating asbestosis

Asbestosis can't be cured. The goal of treatment is to relieve symptoms and control complications. Measures include the following:

- Chest physiotherapy, such as controlled coughing and postural drainage with chest percussion and vibration, helps relieve respiratory signs and symptoms. In advanced disease, it's used to manage hypoxia and cor pulmonale.

- Aerosol therapy, inhaled mucolytics, and fluid intake of at least 3 L daily helps relieve respiratory symptoms.

- Antibiotics should be given promptly for respiratory tract infections.

- Oxygen administration relieves hypoxia. It's given by cannula, mask, or mechanical ventilation.

- Diuretic agents, digitalis glycoside preparations, and salt restriction may be necessary for patients with cor pulmonale.

Extrinsic asthma is caused by sensitivity to specific external allergens.

Extrinsic asthma

Intrinsic asthma

all patients share the disease with at least one immediate family member.

Pathophysiology

In asthma, bronchial linings overreact to various stimuli, causing episodic smooth-muscle spasms that severely constrict the airways. Mucosal edema and thickened secretions further block the airways. (See *Understanding asthma.*)

Asthma can be described as extrinsic or intrinsic:

☝ Extrinsic, or atopic, asthma is sensitive to specific external allergens.

✌ Intrinsic, or nonatopic, asthma is a reaction to internal, nonallergenic factors.

Allergens that cause extrinsic asthma include pollen, animal dander, house dust or mold, kapok or feather pillows, food additives containing sulfites, and any other sensitizing substance. Extrinsic asthma begins in childhood and is commonly accompanied by other hereditary allergies, such as eczema and allergic rhinitis.

No external substance can be implicated in intrinsic asthma. Most episodes occur after a severe respiratory tract infection, especially in adults. Other predisposing conditions include irritants, emotional stress, fatigue, endocrine changes, temperature and humidity variations, and exposure to noxious fumes. Many asthmatics, especially children, have both intrinsic and extrinsic asthma.

What to look for

Patients with mild asthma have adequate air exchange and are asymptomatic between attacks. Signs and symptoms include the following:
• brief wheezing, coughing, and dyspnea on exertion
• intermittent, brief (less than 1 hour) wheezing, coughing, or dyspnea once or twice a week.

Patients with moderate asthma have normal or below-normal air exchange and the following signs and symptoms:

Understanding COPD

Chronic obstructive pulmonary disease (COPD) refers to long-term pulmonary disorders characterized by air flow resistance. Such disorders include asthma, chronic bronchitis, and emphysema. Chronic bronchitis and emphysema are more closely related in cause, pathogenesis, and treatment and are more likely to occur together. Asthma is more acute and intermittent than chronic bronchitis or emphysema.

Predisposing factors
Factors that predispose a patient to COPD include:
• recurrent or chronic respiratory infections
• allergies
• hereditary factors such as an inherited deficiency in $alpha_1$-antitrypsin, an inhibitor to the enzyme trypsin.

Smoking ranks first
Smoking is the most important predisposing factor of COPD. It impairs ciliary action and macrophage function, causing inflammation in the airway, increased mucus production, alveolar destruction, and peribronchiolar fibrosis.

Now I get it!

Understanding asthma

Asthma is an inflammatory disease characterized by hyperresponsiveness of the airway and bronchospasm. These illustrations show the progression of an asthma attack.

When the patient inhales a substance he's hypersensitive to, abnormal antibodies stimulate mast cells in the lung interstitium to release both histamine and the slow-reacting substance of anaphylaxis.

1. Histamine (H) attaches to receptor sites in the larger bronchi, where it causes swelling in smooth muscles.

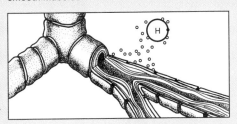

2. Slow-reacting, substance of anaphylaxis (SRS-A) attaches to receptor sites in the smaller bronchi and causes swelling of smooth muscle there. SRS-A also causes fatty acids called prostaglandins to travel by way of the bloodstream to the lungs, where they enhance histamine's effects.

3. Histamine stimulates the mucous membranes to secrete excessive mucus, further narrowing the bronchial lumen, as show below.

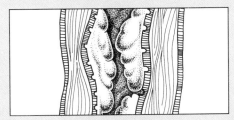

4. On inhalation, the narrowed bronchial lumen can still expand slightly, allowing air to reach the alveoli. On exhalation, increased intrathoracic pressure closes the bronchial lumen completely.

Bronchial lumen on inhalation

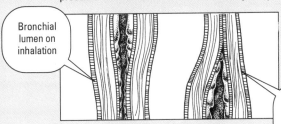

Bronchial lumen on exhalation

5. Mucus fills the lung bases, inhibiting alveolar ventilation, as shown below. Blood, shunted to alveoli in other lung parts, still can't compensate for diminished ventilation.

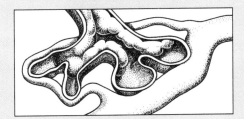

- respiratory distress at rest
- hyperpnea, or an abnormal increase in the depth and rate of respiration
- exacerbations that last several days.
  Patients with severe asthma have continuous signs and symptoms that include the following:
- marked respiratory distress
- marked wheezing or absent breath sounds
- pulsus paradoxus greater than 10 mm Hg
- chest wall contractions. (See *Treating asthma*.)

Battling illness

Treating asthma

The best treatment for asthma is prevention by identifying and avoiding precipitating factors, such as environmental allergens or irritants. Usually, such stimuli can't be removed entirely, so desensitization to specific antigens may be helpful, especially in children. Other common treatments are medication and oxygen.

Medication
Three types of drugs are usually given:

Bronchodilators (theophylline, aminophylline, epinephrine, albuterol, metaproterenol, and terbutaline) decrease bronchoconstriction, reduce bronchial airway edema, and increase pulmonary ventilation. Albuterol and terbutaline cause fewer adverse reactions than epinephrine.

Corticosteroids (hydrocortisone and methylprednisolone) have the same effects as bronchodilators as well as anti-inflammatory and immunosuppressive effects.

Mast cell stabilizers (cromolyn sodium and nedocromil sodium) are effective in patients with atopic asthma who have seasonal disease. When given prophylactically, they block the acute obstructive effects of antigen exposure by inhibiting the degranulation of mast cells, thereby preventing the release of the chemical mediators responsible for anaphylaxis.

Oxygen
Low-flow humidified oxygen may be needed to treat dyspnea, cyanosis, and hypoxemia. The amount delivered is designed to maintain the Pao_2 between 65 and 85 mm Hg, as determined by arterial blood gas studies. Mechanical ventilation is necessary if the patient doesn't respond to initial ventilatory support and drugs or develops respiratory failure.

Alternative therapy
Relaxation exercises, such as yoga, may help increase circulation and help a patient recover from an asthma attack.

What tests tell you

The following tests are used to diagnose asthma:
• Pulmonary function studies reveal signs of airway obstructive disease, low-normal or decreased vital capacity, and increased total lung and residual capacities. Pulmonary function may be normal between attacks. PaO_2 and $PaCO_2$ are usually decreased, except in severe asthma, when $PaCO_2$ may be normal or increased, indicating severe bronchial obstruction.
• Serum immunoglobulin E levels may increase from an allergic reaction.
• Complete blood count with differential reveals increased eosinophil count.
• Chest X-rays can be used to diagnose or monitor asthma's progress and may show hyperinflation with areas of atelectasis.
• ABG analysis is used to detect hypoxemia and guide treatment.
• Skin testing may be used to identify specific allergens. Test results are read in 1 to 2 days to detect an early reaction and again after 4 or 5 days to reveal a late reaction.
• Bronchial challenge testing is used to evaluate the clinical significance of allergens identified by skin testing.

Chronic bronchitis

Chronic bronchitis is inflammation of the bronchi caused by irritants or infection. A form of COPD, chronic bronchitis may be acute or chronic.

In chronic bronchitis, hypersecretion of mucus and chronic productive cough last for 3 months of the year and occur for at least 2 consecutive years. The distinguishing characteristic of bronchitis is obstruction of airflow caused by mucus.

Pathophysiology

Chronic bronchitis occurs when irritants are inhaled for a prolonged period. The result is resistance in the small airways and severe $\dot{V}/\dot{Q}$ imbalance that decreases arterial oxygenation.

Patients have a diminished respiratory drive, so they usually hypoventilate. Chronic hypoxia causes the kidneys to produce erythropoietin. This stimulates excessive RBC production, leading to polycythemia. Hemoglobin levels are high, but the amount of reduced hemoglobin that comes in contact with O_2 is low; therefore, cyanosis is evident.

What to look for

Signs and symptoms of advanced chronic bronchitis include:
- productive cough
- dyspnea
- cyanosis
- use of accessory muscles for breathing
- pulmonary hypertension caused by involvement of small pulmonary arteries (due to inflammation in the bronchial walls and spasms of pulmonary blood vessels from hypoxia).

As pulmonary hypertension continues, right ventricular end-diastolic pressure increases. This leads to cor pulmonale (right ventricular hypertrophy with right-sided heart failure). Heart failure results in increased venous pressure, liver engorgement, and dependent edema. (See *Treating chronic bronchitis*.)

What tests tell you

The following tests are used to diagnose chronic bronchitis:
- Chest X-rays may show hyperinflation and increased bronchovascular markings.
- Pulmonary function tests indicate increased residual volume, decreased vital capacity and forced expiratory flow, and normal static compliance and diffusing capacity.
- ABG analysis displays decreased PaO_2 and normal or increased $PaCO_2$.
- Sputum culture may reveal many microorganisms and neutrophils.
- ECG may show atrial arrhythmias; peaked P waves in leads II, III, and aV_F; and, occasionally, right ventricular hypertrophy.

Battling illness

Treating chronic bronchitis

The most effective treatment for chronic bronchitis is to avoid air pollutants and, if the patient is a smoker, to stop. Other treatments include the following:

- antibiotics to treat recurring infections
- bronchodilators to relieve bronchospasm and facilitate mucus clearance
- adequate hydration
- chest physiotherapy to mobilize secretions
- ultrasonic or mechanical nebulizer treatments to loosen and mobilize secretions
- corticosteroids to combat inflammation
- diuretics for edema
- oxygen for hypoxia.

Cor pulmonale

In this condition, hypertrophy and dilation of the right ventricle develop secondary to a disease affecting the structure or function of the lungs or its vasculature.

Cor pulmonale occurs at the end stage of various chronic disorders of the lungs, pulmonary vessels, chest wall, and respiratory control center. It doesn't occur with disorders stemming from congenital heart disease or those affecting the left side of the heart.

Cor pulmonale causes about 25% of all types of heart failure.

A statement of stats

About 85% of patients with cor pulmonale also have COPD, and about 25% of patients with bronchial COPD eventually develop cor pulmonale. It's most common in smokers and in middle-aged and elderly men; however, the incidence in women is rising.

Pathophysiology

Cor pulmonale results from the following:

☝ disorders that affect the pulmonary parenchyma

✌ pulmonary diseases that affect the airways, such as COPD and bronchial asthma

🤟 vascular diseases, such as vasculitis, pulmonary emboli, or external vascular obstruction resulting from a tumor or aneurysm

🖖 chest wall abnormalities, including thoracic deformities such as kyphoscoliosis and pectus excavatum (funnel chest)

🖐 neuromuscular disorders, such as muscular dystrophy and poliomyelitis

🖐☝ external factors, such as obesity or living at a high altitude.

As long as the heart can compensate

In cor pulmonale, pulmonary hypertension increases the heart's workload. To compensate, the right ventricle hypertrophies to force blood through the lungs. As long as the heart can compensate for the increased pulmonary vascular resistance, signs and symptoms reflect only the underlying disorder.

What to look for

In early stages of cor pulmonale, patients are most likely to report the following:
- chronic productive cough
- exertional dyspnea
- wheezing respirations
- fatigue and weakness.

As the compensatory mechanism begins to fail, larger amounts of blood remain in the right ventricle at the end of diastole, causing ventricular dilation. As cor pulmonale progresses, these additional symptoms occur:
- dyspnea even at rest
- tachypnea
- orthopnea
- dependent edema
- distended neck veins
- enlarged, tender liver
- hepatojugular reflux (distention of the jugular vein induced by pressing over the liver)
- right upper quadrant discomfort
- tachycardia
- decreased cardiac output.

Chest examination reveals characteristics of the underlying lung disease.

Complications

In response to hypoxia, the bone marrow produces more RBCs, causing polycythemia. The blood's viscosity increases, which further aggravates pulmonary hypertension. This increases the right ventricle's workload, causing heart failure.

Eventually, cor pulmonale may lead to biventricular failure, hepatomegaly, edema, ascites, and pleural effusions. Polycythemia increases the risk of thromboembolism. (See *Understanding cor pulmonale.*)

Because cor pulmonale occurs late in the course of COPD and other irreversible diseases, the prognosis is poor. (See *Treating cor pulmonale,* page 180.)

What tests tell you

The following tests are used to diagnose cor pulmonale:
- Pulmonary artery catheterization shows increased right ventricular and pulmonary artery pressures, resulting from increased pulmonary vascular resistance. Both right

Now I get it!

Understanding cor pulmonale

Three types of disorders are responsible for cor pulmonale:
• pulmonary restrictive disorders, such as fibrosis or obesity
• pulmonary obstructive disorders such as bronchitis
• primary vascular disorders such as recurrent pulmonary emboli.

These disorders share a common pathway to the formation of cor pulmonale. Hypoxic constriction of pulmonary blood vessels and obstruction of pulmonary blood flow lead to increased pulmonary resistance, which progresses to cor pulmonale.

Three different types of disorders may cause cor pulmonale but all share a common pathway.

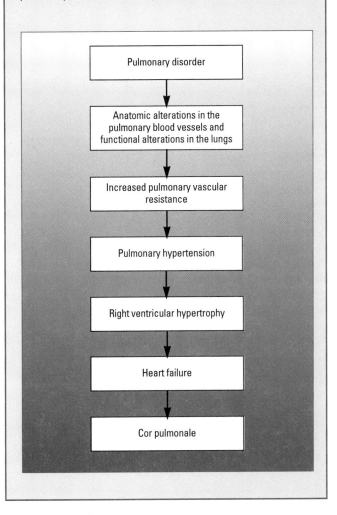

Pulmonary disorder
↓
Anatomic alterations in the pulmonary blood vessels and functional alterations in the lungs
↓
Increased pulmonary vascular resistance
↓
Pulmonary hypertension
↓
Right ventricular hypertrophy
↓
Heart failure
↓
Cor pulmonale

Battling illness

Treating cor pulmonale

Therapy for the patient with cor pulmonale has three aims:

 reducing hypoxemia and pulmonary vasoconstriction

 increasing exercise tolerance

 correcting the underlying condition when possible.

Bed rest, drug therapy, and more
Treatment includes the following:
• bed rest
• digitalis glycosides such as digoxin
• antibiotics for an underlying respiratory tract infection
• a potent pulmonary artery vasodilator, such as diazoxide, nitroprusside, or hydralazine, to treat primary pulmonary hypertension
• continuous administration of low concentrations of oxygen to decrease pulmonary hypertension, polycythemia, and tachypnea
• mechanical ventilation in acute disease
• a low-sodium diet with restricted fluid
• phlebotomy to decrease red blood cell mass and anticoagulation with small doses of heparin to decrease the risk for thromboembolism.

   Treatment may vary, depending on the underlying cause. For example, the patient may need a tracheotomy if he has an upper airway obstruction. He may require corticosteroids if he has vasculitis or an autoimmune disorder.

ventricular systolic and pulmonary artery systolic pressures are over 30 mm Hg, and pulmonary artery diastolic pressure is higher than 15 mm Hg.
• Echocardiography or angiography demonstrates right ventricular enlargement.
• Chest X-rays reveal large central pulmonary arteries and right ventricular enlargement.
• ABG analysis detects decreased PaO_2 (usually less than 70 mm Hg and never more than 90 mm Hg).
• ECG discloses arrhythmias, such as premature atrial and ventricular contractions and atrial fibrillation during severe hypoxia. It may also show right bundle-branch block, right axis deviation, prominent P waves and an inverted T wave in right precordial leads.

• Pulmonary function tests reflect underlying pulmonary disease.
• Magnetic resonance imaging (MRI) measures right ventricular mass, wall thickness, and ejection fraction.
• Cardiac catheterization measures pulmonary vascular pressures.
• Hematocrit is typically over 50%.
• Serum hepatic enzyme levels show an elevated level of aspartate aminotransferase with hepatic congestion and decreased liver function.
• Serum bilirubin levels may be elevated if liver dysfunction and hepatomegaly are present.

Emphysema

A form of COPD, emphysema is the abnormal, permanent enlargement of the acini accompanied by destruction of the alveolar walls. Obstruction results from tissue changes, rather than mucus production, which is the case in asthma and chronic bronchitis. The distinguishing characteristic of this disorder is airflow limitation caused by lack of elastic recoil in the lungs.

Pathophysiology

Emphysema may be caused by a deficiency of alpha$_1$-antitrypsin or cigarette smoking.

In emphysema, recurrent inflammation is associated with the release of proteolytic enzymes (enzymes that promote splitting of proteins by hydrolysis of peptide bonds) from lung cells. This causes irreversible enlargement of the air spaces distal to the terminal bronchioles. Enlargement of air spaces destroys the alveolar walls, which results in a breakdown of elasticity and loss of fibrous and muscle tissues, making the lungs less compliant. (See *Understanding emphysema,* page 182.)

 ## What to look for

Signs and symptoms of emphysema include:
• dyspnea on exertion (initial symptom)
• barrel-shaped chest from lung overdistention

Now I get it!

Understanding emphysema

In normal, healthy breathing, air moves in and out of the lungs to meet metabolic needs. Any change in airway size compromises the lungs' ability to circulate sufficient air.

In a patient with emphysema, recurrent pulmonary inflammation damages and eventually destroys the alveolar walls, creating large air spaces. This breakdown leaves the alveoli unable to recoil normally after expanding and results in bronchiolar collapse on expiration. This traps air in the lungs, leading to overdistention.

No cyanosis
Associated pulmonary capillary destruction usually allows a patient with severe emphysema to match ventilation to perfusion and thereby avoid cyanosis. The lungs are usually enlarged; therefore, total lung capacity and residual volume increase.

Normal alveoli

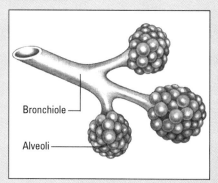

Bronchiole

Alveoli

Abnormal alveoli

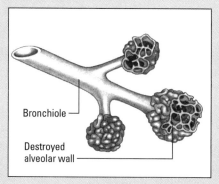

Bronchiole

Destroyed alveolar wall

Treating emphysema

Patients with emphysema need to be counseled about avoiding smoking and air pollution. They also need some or all of the following treatments:

• bronchodilators, such as aminophylline, to promote mucociliary clearance
• antibiotics to treat respiratory tract infections
• immunizations to prevent influenza and pneumococcal pneumonia
• adequate hydration
• chest physiotherapy to mobilize secretions
• oxygen therapy at low settings to correct hypoxia and transtracheal catheterization to receive oxygen at home.

• prolonged expiration because accessory muscles are used for inspiration and abdominal muscles are used to force air out of the lungs
• decreased breath sounds.

Because minimal $\dot{V}/\dot{Q}$ imbalance occurs, hyperventilation keeps blood gases within a normal range until late in the disease. (See *Treating emphysema*.)

What tests tell you

The following tests are used to diagnose emphysema:
• Chest X-rays in advanced disease may show a flattened diaphragm, reduced vascular markings at the lung periph-

ery, overaeration of the lungs, a vertical heart, enlarged anteroposterior chest diameter, and large retrosternal air space.

• Pulmonary function tests indicate increased residual volume and total lung capacity, reduced diffusing capacity, and increased inspiratory flow.

• ABG analysis usually shows reduced Pao_2 and normal $Paco_2$ until late in the disease.

• ECG may reveal tall, symmetrical P waves in leads II, III, and aV_F; vertical QRS axis; and signs of right ventricular hypertrophy late in the disease.

• Complete blood count usually shows an increased hemoglobin level late in the disease when the patient has persistent severe hypoxia.

Pneumonia

If I don't perform gas exchange properly, the blood has too much carbon dioxide and too little oxygen.

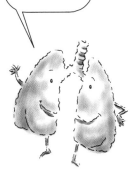

Pneumonia is an acute infection of the lung parenchyma that often impairs gas exchange.

It occurs in both sexes and at all ages. More than 3 million cases of pneumonia occur annually in the United States. It's the leading cause of death from infectious disease.

The prognosis is good for patients with normal lungs and adequate immune systems. However, bacterial pneumonia is the leading cause of death in debilitated patients. (See *Treating pneumonia* and *Distinguishing among types of pneumonia,* page 184.)

Pathophysiology

Pneumonia is classified three ways:

☞ *by origin.* Pneumonia may be viral, bacterial, fungal, or protozoal in origin.

✌ *by location.* Bronchopneumonia involves distal airways and alveoli; lobular pneumonia, part of a lobe; and lobar pneumonia, an entire lobe.

☝ *by type.* Primary pneumonia results from inhalation or aspiration of a pathogen, such as bacteria or a virus, and includes pneumococcal and viral pneumonia. Secondary pneumonia may follow lung damage from a noxious chemical or other insult or may result from hematogenous spread of bacteria. Aspiration pneumonia results from in-

Battling illness

Treating pneumonia

The patient with pneumonia needs antimicrobial therapy based on the causative agent. Reevaluation should be done early in treatment.

Supportive measures include the following:

• humidified oxygen therapy for hypoxia
• bronchodilator therapy
• antitussives
• mechanical ventilation for respiratory failure
• a high-calorie diet and adequate fluid intake
• bed rest
• an analgesic to relieve pleuritic chest pain
• positive end-expiratory pressure ventilation to maintain adequate oxygenation for patients with severe pneumonia on mechanical ventilation.

Distinguishing among types of pneumonia

Type	Characteristics
Viral	
Influenza	• Prognosis poor even with treatment • 50% mortality from cardiopulmonary collapse • Cough (initially nonproductive; later, purulent sputum), marked cyanosis, dyspnea, high fever, chills, substernal pain and discomfort, moist crackles, frontal headache, myalgia
Adenovirus	• Insidious onset • Generally affects young adults • Good prognosis; usually clears with no residual effects • Sore throat, fever, cough, chills, malaise, small amounts of mucoid sputum, retrosternal chest pain, anorexia, rhinitis, adenopathy, scattered crackles, and rhonchi
Respiratory syncytial virus	• Most prevalent in infants and children • Complete recovery in 1 to 3 weeks • Listlessness, irritability, tachypnea with retraction of intercostal muscles, slight sputum production, fever, severe malaise, possible cough or croup, and fine, moist crackles
Measles (rubeola)	• Fever, dyspnea, cough, small amounts of sputum, rash, cervical adenopathy, and profusely runny nose
Chickenpox (varicella pneumonia)	• Uncommon in children, but present in 30% of adults with varicella • Characteristic rash, cough, dyspnea, cyanosis, tachypnea, pleuritic chest pain, and hemoptysis and rhonchi 1 to 6 days after onset of rash
Cytomegalovirus	• Difficult to distinguish from other nonbacterial pneumonias • In adults with healthy lung tissue, resembles mononucleosis and is generally benign; in neonates, occurs as devastating multisystemic infection; in immunocompromised hosts, varies from clinically inapparent to fatal infection • Fever, cough, shaking chills, dyspnea, cyanosis, weakness, and diffuse crackles
Bacterial	
Streptococcus	• Sudden onset of a single, shaking chill, and sustained temperature of 102° to 104° F (38.9° to 40° C); often preceded by upper respiratory tract infection
Klebsiella	• More likely in patients with chronic alcoholism, pulmonary disease, and diabetes • Fever and recurrent chills; cough producing rusty, bloody, viscous sputum (currant jelly); cyanosis of lips and nail beds from hypoxemia; shallow, grunting respirations

Clinical manifestations of different types of pneumonia vary.

Distinguishing among types of pneumonia (continued)

Type	Characteristics
Bacterial (continued)	
Staphylococcus	• Commonly occurs in patients with viral illness, such as influenza or measles, and in those with cystic fibrosis. • Temperature of 102° to 104° F, recurrent shaking chills, bloody sputum, dyspnea, tachypnea, and hypoxemia.
Aspiration	
	• Results from vomiting and aspiration of gastric or oropharyngeal contents into trachea and lungs • Noncardiogenic pulmonary edema possible with damage to respiratory epithelium from contact with gastric acid • Subacute pneumonia possible with cavity formation • Lung abscess possible if foreign body present • Crackles, dyspnea, cyanosis, hypotension, and tachycardia

halation of foreign matter, such as vomitus or food particles, into the bronchi.

Letting in a pathogen

In general, the lower respiratory tract can be exposed to pathogens by inhalation, aspiration, vascular dissemination, or direct contact with contaminated equipment such as suction catheters. Once inside, pathogens begin to colonize and infection develops.

How bacterial pneumonia develops

In bacterial pneumonia, which can occur in any part of the lungs, an infection initially triggers alveolar inflammation and edema. This produces an area of low ventilation with normal perfusion. Capillaries become engorged with blood, causing stasis. As the alveolocapillary membrane breaks down, alveoli fill with blood and exudate, resulting in atelectasis, or lung collapse. (See *A close look at atelectatic alveoli*, page 186.)

In severe bacterial infections, the lungs look heavy and liverlike, reminiscent of ARDS.

I can sneak into the lower respiratory tract through inhalation, aspiration, or vascular dissemination or via contaminated equipment.

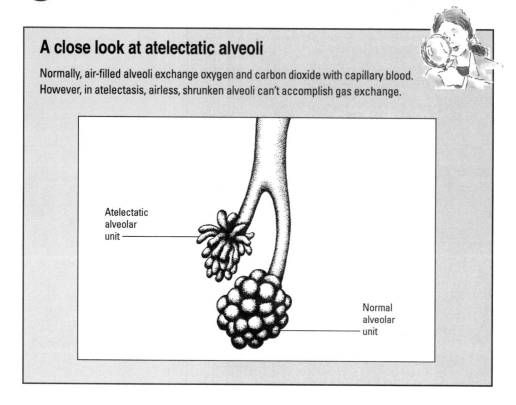

A close look at atelectatic alveoli

Normally, air-filled alveoli exchange oxygen and carbon dioxide with capillary blood. However, in atelectasis, airless, shrunken alveoli can't accomplish gas exchange.

Atelectatic alveolar unit

Normal alveolar unit

How viral pneumonia develops

In viral pneumonia, the virus first attacks bronchiolar epithelial cells. This causes interstitial inflammation and desquamation. The virus also invades bronchial mucous glands and goblet cells. It then spreads to the alveoli, which fill with blood and fluid. In advanced infection, a hyaline membrane may form. Like bacterial infections, viral pneumonia clinically resembles ARDS.

How aspiration pneumonia develops

In aspiration pneumonia, inhalation of gastric juices or hydrocarbons triggers inflammatory changes and also inactivates surfactant over a large area. Decreased surfactant leads to alveolar collapse. Acidic gastric juices may damage the airways and alveoli. Particles containing aspirated gastric juices may obstruct the airways and reduce airflow, leading to secondary bacterial pneumonia.

Risk raisers

Certain predisposing factors increase the risk of pneumonia. For bacterial and viral pneumonia, these include:

(Text continues on page 191.)

Incredibly Easy miniguide: Asthma

Remember, asthma is characterized by bronchospasm, increased mucus secretion, and mucosal edema. All three cause obstruction.

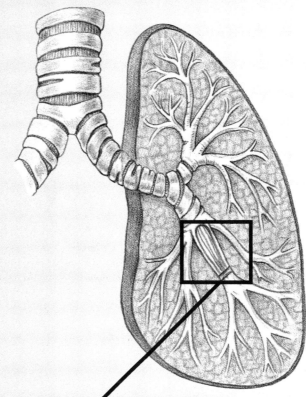

Normal bronchiole

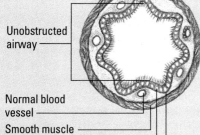

Unobstructed airway

Normal blood vessel

Smooth muscle

Epithelial cells

Normal basement membrane

Obstructed bronchiole

Muscle spasm

Airway obstructed with mucus plug

Epithelial cells

Engorged blood vessel

Thickening of basement membrane

Notice the difference between the obstructed and normal bronchiole.

Incredibly Easy miniguide: Adult respiratory distress syndrome

Adult respiratory distress syndrome is a form of pulmonary edema that can quickly lead to acute respiratory failure.

There are no signs or symptoms yet.

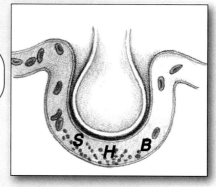

In phase 1, injury reduces normal blood flow to the lungs. Platelets aggregate and release histamine (H), serotonin (S), and bradykinin (B).

Oh my, now the patient develops tachypnea, dyspnea, and tachycardia.

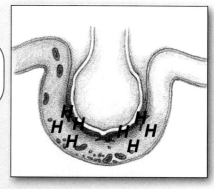

In phase 2, those substances, especially histamine, inflame and damage the alveolocapillary membrane, increasing capillary permeability. Fluids then shift into the interstitial space.

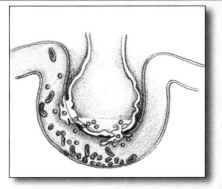

In phase 3, as capillary permeability increases, proteins and fluids leak out, increasing interstitial osmotic pressure and causing pulmonary edema.

Phase 3 next! Look out for increased tachypnea, dyspnea, cyanosis. hypoxemia, decreased pulmonary compliance, crackles, and rhonchi.

Stay alert! Time for phase 4. Look for thick, frothy, sticky sputum and marked hypoxemia with increased respiratory distress.

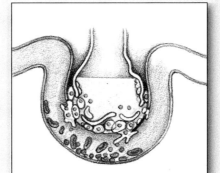

In phase 4, decreased blood flow and fluids in the alveoli damage surfactant and impair the cell's ability to produce more. As a result, alveoli collapse, impairing gas exchange.

Next, it's phase 5! The patient may develop increased tachypnea, hypoxemia, and hypocapnia.

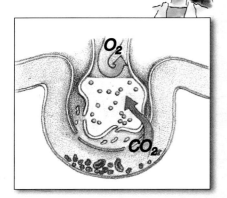

In phase 5, sufficient oxygen can't cross the alveolocapillary membrane, but carbon dioxide (CO_2) can and is lost with every exhalation. Oxygen (O_2) and carbon dioxide levels decrease in the blood.

Uh-oh! Now metabolic acidosis develops.

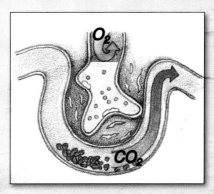

In phase 6, pulmonary edema worsens, inflammation leads to fibrosis, and gas exchange is further impeded.

Incredibly Easy miniguide: Pneumothorax

Traumatic open pneumothorax

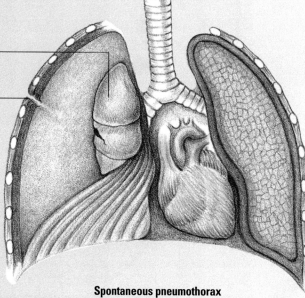

Complete collapse

Knife wound

In traumatic open pneumothorax, air flows between the pleural space and the outside of the body. Here pneumothorax occurs as the result of a knife wound.

In pneumothorax, air in the pleural space prohibits complete lung expansion.

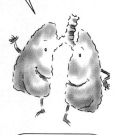

Spontaneous pneumothorax

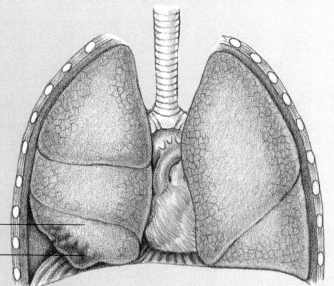

Partial collapse

Ruptured bleb

In spontaneous pneumothorax, a type of closed pneumothorax, air enters the pleural space from within the lung. The usual cause is a ruptured bleb on the surface of the lung.

> A long list of factors can predispose a patient to bacterial and viral pneumonia.

- chronic illness and debilitation
- cancer (particularly lung cancer)
- abdominal and thoracic surgery
- atelectasis
- colds or other viral respiratory infections
- chronic respiratory disease, such as COPD, asthma, bronchiectasis, or cystic fibrosis
- influenza
- smoking
- malnutrition
- alcoholism
- sickle cell disease
- tracheostomy
- exposure to noxious gases
- aspiration
- immunosuppressive therapy.

Aspiration pneumonia is more likely to occur in elderly or debilitated patients, those receiving nasogastric tube feedings, and those with an impaired gag reflex, poor oral hygiene, or a decreased level of consciousness.

What to look for

The clinical manifestations of different types of pneumonia vary.

What tests tell you

The following tests are used to diagnose pneumonia:
- Chest X-rays confirm the diagnosis by disclosing infiltrates.
- Sputum specimen, Gram stain, and culture and sensitivity tests help differentiate the type of infection and the drugs that are effective in treatment.
- White blood cell count indicates leukocytosis in bacterial pneumonia and a normal or low count in viral or mycoplasmal pneumonia.
- Blood cultures reflect bacteremia and are used to determine the causative organism.
- ABG levels vary, depending on the severity of pneumonia and the underlying lung state.
- Bronchoscopy or transtracheal aspiration allows the collection of material for culture.

• Pulse oximetry may show a reduced oxygen saturation (SaO_2) level.

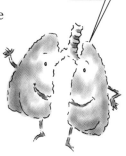

When air accumulates in my pleural cavity, I begin to collapse.

Pneumothorax

Pneumothorax is an accumulation of air in the pleural cavity that leads to partial or complete lung collapse. When the amount of air between the visceral and parietal pleurae increases, increasing tension in the pleural cavity can cause the lung to progressively collapse. In some cases, venous return to the heart is impeded, causing a life-threatening condition called tension pneumothorax.

Are you spontaneous or traumatic?

Pneumothorax is classified as either traumatic or spontaneous. A traumatic pneumothorax may be further classified as open (sucking chest wound) or closed (blunt or penetrating trauma). Note that an open (penetrating) wound may cause a closed pneumothorax if communication between the atmosphere and the pleural space seals itself off. A spontaneous pneumothorax, which is also considered closed, can be further classified as primary (idiopathic) or secondary (related to a specific disease).

Pathophysiology

Spontaneous pneumothorax may occur from unknown causes or from an underlying pulmonary disease such as COPD.

Yikes! Injury, pneumothorax, hemothorax

A penetrating injury, such as a stab wound, gunshot wound, or impaled object may cause traumatic open pneumothorax (sucking chest wound), traumatic closed pneumothorax, or hemothorax (accumulation of blood in the pleural cavity).

Blunt trauma from a car crash, a fall, or a crushing chest injury may also cause a traumatic closed pneumothorax or hemothorax.

Traumatic pneumothorax also may result from the following:

 insertion of a central line

 thoracic surgery

thoracentesis

 pleural or transbronchial biopsy.

Tension pneumothorax can develop from either spontaneous or traumatic pneumothorax. (See *Understanding tension pneumothorax,* page 194.)

How open pneumothorax occurs

Open pneumothorax results when atmospheric air (positive pressure) flows directly into the pleural cavity (negative pressure). As the air pressure in the pleural cavity becomes positive, the lung collapses on the affected side. Lung collapse leads to decreased total lung capacity. The patient then develops $\dot{V}/\dot{Q}$ imbalance leading to hypoxia.

How closed pneumothorax occurs

Closed pneumothorax occurs when air enters the pleural space from within the lung. This causes increased pleural pressure and prevents lung expansion during inspiration. It may be called traumatic pneumothorax when blunt chest trauma causes lung tissue to rupture, resulting in air leakage.

How spontaneous pneumothorax occurs

Spontaneous pneumothorax is a type of closed pneumothorax. It's more common in men and in older patients with chronic pulmonary disease. However, it also occurs in healthy young adults. The usual cause is rupture of a subpleural bleb (a small cystic space) at the surface of the lung. This causes air leakage into the pleural spaces; then the lung collapses, causing hypoxia and decreased total lung capacity, vital capacity, and lung compliance. The total amount of lung collapse can range from 5% to 95%.

Don't forget to see our full-color illustration of pneumothorax on page 190.

 ## What to look for

Although the causes of traumatic and spontaneous pneumothorax vary greatly, the effects are similar. The cardinal signs and symptoms of pneumothorax include:
• sudden, sharp, pleuritic pain exacerbated by chest movement, breathing, and coughing
• asymmetric chest wall movement
• shortness of breath
• cyanosis

Now I get it!

Understanding tension pneumothorax

In tension pneumothorax, air accumulates intrapleurally and can't escape. Intrapleural pressure rises, collapsing the ipsilateral lung.

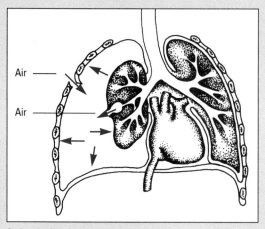

On inspiration, the mediastinum shifts toward the unaffected lung, impairing ventilation.

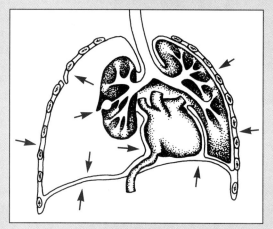

On expiration, the mediastinal shift distorts the vena cava and reduces venous return.

• respiratory distress.

The signs and symptoms of open pneumothorax also include:
• absent breath sounds on affected side
• chest rigidity on affected side
• tachycardia
• crackling beneath the skin on palpation, indicating subcutaneous emphysema (air in the tissues).

Tension pneumothorax produces the most severe respiratory symptoms, including:
• decreased cardiac output
• hypotension
• compensatory tachycardia
• tachypnea
• lung collapse due to air or blood in the intrapleural space

Battling illness

Treating pneumothorax

Treatment depends on the type of pneumothorax.

Spontaneous
Treatment is usually conservative for spontaneous pneumothorax under these conditions:
• no signs of increased pleural pressure
• lung collapse less than 30%
• no dyspnea or indications of physiologic compromise.
   Such treatment consists of bed rest; careful monitoring of blood pressure, pulse, and respiratory rate; oxygen administration; and, possibly, aspiration of air with a large-bore needle attached to a syringe.
   If more than 30% of the lung collapses, a thoracostomy tube is placed in the second or third intercostal space in the midclavicular line to try to reexpand the lung. The tube then connects to an underwater seal or to low-pressure suction. Treatment for recurring spontaneous pneumothorax is thoracotomy and pleurectomy, which causes the lung to adhere to the parietal pleura.

Traumatic
Traumatic pneumothorax requires chest tube drainage and may also require surgical repair.

Tension
Tension pneumothorax is a medical emergency. If the tension in the pleural space isn't relieved, the patient will die from inadequate cardiac output or hypoxemia. A large-bore needle is inserted into the pleural space through the second intercostal space. If large amounts of air escape through the needle after insertion, it's left in place until a thoracostomy tube can be inserted.

• mediastinal shift and tracheal deviation to the opposite side
• cardiac arrest. (See *Treating pneumothorax.*)

What tests tell you

The following tests are used to diagnose pneumothorax:
• Chest X-rays confirm the diagnosis by revealing air in the pleural space and, possibly, a mediastinal shift.
•ABG studies may show hypoxemia, possibly with respiratory acidosis and hypercapnia. Sao_2 levels may decrease at first, but typically return to normal within 24 hours.

Pulmonary edema

Pulmonary edema is a common complication of cardiac disorders. It's marked by accumulated fluid in the extravascular spaces of the lung. It may occur as a chronic condition or develop quickly and rapidly become fatal.

Pathophysiology

Pulmonary edema may result from left-sided heart failure caused by arteriosclerotic, cardiomyopathic, hypertensive, or valvular heart disease.

Off balance

Normally, pulmonary capillary hydrostatic pressure, capillary oncotic pressure, capillary permeability, and lymphatic drainage are in balance. This prevents fluid infiltration to the lungs. When this balance changes, or the lymphatic drainage system is obstructed, pulmonary edema results.

If colloid osmotic pressure decreases, the hydrostatic force that regulates intravascular fluids is lost because nothing opposes it. Fluid flows freely into the interstitium and alveoli, impairing gas exchange and leading to pulmonary edema. (See *Understanding pulmonary edema.*)

Let's face it. We have to depend on each other.

What to look for

Signs and symptoms vary with the stage of pulmonary edema. In the early stages, the following occur:
- dyspnea on exertion
- paroxysmal nocturnal dyspnea
- orthopnea
- cough
- mild tachypnea
- increased blood pressure
- dependent crackles
- neck vein distention
- diastolic S_3 gallop
- tachycardia.

As tissue hypoxia and decreased cardiac output occur, you'll see these signs and symptoms:
- labored, rapid respiration
- more diffuse crackles
- cough producing frothy, bloody sputum
- increased tachycardia
- arrhythmias
- cold, clammy skin
- diaphoresis
- cyanosis
- falling blood pressure

Now I get it!

Understanding pulmonary edema

In pulmonary edema, diminished function of the left ventricle causes blood to pool there and in the left atrium. Eventually, blood backs up into the pulmonary veins and capillaries.

Increasing capillary hydrostatic pressure pushes fluid into the interstitial spaces and alveoli. The illustrations below show a normal alveolus and the effects of pulmonary edema.

Normal alveolus

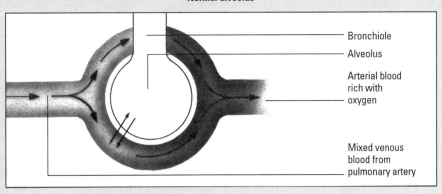

Bronchiole

Alveolus

Arterial blood rich with oxygen

Mixed venous blood from pulmonary artery

Alveolus in patient with pulmonary edema

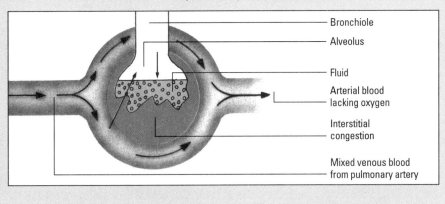

Bronchiole

Alveolus

Fluid

Arterial blood lacking oxygen

Interstitial congestion

Mixed venous blood from pulmonary artery

Note that gas exchange is impaired. Oxygen can't get to tissues.

• thready pulse. (See *Treating pulmonary edema*, page 198.)

What tests tell you

Clinical features of pulmonary edema permit a working diagnosis. The following diagnostic tests are used to confirm the disease:

Battling illness

Treating pulmonary edema

Treatment for pulmonary edema has three aims:

 reducing extravascular fluid

 improving gas exchange and myocardial function

 correcting the underlying disease, if possible.

The following treatments are used:

• high concentrations of oxygen administered by nasal cannula (the patient usually can't tolerate a mask)
• in persistently low arterial oxygen levels, assisted ventilation to improve oxygen delivery to the tissues and acid-base balance
• diuretics, such as furosemide, ethacrynic acid, and bumetanide, to increase urination, which helps mobilize extravascular fluid
• positive inotropic agents, such as digitalis glycosides and amrinone, to enhance contractility in myocardial dysfunction
• pressor agents to enhance contractility and promote vasoconstriction in peripheral vessels
• antiarrhythmics for arrhythmias related to decreased cardiac output
• arterial vasodilators such as nitroprusside to decrease peripheral vascular resistance, preload, and afterload
• morphine to reduce anxiety and dyspnea and dilate the systemic venous bed, promoting blood flow from pulmonary circulation to the periphery.

• ABG analysis usually shows hypoxia with variable $Paco_2$, depending on the patient's degree of fatigue. Metabolic acidosis may be revealed.
• Chest X-rays show diffuse haziness of the lung fields and, usually, cardiomegaly and pleural effusion.
• Pulse oximetry may reveal decreasing Sao_2 levels.
• Pulmonary artery catheterization identifies left-sided heart failure and helps rule out ARDS.
• ECG may show previous or current myocardial infarction.

Tuberculosis

Tuberculosis is an infectious disease that primarily affects the lungs but can invade other body systems. In tuberculosis, pulmonary infiltrates accumulate, cavities develop, and masses of granulated tissue form within the lungs. Tuberculosis may occur as an acute or chronic infection.

The American Lung Association estimates that active tuberculosis afflicts nearly 14 of every 100,000 people in the United States. Tuberculosis is twice as common in men and four times as common in nonwhites as in whites. Incidence is highest in people who live in crowded, poorly ventilated, unsanitary conditions, such as prisons, tenement houses, and homeless shelters. The typical newly diagnosed tuberculosis patient is a single, homeless, nonwhite man.

Pathophysiology

Tuberculosis results from exposure to *Mycobacterium tuberculosis* and, sometimes, other strains of mycobacteria. Here's what happens:

I'm flying on a droplet and hoping to spread.

• *Transmission.* An infected person coughs or sneezes, spreading infected droplets. When someone without immunity inhales these droplets, the bacilli are deposited in the lungs.

• *Immune response.* The immune system responds by sending leukocytes, and inflammation results. After a few days, leukocytes are replaced by macrophages. Bacilli are then ingested by the macrophages and carried off by the lymphatics to the lymph nodes.

No doubt, the immune system will try to wipe me out.

• *Tubercle formation.* Machrophages that ingest the bacilli fuse to form epithelioid cell tubercles, tiny nodules surrounded by lymphocytes. Within the lesion, caseous necrosis develops and scar tissue encapsulates the tubercle. The organism may or may not be killed in the process.

• *Dissemination.* If the tubercles and inflamed nodes rupture, the infection contaminates the surrounding tissue and may spread through the blood and lymphatic circulation to distant sites. This process is called hematogenous dissemination. (See *Understanding tuberculosis invasion,* page 200.)

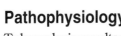

What to look for

After exposure to *M. tuberculosis,* roughly 5% of infected people develop active tuberculosis within 1 year. They may complain of a low-grade fever at night, a productive cough lasting longer than 3 weeks, and symptoms of airway obstruction from lymph node involvement.

In other infected people, microorganisms cause a latent infection. The host's immunologic defense system may destroy the bacillus. Alternatively, the encapsulated bacilli may live within the tubercle. It may lie dormant for years, reactivating later to cause active infection.

Now I get it!

Understanding tuberculosis invasion

After infected droplets are inhaled, they enter the lungs and are deposited either in the lower part of the upper lobe or the upper part of the lower lobe. Leukocytes surround the droplets, which leads to inflammation. As part of the inflammatory response, some mycobacteria are carried off in the lymphatic circulation by the lymph nodes.

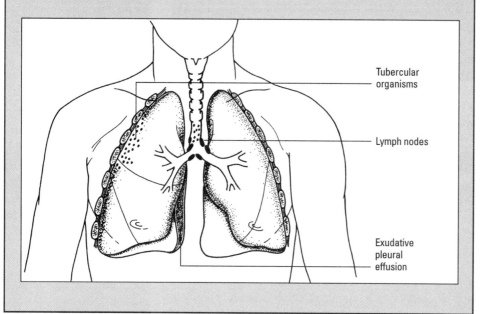

Tubercular organisms

Lymph nodes

Exudative pleural effusion

Not just damage to the lungs

Tuberculosis can cause massive pulmonary tissue damage, with inflammation and tissue necrosis eventually leading to respiratory failure. Bronchopleural fistulas can develop from lung tissue damage, resulting in pneumothorax. The disease can also lead to hemorrhage, pleural effusion, and pneumonia. Small mycobacterial foci can infect other body organs, including the kidneys, the skeleton, and the central nervous system.

With proper treatment, the prognosis for a patient with tuberculosis is usually excellent. (See *Treating tuberculosis.*)

I may harm other organs and body systems in addition to the lungs.

What tests tell you

The following tests are used to diagnose tuberculosis:
• Chest X-rays show nodular lesions, patchy infiltrates (mainly in upper lobes), cavity formation, scar tissue, and calcium deposits.
• Tuberculin skin test reveals infection at some point, but doesn't indicate active disease.
• Stains and cultures of sputum, CSF, urine, drainage from abscesses, or pleural fluid show heat-sensitive, non-motile, aerobic, acid-fast bacilli.
• CT or MRI scans allow the evaluation of lung damage and may confirm a difficult diagnosis.
• Bronchoscopy shows inflammation and altered lung tissue. It also may be performed to obtain sputum if the patient can't produce an adequate sputum specimen.

Several of these tests may be needed to distinguish tuberculosis from other diseases that mimic it, such as lung carcinoma, lung abscess, pneumoconiosis, and bronchiectasis.

Battling illness

Treating tuberculosis

The usual treatment is daily oral doses of isoniazid or rifampin, with ethambutol added in some cases, for at least 9 months. After 2 to 4 weeks, the disease is no longer infectious, and the patient can resume normal activities while continuing to take medication.

The patient with atypical mycobacterial disease or drug-resistant tuberculosis may require second-line drugs, such as capreomycin, streptomycin, para-aminosalicylic acid, pyrazinamide, and cycloserine.

All done. Now catch your breath, and then try our quick quiz.

Quick quiz

1. One of the hallmark signs of ARDS is a pulmonary capillary wedge pressure of:
    A. 16 mm Hg.
    B. 12 mm Hg.
    C. 18 mm Hg.

Answer: B. Patients with ARDS have a pulmonary capillary wedge pressure of 12 mm Hg or less because their pulmonary edema isn't cardiac in nature. An elevated pressure would indicate pulmonary edema related to heart failure.

2. Asthma is caused by:
    A. sensitivity to specific allergens.
    B. a severe respiratory tract infection.
    C. emotional stress.

Answer: A. Asthma may result from sensitivity to specific external allergens, including pollen, animal dander, house dust, and mold. Although a severe respiratory tract infection may precede an attack and emotional stress may aggravate asthma, these things don't cause the disease.

3. Tuberculosis is transmitted through:
    A. the fecal-oral route.
    B. contact with blood.
    C. inhalation of infected droplets.

Answer: C. Transmission occurs when an infected person coughs or sneezes, spreading infected droplets.

Scoring

   If you answered two or three correctly, excellent! When it comes to understanding the respiratory system, no one can accuse you of being full of hot air.

    If you answered less than two correctly, that's life. Take a deep breath and get ready for Chapter 8 on the cardiovascular system. Remember, the beat goes on.

Cardiovascular system

Just the facts

In this chapter you'll learn:

♦ the relationship between oxygen supply and demand

♦ the risk factors for cardiovascular disease

♦ the causes, pathophysiology, diagnostic tests, and treatments for several common cardiovascular disorders.

Understanding the cardiovascular system

The cardiovascular system begins its activity when the fetus is barely a month old. It's the last system to cease activity at the end of life.

The heart, arteries, veins, and lymphatics make up the cardiovascular system. These structures transport life-supporting oxygen and nutrients to cells, remove metabolic waste products, and carry hormones from one part of the body to another. Circulation requires normal heart function, which propels blood through the system by continuous rhythmic contractions.

Despite advances in disease detection and treatment, cardiovascular disease remains the leading cause of death in the United States. Heart attack, or myocardial infarction (MI), is the primary cause of cardiovascular related deaths. MI often occurs with little or no warning.

My function is so vital that it defines the very presence of life.

Oxygen balancing act

A critical balance exists between myocardial oxygen supply and demand. A decrease in oxygen supply or an increase in oxygen demand can disturb this balance and threaten myocardial function.

The four major determinants of myocardial oxygen demand are:

 heart rate

 contractile force

 muscle mass

ventricular wall tension.

Cardiac workload and oxygen demand increase if the heart rate speeds up or if the force of contractions becomes elevated. This can occur in hypertension, ventricular dilation, or heart muscle hypertrophy.

The heart's law of supply and demand

If myocardial oxygen demand increases, so must oxygen supply. To effectively increase oxygen supply, coronary perfusion must also increase. Tissue hypoxia — the most potent stimulus — causes coronary arteries to dilate and increases coronary flow. Normal coronary vessels can dilate and increase flow five to six times above resting levels. However, stenotic, diseased vessels can't dilate, so oxygen deficit may result.

If myocardial oxygen demand increases, so must oxygen supply.

One-way ticket

Normally, blood flows unimpeded across the valves in one direction. The valves open and close in response to a pressure gradient. When the pressure in the chamber proximal to the valve exceeds the pressure in the chamber beyond the valve, the valves open. When the pressure beyond the valve exceeds the pressure in the proximal chamber, the valves close. The valve leaflets, or cusps, are so responsive that even a pressure difference of less than 1 mm Hg between chambers will open and close them.

Why do you flow so low?

Valvular disease is the major cause of low blood flow. A diseased valve allows blood to flow backward across leaflets that haven't closed securely. This phenomenon is called regurgitation. The backflow of blood through the valves forces the heart to pump more blood, increasing cardiac workload. The valve opening may also become restricted and impede the forward flow of blood.

Another cause of low flow

The heart may fail to meet the tissues' metabolic requirements for blood and fail to function as a pump. Eventually, the circulatory system may fail to perfuse body tissues, and blood volume and vascular tone may be altered.

Inner awareness

The body closely monitors both blood volume and vascular tone. Blood flow to each tissue is monitored by microvessels, which measure how much blood each tissue needs and control the local blood flow. The nerves that control circulation also help direct blood flow to tissues.

If my valves don't function properly, I have to work even harder.

How the heart listens

The heart pays attention to the tissues' demands. It responds to the return of blood through the veins and to nerve signals that make it pump the required amounts of blood.

Paying attention to arterial pressure

Arterial pressure is carefully regulated by the body: If it falls below or rises above its normal mean level, immediate circulatory changes occur.

If arterial pressure falls below normal, the following changes occur:
- increased heart rate
- increased force of contraction
- increased constriction of arterioles.

If arterial pressure rises above normal, the following changes occur:
- reflex slowing of heart rate
- decreased force of contraction
- vasodilation.

Risk factors

Some risk factors for cardiovascular disease are controllable and some aren't.

Most risk factors for heart disease are controllable.

Controllable risk factors

Some risk factors can be avoided or altered, potentially slowing the disease process or even reversing it. These factors include:
- elevated serum lipid levels
- hypertension
- cigarette smoking
- diabetes mellitus
- sedentary lifestyle
- stress
- obesity
- excessive intake of saturated fats, carbohydrates, and salt.

Uncontrollable risk factors

Four uncontrollable factors increase a person's risk of cardiovascular disease. They are:

   age

   male gender

   family history

   race.

A few you just have to live with.

Safer before 40

Susceptibility to cardiovascular disease increases with age; disease before age 40 is unusual. However, the age-disease correlation may simply reflect the longer duration of exposure to other risk factors.

Gender bias?

Women are less susceptible than men to heart disease until after menopause; then they become as susceptible as men. One theory proposes that estrogen has a protective effect.

Nature vs. nurture

A positive family history also increases a person's chances of developing premature cardiovascular disease. For example, genetic factors can cause some pronounced, accelerated forms of atherosclerosis such as lipid disease. However, family history of cardiovascular disease may reflect a strong environmental component. Risk factors — such as obesity or a lifestyle that causes tension — may recur in families.

Race as a risk factor

Although cardiovascular disease affects all races, blacks are more susceptible to disease than whites.

Cardiovascular disorders

The disorders discussed below include cardiac tamponade, coronary artery disease (CAD), dilated cardiomyopathy, heart failure, hypertension, hypertrophic cardiomyopathy, MI, pericarditis, and rheumatic heart disease.

Cardiac tamponade

In cardiac tamponade, a rapid rise in intrapericardial pressure impairs diastolic filling of the heart. The rise in pressure usually results from blood or fluid accumulation in the pericardial sac. As little as 200 ml of fluid can create an emergency if it accumulates rapidly. If untreated, cardiogenic shock and death can occur.

If fluid accumulates slowly and pressure rises — such as in pericardial effusion caused by cancer — signs and symptoms may not be evident immediately. This is because the fibrous wall of the pericardial sac can stretch to accommodate up to 2 L of fluid.

As little as 200 ml of fluid may cause cardiac tamponade.

Don't forget to check out our full-color illustration of cardiac tamponade on page 223.

Pathophysiology

Cardiac tamponade may result from the following:
• effusion, such as in cancer, bacterial infections, tuberculosis and, rarely, acute rheumatic fever
• hemorrhage caused by trauma, such as a gunshot or stab wound in the chest, cardiac surgery, or perforation

by a catheter during cardiac or central venous catheterization

• hemorrhage from nontraumatic causes, such as rupture of the heart or great vessels and anticoagulant therapy in a patient with pericarditis

• viral, postirradiation, or idiopathic pericarditis

• acute MI

• chronic renal failure during dialysis

• drug reaction from procainamide, hydralazine, minoxidil, isoniazid, penicillin, methysergide, or daunorubicin

• connective tissue disorders, such as rheumatoid arthritis, systemic lupus erythematosus, rheumatic fever, vasculitis, and scleroderma.

If cardiac tamponade has no known cause, it's called Dressler's syndrome.

Too much fluid, not enough blood

In cardiac tamponade, the progressive accumulation of fluid in the pericardium causes compression of the heart chambers. This obstructs blood flow into the ventricles and reduces the amount of blood that can be pumped out of the heart with each contraction.

Every time the ventricles contract, more fluid accumulates in the pericardial sac. This further limits the amount of blood that can fill the chamber during the next cardiac cycle. Reduced cardiac output may be fatal without prompt treatment.

The amount of fluid necessary to cause cardiac tamponade varies greatly. It may be as small as 200 ml when the fluid accumulates rapidly or more than 2,000 ml if the fluid accumulates slowly and the pericardium stretches to adapt.

 ## What to look for

Cardiac tamponade has three classic features known as Beck's triad. They are:

 elevated central venous pressure (CVP) with neck vein distention

muffled heart sounds

pulsus paradoxus (inspiratory drop in systemic blood pressure greater than 15 mm Hg).

Other signs and symptoms include:
• orthopnea
• diaphoresis
• anxiety
• restlessness
• cyanosis
• weak, rapid peripheral pulse. (See *Treating cardiac tamponade.*)

What tests tell you

The following tests are used to diagnose cardiac tamponade:
• Chest X-ray shows a slightly widened mediastinum and enlargement of the cardiac silhouette.
• Electrocardiography (ECG) rules out other cardiac dis-

Battling illness

Treating cardiac tamponade

The goal of treatment for cardiac tamponade is to relieve intrapericardial pressure and cardiac compression by removing the accumulated blood or fluid. This can be done three different ways:
• pericardiocentesis, or needle aspiration of the pericardial cavity
• surgical creation of an opening
• insertion of a drain into the pericardial sac to drain the effusion.

More measures
In the hypotensive patient, cardiac output is maintained through trial volume loading with I.V. normal saline solution with albumin and, perhaps, an inotropic drug such as dopamine.

Depending on the cause of tamponade, additional treatment may include the following:
• in traumatic injury, blood transfusion or a thoracotomy to drain reaccumulating fluid or to repair bleeding sites
• in heparin-induced tamponade, the heparin antagonist protamine
• in warfarin-induced tamponade, vitamin K.

orders. The QRS amplitude may be reduced, and electrical alternans of the P wave, QRS complex, and T wave may be present. Generalized ST-segment elevation is noted in all leads.

• Pulmonary artery pressure monitoring reveals increased right atrial pressure or CVP and right ventricular diastolic pressure.

• Echocardiography records pericardial effusion with signs of right ventricular and atrial compression.

Coronary artery disease

CAD causes the loss of oxygen and nutrients to myocardial tissue because of poor coronary blood flow. This disease is nearly epidemic in the Western world. It's most prevalent in white, middle-aged men and in the elderly, with more than 50% of men age 60 or older showing signs of CAD on autopsy.

Pathophysiology

Atherosclerosis is the most common cause of CAD. In this condition, fatty, fibrous plaques, possibly including calcium deposits, progressively narrow the coronary artery lumens, which reduces the volume of blood that can flow through them. This can lead to myocardial ischemia (heart tissue death).

What you can and can't control

Many risk factors are associated with atherosclerosis and CAD. Some are controllable and some not.

Uncontrollable risk factors include being over age 40, being male, being white, and having a family history of CAD. Controllable risk factors include the following:

• systolic blood pressure greater than 160 mm Hg or diastolic blood pressure greater than 95 mm Hg

• increased low-density and decreased high-density lipoprotein levels

• smoking (risk dramatically drops within 1 year of quitting)

• stress or being a type A personality (aggressive, competitive, addicted to work, chronically impatient)

• obesity, which increases the risk of diabetes mellitus, hypertension, and high cholesterol

• inactivity

Don't forget to see the full-color illustration of atherosclerosis on page 226.

• diabetes mellitus, especially in women.

Other risk factors that can be modified include increased levels of serum fibrinogen and uric acid, elevated hematocrit, reduced vital capacity, high resting heart rate, thyrotoxicosis, and the use of oral contraceptives.

Less common causes

Other factors that can reduce blood flow include:
• dissecting aneurysm
• infectious vasculitis
• syphilis
• congenital defects in the coronary vascular system.

Coronary artery spasms can also impede blood flow. These spontaneous, sustained contractions of one or more coronary arteries occlude the vessel, reduce blood flow to the myocardium, and cause angina pectoris (chest pain) Without treatment, ischemia and, eventually, MI result.

A precarious balance

As atherosclerosis progresses, luminal narrowing is accompanied by vascular changes that impair the diseased vessel's ability to dilate. This causes a precarious balance between myocardial oxygen supply and demand, threatening the myocardium beyond the lesion. When oxygen demand exceeds what the diseased vessels can supply, localized myocardial ischemia results.

From aerobic to anaerobic

Transient ischemia causes reversible changes at the cellular and tissue levels, depressing myocardial function. Untreated, it can lead to tissue injury or necrosis. Oxygen deprivation forces the myocardium to shift from aerobic to anaerobic metabolism. As a result, lactic acid (the end product of anaerobic metabolism) accumulates. This reduces cellular pH.

With each contraction, less blood

The combination of hypoxia, reduced energy availability, and acidosis rapidly impairs left ventricular function. The strength of contractions in the affected myocardial region is reduced as the fibers shorten inadequately with less force and velocity. In addition, the ischemic section's wall motion is abnormal. This generally results in less blood being ejected from the heart with each contraction.

Haphazard hemodynamics

Because of reduced contractility and impaired wall motion, the hemodynamic response becomes variable. It depends on the ischemic segment's size and the degree of reflex compensatory response by the autonomic nervous system.

Depression of left ventricular function may reduce stroke volume and thereby lower cardiac output. Reduction in systolic emptying increases ventricular volumes. As a result, left-sided heart pressures and pulmonary wedge pressure increase.

Compliance counts

These increases in left-sided heart pressures and pulmonary wedge pressure are magnified by changes in wall compliance induced by ischemia. Compliance is reduced, magnifying the elevation in pressure.

A sympathetic response

During ischemia, sympathetic nervous system response leads to slight elevations in blood pressure and heart rate before the onset of pain. With the onset of pain, further sympathetic activation occurs.

What to look for

Angina is the classic sign of CAD. The patient may describe a burning, squeezing, or crushing tightness in the substernal or precordial area that radiates to the left arm, neck, jaw, or shoulder blade. He may clench his fist over his chest or rub his left arm when describing it. Pain is often accompanied by nausea, vomiting, fainting, sweating, and cool extremities.

Angina commonly occurs after physical exertion but may also follow emotional excitement, exposure to cold, or eating a large meal. Sometimes, it develops during sleep and awakens the patient.

Stable or unstable?

If the pain is predictable and relieved by rest or nitrates, it's called stable angina. If it increases in frequency and duration and is more easily induced, it's called unstable or unpredictable angina. Left untreated, unstable angina may progress to MI. (See *Treating CAD*.)

What tests tell you

The following diagnostic tests confirm CAD:
• ECG during an episode of angina shows ischemia, as demonstrated by T-wave inversion, ST-segment depres-

Battling illness

Treating CAD

Because coronary artery disease (CAD) is so widespread, controlling risk factors is important. Other treatment may focus on one of two goals: reducing myocardial oxygen demand or increasing the oxygen supply and alleviating pain. Interventions may be noninvasive or invasive.

Controlling risk

Because CAD is so widespread, overweight patients should limit calories, and all patients should limit their intake of salt, fats, and cholesterol and stop smoking. Regular exercise is important, although it may need to be done more slowly to alleviate pain. If stress is a known pain trigger, patients should learn stress reduction techniques.

Other preventive actions include controlling hypertension with diuretics or beta blockers, controlling elevated serum cholesterol or triglyceride levels with antilipemics, and minimizing platelet aggregation and blood clot formation with aspirin.

Noninvasive measures

Drug therapy consists mainly of nitrates, such as nitroglycerin, isosorbide dinitrate, or beta-adrenergic blockers that dilate vessels. Aspirin is also given to reduce arterial inflammation and occlusion.

Invasive measures

Three invasive treatments are commonly used: coronary artery bypass graft (CABG) surgery, percutaneous transluminal coronary angioplasty (PTCA), and laser angioplasty.

CABG

Critically narrowed or blocked arteries may need CABG surgery to alleviate uncontrollable angina and prevent myocardial infarction (MI). In this procedure, a part of the saphenous vein in the leg or the internal mammary artery in the chest are grafted between the aorta and the affected artery beyond the obstruction.

PTCA

PTCA may be performed during cardiac catheterization to compress fatty deposits and relieve occlusion. In patients with calcification, this procedure may reduce the obstruction by fracturing the plaque.

PTCA causes fewer complications than surgery, but it does have these risks:

• circulatory insufficiency
• MI
• restenosis of the vessels
• retroperitoneal bleeding
• sudden coronary reocclusions
• vasovagal response and arrhythmias
• death (rare).

PTCA is a good alternative to grafting in elderly patients and others who can't tolerate cardiac surgery. However, patients with a left main coronary artery occlusion or lesions in extremely tortuous vessels aren't candidates for PTCA.

Laser angioplasty

Laser angioplasty corrects occlusion by vaporizing fatty deposits with a hot-tip laser device. Rotational ablation, or rotational atherectomy, removes plaque with a high-speed, rotating burr covered with diamond crystals.

sion and, possibly, arrhythmias such as premature ventricular contractions. Results may be normal during pain-free periods. Arrhythmias may occur without infarction, secondary to ischemia.

• Treadmill or bicycle exercise stress test may provoke chest pain and ECG signs of myocardial ischemia. Monitoring of electrical rhythm may demonstrate T-wave inversion or ST-segment depression in the ischemic areas.

• Coronary angiography reveals the location and extent of coronary artery stenosis or obstruction, collateral circulation, and the arteries' condition beyond the narrowing.

• Myocardial perfusion imaging with thallium-201 during treadmill exercise detects ischemic areas of the myocardium, visualized as "cold spots."

Dilated cardiomyopathy

Also called congestive cardiomyopathy, dilated cardiomyopathy results from extensively damaged myocardial muscle fibers. This disorder interferes with myocardial metabolism and grossly dilates every heart chamber, giving the heart a globular shape.

This disease usually affects middle-aged men but can occur in any age-group. Because it usually isn't diagnosed until the advanced stages, the prognosis is generally poor. Most patients, especially those over age 55, die within 2 years of symptom onset.

Pathophysiology

The exact cause of dilated cardiomyopathy is unknown. It may be linked to myocardial destruction caused by the following:

• infectious agents, as in viral myocarditis (especially after infection with coxsackievirus B, poliovirus, or influenza virus) and acquired immunodeficiency syndrome

• metabolic agents that cause endocrine and electrolyte disorders and nutritional deficiencies, as in hyperthyroidism, pheochromocytoma, beriberi, and kwashiorkor

• muscle disorders, such as myasthenia gravis, muscular dystrophy, and myotonic dystrophy

• infiltrative disorders, such as hemochromatosis and amyloidosis

• sarcoidosis

• rheumatic fever, especially in children with myocarditis

Just my luck. Damage my muscle fibers and I lose my beautiful figure.

- alcoholism
- use of doxorubicin, cyclophosphamide, cocaine, and flu- orouracil
- X-linked inheritance patterns.

The pregnancy link

In addition, dilated cardiomyopathy may develop during the last trimester of pregnancy or a few months after delivery. Its cause is unknown, but it occurs most often in multiparous women over age 30, particularly those with malnutrition or preeclampsia. In some patients, cardiomegaly and heart failure reverse with treatment, allowing a subsequent normal pregnancy. But if cardiomegaly persists despite treatment, the prognosis is poor.

Complications

Dilated cardiomyopathy can lead to intractable heart failure, arrhythmias, and emboli. Ventricular arrhythmias may lead to syncope and sudden death.

Blame it on a sluggish ventricle

Cardiomyopathy involves the ventricular myocardium, as opposed to other heart structures, such as the valves or coronary arteries. Dilated cardiomyopathy is characterized by a grossly dilated, hypodynamic ventricle that contracts poorly and, to a lesser degree, myocardial hypertrophy. All four chambers become dilated as a result of increased volumes and pressures. Thrombi frequently develop within these chambers due to blood pooling and stasis, which may lead to embolization.

If hypertrophy coexists, the heart ejects blood less efficiently. A large volume remains in the left ventricle after systole, causing heart failure.

The onset of the disease is usually insidious. It may progress to end-stage refractory heart failure. If so, the patient's prognosis is poor. He may need heart transplantation.

What to look for

Signs and symptoms of dilated cardiomyopathy include:
- shortness of breath
- orthopnea
- dyspnea on exertion and paroxysmal nocturnal dyspnea
- fatigue

- dry cough at night
- palpitations
- vague chest pain
- narrow pulse pressure
- irregular rhythms
- S_3 and S_4 gallop rhythms
- pansystolic murmur. (See *Treating dilated cardiomyopathy.*)

What tests tell you

No single test confirms dilated cardiomyopathy. Diagnosis requires elimination of other possible causes of heart failure and arrhythmias. The following tests are used:

- ECG and angiography rule out ischemic heart disease. ECG may also show biventricular hypertrophy, sinus tachycardia, atrial enlargement, ST-segment and T-wave abnormalities and, in 20% of patients, atrial fibrillation or

Battling illness

Treating dilated cardiomyopathy

The goals of treatment for dilated cardiomyopathy are to correct the underlying causes and to improve the heart's pumping ability. The second goal is achieved with digitalis glycosides, diuretics, oxygen, anticoagulants, vasodilators, and a low-sodium diet supplemented by vitamin therapy. Antiarrhythmics may be used to treat arrhythmias.

Therapy may also include prolonged bed rest and selective use of corticosteroids, particularly when myocardial inflammation is present. Vasodilators reduce preload and afterload, decreasing congestion and increasing cardiac output.

Therapy: Acute, long-term, and final options
Acute heart failure necessitates vasodilation with I.V. nitroprusside or nitroglycerin. Long-term treatment may include prazosin, hydralazine, isosorbide dinitrate, and anticoagulants if the patient is on bed rest. Dopamine, dobutamine, and amrinone may be useful during the acute stage.

When these treatments fail, heart transplantation may be the only option for carefully selected patients.

left bundle-branch block. QRS complexes are decreased in amplitude.
• Chest X-ray demonstrates moderate to marked cardiomegaly, usually affecting all heart chambers, along with pulmonary congestion, pulmonary venous hypertension, and pleural effusion. Pericardial effusion may give the heart a globular shape.
• Echocardiography may identify ventricular thrombi, global hypokinesis, and the degrees of left ventricular dilation and dysfunction.
• Cardiac catheterization can show left ventricular dilation and dysfunction, elevated left ventricular and right ventricular filling pressures, and diminished cardiac output.
• Gallium scans may identify patients with dilated cardiomyopathy and myocarditis.
• Transvenous endomyocardial biopsy may be useful in some patients to determine the underlying disorder, such as amyloidosis or myocarditis.

Heart failure

When the myocardium can't pump effectively enough to meet the body's metabolic needs, heart failure occurs. Pump failure usually occurs in a damaged left ventricle, but it may also happen in the right ventricle. Usually, left-sided heart failure develops first.

Heart failure is classified in the following ways:

 high output or low output

 acute or chronic

left-sided or right-sided (See *Understanding left-sided and right-sided heart failure,* pages 218 and 219.)

forward or backward. (See *Classifying heart failure,* page 220.)

Symptoms of heart failure may restrict a person's ability to perform activities of daily living and severely affect quality of life. Advances in diagnostic and therapeutic techniques have greatly improved the outlook for these patients. However, the prognosis still depends on the underlying cause and its response to treatment.

Now I get it!

Understanding left-sided and right-sided heart failure

These illustrations show how myocardial damage leads to heart failure.

Left-sided heart failure

1. Increased workload and end-diastolic volume enlarge the left ventricle (see illustration below). Because of lack of oxygen, the ventricle enlarges with stretched tissue rather than functional tissue. The patient may experience increased heart rate, pale and cool skin, tingling in the extremities, decreased cardiac output, and arrhythmias.

2. Diminished left ventricular function allows blood to pool in the ventricle and the atrium and eventually back up into the pulmonary veins and capillaries, as shown below. At this stage, the patient may experience dyspnea on exertion, confusion, dizziness, postural hypotension, decreased peripheral pulses and pulse pressure, cyanosis, and an S_3 gallop.

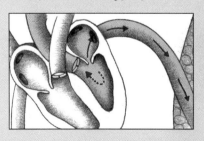

3. As the pulmonary circulation becomes engorged, rising capillary pressure pushes sodium and water into the interstitial space (as shown below), causing pulmonary edema. You'll note coughing, subclavian retractions, crackles, tachypnea, elevated pulmonary artery pressure, diminished pulmonary compliance, and increased partial pressure of carbon dioxide.

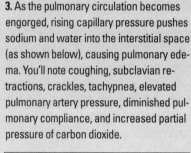
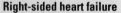

4. When the patient lies down, fluid in the extremities moves into the systemic circulation. Because the left ventricle can't handle the increased venous return, fluid pools in the pulmonary circulation, worsening pulmonary edema. You may note decreased breath sounds, dullness on percussion, crackles, and orthopnea.

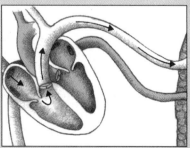

5. The right ventricle may now become stressed because it's pumping against greater pulmonary vascular resistance and left ventricular pressure (see illustration below). When this occurs, the patient's symptoms worsen.

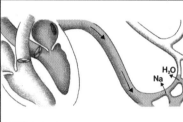

Right-sided heart failure

6. The stressed right ventricle enlarges with the formation of stretched tissue, as shown below. Increasing conduction time and deviation of the heart from its normal axis can cause arrhythmias. If the patient doesn't already have left-sided heart failure, he may experience increased heart rate, cool skin, cyanosis, decreased cardiac output, palpitations, and dyspnea.

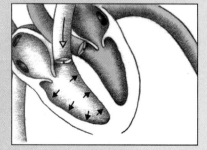

Understanding left-sided and right-sided heart failure *(continued)*

7. Blood pools in the right ventricle and right atrium. The backed-up blood causes pressure and congestion in the vena cava and systemic circulation, as shown below. The patient will have elevated central venous pressure, jugular vein distention, and hepatojugular reflux.

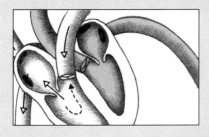

8. Backed-up blood also distends the visceral veins, especially the hepatic vein. As the liver and spleen become engorged (see illustration below), their function is impaired. The patient may develop anorexia, nausea, abdominal pain, palpable liver and spleen, weakness, and dyspnea secondary to abdominal distention.

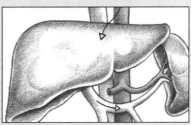

9. Rising capillary pressure forces excess fluid from the capillaries into the interstitial space, as shown below. This causes tissue edema, especially in the lower extremities and abdomen. The patient may experience weight gain, pitting edema, and nocturia.

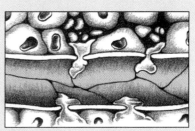

Pathophysiology

Heart failure may result from a primary abnormality of the heart muscle, such as an infarction, that impairs ventricular function and prevents the heart from pumping enough blood. It's also caused by problems unrelated to MI, including:

• mechanical disturbances in ventricular filling during diastole, which occur because blood volume is too low for the ventricle to pump. This problem occurs in mitral stenosis secondary to rheumatic heart disease or constrictive pericarditis and in atrial fibrillation.

• systolic hemodynamic disturbances — such as excessive cardiac workload caused by volume overload or pressure overload — that limit the heart's pumping ability. This problem can result from mitral or aortic insufficiency, which leads to volume overload. It can also result from aortic stenosis or systemic hypertension, which cause increased resistance to ventricular emptying and decreased cardiac output.

Classifying heart failure

Heart failure may be classified different ways according to its pathophysiology.

Right-sided or left-sided

Right-sided heart failure is a result of ineffective right ventricular contractile function. It may be due to an acute right ventricular infarction or pulmonary embolus. However, the most common cause is profound backward flow due to left-sided heart failure.

Left-sided heart failure is the result of ineffective left ventricular contractile function. It may lead to pulmonary congestion or pulmonary edema and decreased cardiac output. Left ventricular myocardial infarction (MI), hypertension, and aortic and mitral valve stenosis or regurgitation are common causes.

As the decreased pumping ability of the left ventricle persists, fluid accumulates, backing up into the left atrium and then into the lungs. If this worsens, pulmonary edema and right-sided heart failure may also result.

Forward or backward

Forward heart failure is caused by inadequate delivery of blood to the arterial system. It's a direct result of increased afterload, which causes less blood to be ejected from the left ventricle and decreased perfusion to vital organs. It's usually a result of hypertension or aortic stenosis.

Backward heart failure is caused when the left ventricle fails to empty. It results in the progressive accumulation of fluid on the left side of the heart and can lead to right-sided heart failure. It's associated with MI and cardiomyopathy.

Acute or chronic

"Acute" refers to the timing of the onset of symptoms and whether compensatory mechanisms kick in. Typically, fluid status is normal or low, and sodium and water retention don't occur.

In chronic heart failure, signs and symptoms have been present for a period of time, compensatory mechanisms have taken effect, and fluid volume overload persists. Drugs, diet changes, and activity restrictions usually control symptoms.

Low or high ventricular output

Low output failure results from MI, hypotension, cardiomyopathies, or hemorrhage. High output failure may be caused by pregnancy, thyrotoxicosis, and anemia. In both types, patients fully compensate for decreased ventricular function.

Factors favorable to failure

Certain conditions can predispose a patient to heart failure, especially if he has underlying heart disease. These include:

• arrhythmias, such as tachyarrhythmias, which can reduce ventricular filling time; arrhythmias that disrupt the normal atrial and ventricular filling synchrony; and bradycardia, which can reduce cardiac output

• pregnancy and thyrotoxicosis, which increase cardiac output

• pulmonary embolism, which elevates pulmonary arterial pressures, causing right ventricular failure

• infections, which increase metabolic demands and further burden the heart

• anemia, which leads to increased cardiac output to meet the oxygen needs of the tissues

• increased physical activity, increased salt or water intake, emotional stress, or failure to comply with the prescribed treatment regimen for the underlying heart disease.

Complications

Eventually sodium and water may enter the lungs, causing pulmonary edema, a life-threatening condition. Decreased perfusion to the brain, kidneys, and other major organs can cause them to fail. MI can occur because the oxygen demands of the overworked heart can't be met.

How the body reacts

The patient's underlying condition determines whether heart failure is acute or insidious.

Heart failure is often associated with systolic or diastolic overloading and myocardial weakness. As stress on the heart muscle reaches a critical level, the muscle's contractility is reduced and cardiac output declines. Venous input to the ventricle remains the same, however.

The body's responses to decreased cardiac output include:
• reflex increase in sympathetic activity
• release of renin from the juxtaglomerular cells of the kidney
• anaerobic metabolism by affected cells
• increased extraction of oxygen by the peripheral cells.

Making some adjustments

When blood in the ventricles increases, the heart makes the following adaptations (also known as compensation):
• *short-term adaptations.* As the end-diastolic fiber length increases, the ventricular muscle responds by dilating and increasing the force of contraction. (This is called the Frank-Starling curve.)
• *long-term adaptation.* Ventricular hypertrophy increases the heart muscle's ability to contract and push its volume of blood into the circulation.

Compensation may occur for long periods of time before signs and symptoms develop. (See *Treating heart failure,* page 222.)

I'm an incredibly adaptable organ, if I do say so myself.

Battling illness

Treating heart failure

The goal of treatment for heart failure is to improve pump function, thereby reversing the compensatory mechanisms that produce or intensify the clinical effects.

Heart failure can usually be controlled quickly with the following treatments:
• administration of diuretics, such as furosemide, hydrochlorothiazide, ethacrynic acid, bumetanide, spironolactone, or triamterene, to reduce total blood volume and circulatory congestion
• prolonged bed rest
• oxygen administration to increase oxygen delivery to the myocardium and other vital organs
• administration of inotropic drugs, such as digoxin, to strengthen myocardial contractility; sympathomimetics, such as dopamine and dobutamine, in acute situations; or amrinone, to increase contractility and cause arterial vasodilation
• administration of vasodilators to increase cardiac output or angiotensin-converting enzyme inhibitors to decrease afterload
• antiembolism stockings to prevent venostasis and thromboembolism formation.

Acute pulmonary edema

As a result of decreased contractility and elevated fluid volume and pressure, fluid may be driven from the pulmonary capillary beds into the alveoli, causing pulmonary edema. Treatment for acute pulmonary edema includes the following:
• administration of morphine
• administration of nitroglycerin or nitroprusside to diminish blood return to the heart
• administration of dobutamine, dopamine, or amrinone to increase myocardial contractility and cardiac output
• administration of diuretics to reduce fluid volume
• administration of supplemental oxygen
• high Fowler's position.

Continued care

After recovery, the patient must continue medical care and usually must continue taking digitalis glycosides, diuretics, and potassium supplements. The patient with valve dysfunction who has recurrent, acute heart failure may need surgical valve replacement.

What to look for

The early signs and symptoms of heart failure include:
• fatigue
• exertional, paroxysmal, and nocturnal dyspnea
• neck vein engorgement
• hepatomegaly.
    Later signs and symptoms include:
• tachypnea
• palpitations
• dependent edema
• unexplained, steady weight gain
• nausea
• chest tightness
• slowed mental response

Geez, if I can't do my job, look what happens!

(Text continues on page 227.)

Incredibly Easy miniguide: Cardiac tamponade

Heart in cardiac tamponade

In cardiac tamponade, blood or fluid fills the pericardial space, compressing the heart chambers, increasing intracardial pressure, and obstructing venous return.

Crimped superior vena cava

Fibrous pericardium

Pericardial space with fluid

Normal heart

Note the normal heart at left. Unlike a heart with tamponade, this healthy heart is able to deliver oxygen, nutrients, and other substances to all the body's cells.

Normal superior vena cava

Fibrous pericardium without fluid in the pericardial space

Incredibly Easy miniguide: Myocardial infarction

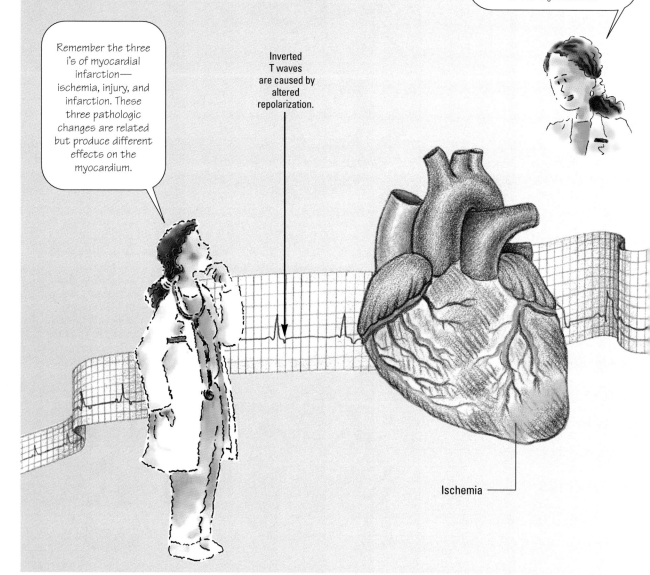

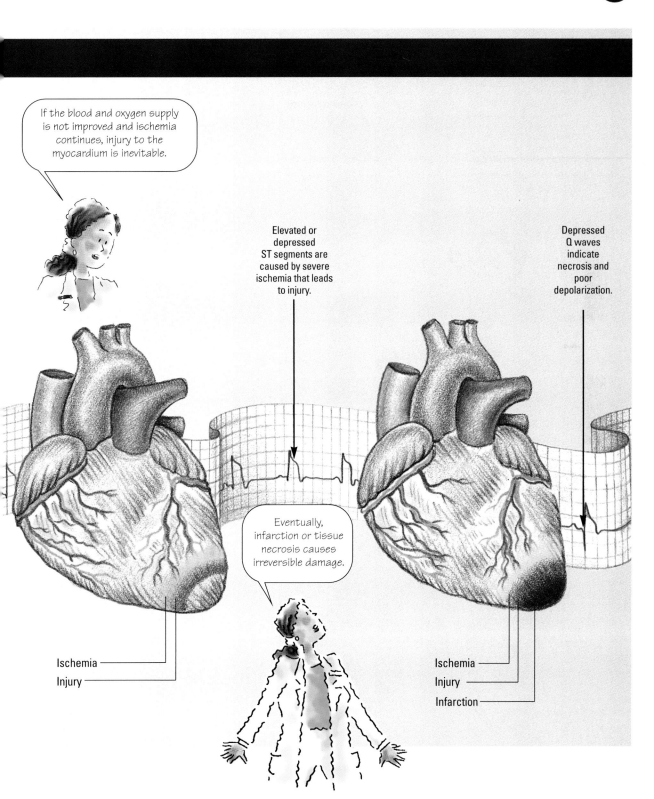

Incredibly Easy miniguide: Atherosclerosis

Coronary artery

Atherosclerosis is the most common cause of coronary heart disease. Stiffening and loss of dilatory response leads to fatty deposits in the vessel.

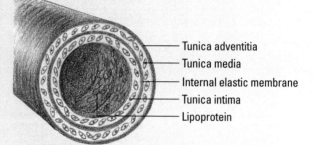

- Tunica adventitia
- Tunica media
- Internal elastic membrane
- Tunica intima
- Lipoprotein

Accumulation of fibrous plaques and lipids progressively narrows the lumen and impedes blood flow to the myocardium.

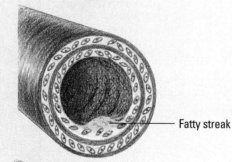

- Fatty streak

In advanced stages, calcification or rupture may occur.

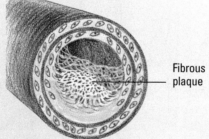

- Fibrous plaque

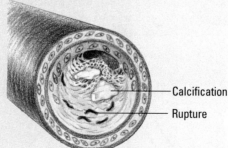

- Calcification
- Rupture

So stay away from red meat, get on that treadmill...

- anorexia
- hypotension
- diaphoresis
- narrow pulse pressure
- pallor
- oliguria
- gallop rhythm and inspiratory crackles on auscultation
- dullness over the lung bases
- hemoptysis
- cyanosis
- marked hepatomegaly
- pitting ankle enema
- sacral edema in bedridden patients.

What tests tell you

The following tests help diagnose heart failure:
- ECG reveals heart strain, heart enlargement, ischemia, atrial enlargement, tachycardia, and extrasystole.
- Chest X-ray shows increased pulmonary vascular markings, interstitial edema, or pleural effusion and cardiomegaly.
- Pulmonary artery pressure monitoring shows elevated pulmonary artery and capillary wedge pressures and elevated left ventricular end-diastolic pressure in left-sided heart failure, and elevated right atrial pressure or elevated CVP in right-sided heart failure.

Hypertension

Hypertension is an intermittent or sustained elevation of diastolic or systolic blood pressure. Generally, a sustained systolic blood pressure of 140 mm Hg or higher or a diastolic pressure of 90 mm Hg indicates hypertension.

Hypertension affects more than 60 million adults in the United States. Blacks are twice as likely as whites to be affected, and they are four times as likely to die of the disorder.

The two major types of hypertension are essential (also called primary or idiopathic) and secondary. The etiology of essential hypertension, the most common type, is complex. It involves several interacting homeostatic mechanisms. Hypertension is classified as secondary if

In hypertension, the blood exerts too much pressure on the arteries.

it's related to a systemic disease that raises peripheral vascular resistance or cardiac output. Malignant hypertension is a severe, fulminant form of the disorder that may arise from either type.

Pathophysiology

Hypertension may be caused by increases in cardiac output, total peripheral resistance, or both. Cardiac output is increased by conditions that increase heart rate or stroke volume. Peripheral resistance is increased by factors that increase blood viscosity or reduce the lumen size of vessels, especially the arterioles.

Family history, race, stress, obesity, a diet high in fat or sodium, use of tobacco or oral contraceptives, a sedentary lifestyle, and aging may all play a role. Their effects continue to be studied.

Essential hypertension usually begins insidiously as a benign disease, slowly progressing to a malignant state. If untreated, even mild cases can cause major complications and death. Carefully managed treatment, which may include lifestyle modifications and drug therapy, improves prognosis.

Why? Why? Why?

Several theories help to explain the development of hypertension. For example, it is thought to arise from:
- changes in the arteriolar bed causing increased resistance
- abnormally increased tone in the sensory nervous system that originates in the vasomotor system centers, causing increased peripheral vascular resistance
- increased blood volume resulting from renal or hormonal dysfunction
- an increase in arteriolar thickening caused by genetic factors, leading to increased peripheral vascular resistance
- abnormal renin release resulting in the formation of angiotensin II, which constricts the arterioles and increases blood volume. (See *Understanding blood pressure regulation.*)

Do you want to know my theory?

Secondary hypertension

Secondary hypertension may be caused by the following conditions:
- renovascular disease

- renal parenchymal disease
- pheochromocytoma
- primary hyperaldosteronism
- Cushing's syndrome
- diabetes mellitus
- dysfunction of the thyroid, pituitary, or parathyroid gland
- coarctation of the aorta
- pregnancy
- neurologic disorders.

What happens in secondary hypertension

The pathophysiology of secondary hypertension is related to the underlying disease. For example, consider the following:
- The most common cause of secondary hypertension is chronic renal disease. Insult to the kidney from chronic glomerulonephritis or renal artery stenosis interferes with sodium excretion, the renin-angiotensin-aldosterone

---

Understanding blood pressure regulation

Hypertension may result from a disturbance in one of the body's intrinsic regulatory mechanisms.

Renin-angiotensin system
Here's how the renin-angiotensin system acts to increase blood pressure:
- Sodium depletion, reduced blood pressure and dehydration stimulate renin release.
- Renin reacts with angiotensinogen, a liver enzyme, and converts it to angiotensin I, which increases preload and afterload.
- Angiotensin I converts to angiotensin II in the lungs. Angiotensin II is a potent vasoconstrictor that targets the arterioles.
- Circulating angiotensin II works to increase preload and afterload by stimulating the adrenal cortex to secrete aldosterone. This increases blood volume by conserving sodium and water.

Autoregulation
Several intrinsic mechanisms work to change an artery's diameter to maintain tissue and organ perfusion despite fluctuations in systemic blood pressure. These mechanisms include stress relaxation and capillary fluid shift.

- In stress relaxation, blood vessels gradually dilate when blood pressure rises to reduce peripheral resistance.
- In capillary fluid shift, plasma moves between vessels and extravascular spaces to maintain intravascular volume.

Get that heart rate going
When blood pressure drops, baroreceptors in the aortic arch and carotid sinuses decrease their inhibition of the medulla's vasomotor center. This action increases sympathetic stimulation of the heart by norepinephrine. This increases cardiac output by strengthening the contractile force, raising the heart rate, and augmenting peripheral resistance by vasoconstriction. Stress can also stimulate the sympathetic nervous system to increase cardiac output and peripheral vascular resistance.

ADH
The release of antidiuretic hormone (ADH) can regulate hypotension by causing reabsorption of water by the kidney. With reabsorption, blood plasma volume increases, raising blood pressure.

system, or renal perfusion. This causes blood pressure to rise.

• In Cushing's syndrome, increased cortisol levels raise blood pressure by increasing renal sodium retention, angiotensin II levels, and vascular response to norepinephrine.

• In primary aldosteronism, increased intravascular volume, altered sodium concentrations in vessel walls, or very high aldosterone levels cause vasoconstriction (increased resistance).

• Pheochromocytoma is a secreting tumor of chromaffin cells, usually of the adrenal medulla. It causes hypertension due to increased secretion of epinephrine and norepinephrine. Epinephrine functions mainly to increase cardiac contractility and rate. Norepinephrine functions mainly to increase peripheral vascular resistance.

Oh no! CAD, MI, heart failure

Complications occur late in the disease and can attack any organ system. Cardiac complications include CAD, angina, MI, heart failure, arrhythmias, and sudden death. Neurologic complications include cerebral infarctions and hypertensive encephalopathy. Hypertensive retinopathy can cause blindness. Renovascular hypertension can lead to renal failure. (See *Treating hypertension* and *A close look at blood vessel damage.*)

What to look for

Hypertension usually doesn't produce signs and symptoms until vascular changes in the heart, brain, or kidneys occur. Severely elevated blood pressure damages the intima of small vessels, resulting in fibrin accumulation in the vessels, local edema and, possibly, intravascular clotting. Symptoms depend on the location of the damaged vessels:

 brain — cerebrovascular accident, transient ischemic attacks

 retina — blindness

heart — MI

kidneys — proteinuria, edema and, eventually, renal failure.

Battling illness

Treating hypertension

Although essential hypertension has no cure, drugs and modifications in diet and lifestyle can control it. Generally, lifestyle modification is the first treatment used, especially in early, mild cases. If this doesn't work, the doctor prescribes various types of antihypertensives.

Treatment of secondary hypertension includes correcting the underlying cause and controlling hypertensive effects. Hypertensive crisis, or severely elevated blood pressure, may not respond to drugs and may be fatal.

Hypertension increases the heart's workload. This causes left ventricular hypertrophy and, later, left-sided heart failure, pulmonary edema, and right-sided heart failure.

What tests tell you

The following tests may reveal predisposing factors for and help identify the cause of hypertension:

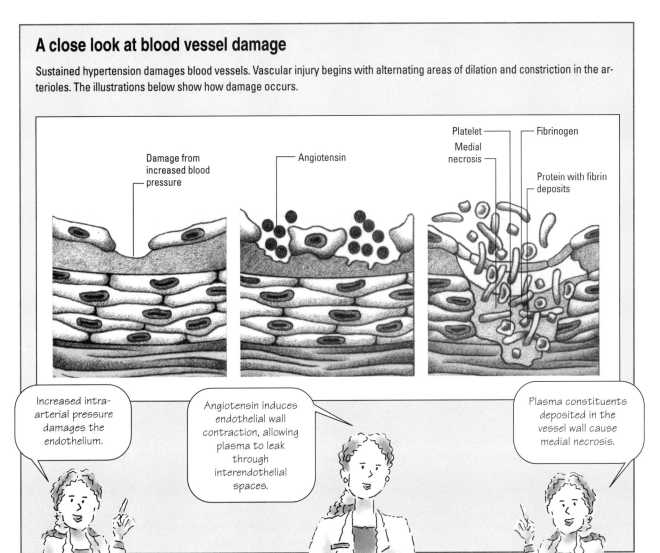

A close look at blood vessel damage

Sustained hypertension damages blood vessels. Vascular injury begins with alternating areas of dilation and constriction in the arterioles. The illustrations below show how damage occurs.

• Urinalysis may show protein, red blood cells, or white blood cells (WBCs), suggesting renal disease; or glucose, suggesting diabetes mellitus.
• Excretory urography may reveal renal atrophy, indicating chronic renal disease. One kidney that's more than ⅝″ (1.6 cm) shorter than the other suggests unilateral renal disease.
• Serum potassium levels less than 3.5 mEq/L may indicate adrenal dysfunction (primary hyperaldosteronism).
• Blood urea nitrogen (BUN) levels that are elevated to more than 20 mg/dl and serum creatinine levels that are elevated to more than 1.5 mg/dl suggest renal disease.

These tests may help detect cardiovascular damage and other complications:
• ECG may show left ventricular hypertrophy or ischemia.
• Chest X-ray may demonstrate cardiomegaly.

Hypertrophic cardiomyopathy

Hypertrophic cardiomyopathy is a primary disease of the cardiac muscle. You may hear it called several other names: idiopathic hypertrophic subaortic stenosis, hypertrophic obstructive cardiomyopathy, or muscular aortic stenosis.

The course of this disorder varies. Some patients have progressive deterioration; others remain stable for years.

Pathophysiology

About half the time, hypertrophic cardiomyopathy is transmitted genetically as an autosomal dominant trait. Other causes aren't known.

A thickening of the septum

Hypertrophic cardiomyopathy is characterized by left ventricular hypertrophy and an unusual cellular hypertrophy of the upper ventricular septum. These changes may result in an outflow tract pressure gradient, a pressure difference that results in an obstruction of blood outflow. The obstruction may change between examinations and even from beat to beat.

Hypertrophy of the intraventricular septum can pull the papillary muscle out of its usual alignment. This causes altered function of the anterior leaflet of the mitral valve

Hypertrophic cardiomyopathy is marked by disproportionate asymmetrical thickening of the intraventricular septum.

and mitral regurgitation. The myocardial wall may stiffen over time, causing increased resistance to blood entering the right atrium and an increase in diastolic filling pressures.

High, low, or normal

Cardiac output may be low, normal, or high, depending on whether the stenosis is obstructive or nonobstructive. Eventually, left ventricular dysfunction — a result of rigidity and decreased compliance — causes pump failure.

Complications

Pulmonary hypertension and heart failure may occur secondary to left ventricular stiffness. Sudden death is also possible and usually results from ventricular arrhythmias, such as ventricular tachycardia and premature ventricular contractions.

The thickened heart muscle becomes inflexible, hindering normal expansion and contraction of the heart chambers.

What to look for

Signs and symptoms of hypertrophic cardiomyopathy include the following:
- atrial fibrillation
- orthopnea
- dyspnea on exertion
- syncope
- angina
- fatigue
- edema
- tachypnea
- migratory joint pain
- abdominal pain. (See *Treating hypertrophic cardiomyopathy,* page 234.)

What tests tell you

The following tests are used to diagnose hypertrophic cardiomyopathy:
- Echocardiography shows left ventricular hypertrophy and a thick, asymmetrical intraventricular septum in obstructive disease. In nonobstructive disease, ventricular areas are hypertrophied, and the septum may have a ground-glass appearance. Poor septal contraction, abnormal motion of the anterior mitral leaflet during systole, and narrowing or occlusion of the left ventricular outflow tract in obstructive disease may also be seen. The left ven-

tricular cavity looks small, with vigorous posterior wall motion but reduced septal excursion.
• Cardiac catheterization reveals elevated left ventricular end-diastolic pressure and, possibly, mitral insufficiency.
• ECG usually shows left ventricular hypertrophy, ST-segment and T-wave abnormalities, Q waves in leads II, III, aV_F, and in V_4 to V_6 (due to hypertrophy, not infarction), left anterior hemiblock, left axis deviation, and ventricular and atrial arrhythmias.
• Chest X-ray may show a mild to moderate increase in heart size.
• Thallium scan usually reveals myocardial perfusion defects.

Battling illness

Treating hypertrophic cardiomyopathy

The goals of treatment for hypertrophic cardiomyopathy are to relax the ventricle and relieve outflow tract obstruction. Drugs are the first line of treatment; surgery is done when all else fails.

Drug therapy

The following drugs are used to treat the disorder:
• Propranolol, a beta-adrenergic blocking agent, is the drug of choice. It slows the heart rate and increases ventricular filling by relaxing the obstructing muscle, thereby reducing angina, syncope, dyspnea, and arrhythmias. However, it may aggravate symptoms of cardiac decompensation.
• Calcium channel blockers may reduce elevated diastolic pressures, decrease the severity of outflow tract gradients, and increase exercise tolerance. Disopyramide can be used to reduce left ventricular hypercontractility and the outflow gradient.
• Heparin is given during episodes of atrial fibrillation. When accompanying hypertrophic cardiomyopathy, atrial fibrillation is a medical emergency that calls for cardioversion. Because of the high risk of systemic embolism, heparin must be administered until fibrillation subsides.
• Amiodarone is given if heart failure occurs, unless an atrioventricular block exists. This drug is also effective in reducing ventricular and supraventricular arrhythmias and improving left ventricular pressure gradients. *Vasodilators, such as nitroglycerin and diuretics, and sympathetic stimulators, such as isoproterenol, are contraindicated. Inotropic drugs are also contraindicated.*

Surgery

Surgery is done if drug therapy fails. Resection of the hypertrophied septum, called ventricular myotomy, may be performed alone or with mitral valve replacement to ease outflow tract obstruction and relieve symptoms. However, this procedure is still experimental because it can cause complete heart block and a ventricular septal defect.

Myocardial infarction

MI results from reduced blood flow through one of the coronary arteries. This causes myocardial ischemia, injury, and necrosis. In transmural (Q wave) MI, tissue damage extends through all myocardial layers. In subendocardial (non-Q wave) MI, usually only the innermost layer is damaged.

In MI, one of the heart's arteries fails to deliver enough blood to the part of the heart muscle it serves.

A statement of stats

In North America and Western Europe, MI is one of the leading causes of death. Death usually results from cardiac damage or complications. Mortality is about 25%. However, more than 50% of sudden deaths occur within 1 hour after the onset of symptoms, before the patient reaches the hospital. Of those who recover, up to 10% die within a year.

Men are more susceptible to MI than premenopausal women, although the incidence is increasing in women who smoke and take oral contraceptives. The incidence in postmenopausal women is similar to that in men.

Pathophysiology

MI results from occlusion of one or more of the coronary arteries. Occlusion can stem from atherosclerosis, thrombosis, platelet aggregation, or coronary artery stenosis or spasm. Predisposing factors include:

- aging
- diabetes mellitus
- elevated serum triglyceride, low-density lipoprotein, and cholesterol levels and decreased serum high-density lipoprotein levels
- excessive intake of saturated fats, carbohydrates, or salt
- hypertension
- obesity
- positive family history of CAD
- sedentary lifestyle
- smoking
- stress or a type A personality
- use of amphetamines or cocaine.

After menopause, women are as susceptible to MI as men.

Complications

Elderly patients are more prone to complications and death. The most common complications after an acute MI include:
- arrhythmias
- cardiogenic shock
- heart failure causing pulmonary edema
- pericarditis.
  Other complications include:
- rupture of the atrial or ventricular septum, ventricular wall, or valves
- ventricular aneurysms
- mural thrombi causing cerebral or pulmonary emboli
- extensions of the original infarction
- post-MI pericarditis (Dressler's syndrome), which occurs days to weeks after an MI and causes residual pain, malaise, and fever
- psychological problems caused by fear of another MI or organic brain disorder from tissue hypoxia
- personality changes.

Reduced, altered, elevated

MI results from prolonged ischemia to the myocardium with irreversible cell damage and muscle death. Functionally, the MI causes:
- reduced contractility with abnormal wall motion
- altered left ventricular compliance
- reduced stroke volume
- reduced ejection fraction
- elevated left ventricular end-diastolic pressure. (See *Understanding MI*.)

Injury surrounded by ischemia

All MIs have a central area of necrosis or infarction surrounded by an area of injury. The area of injury is surrounded by a ring of ischemia. Tissue regeneration does not occur after an MI because the affected myocardial muscle is dead.

Compensatory kick

Scar tissue that forms on the necrotic area may inhibit contractility. When this occurs, the compensatory mechanisms (vascular constriction, increased heart rate, and renal retention of sodium and water) kick in to try to maintain cardiac output. Ventricular dilation also may occur. If a

When scar tissue inhibits contractility, compensatory mechanisms kick in.

Now I get it!

Understanding MI

In myocardial infarction (MI), blood supply to the myocardium is interrupted. Here's what happens:

1. Injury to the endothelial lining of the coronary arteries causes platelets, white blood cells, fibrin, and lipids to converge at the injured site, as shown below. Foam cells, or resident macrophages, congregate under the damaged lining and absorb oxidized cholesterol, forming a fatty streak that narrows the arterial lumen.

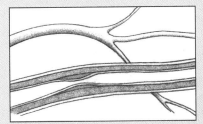

2. Because the arterial lumen narrows gradually, collateral circulation develops and helps maintain myocardial perfusion distal to the obstruction. The illustration below shows collateral circulation.

3. When myocardial demand for oxygen is more than the collateral circulation can supply, myocardial metabolism shifts from aerobic to anaerobic, producing lactic acid, which stimulates nerve end

ings, as shown below.

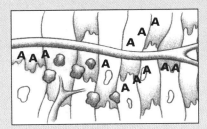

4. Lacking oxygen, the myocardial cells die. This decreases contractility, stroke volume, and blood pressure.

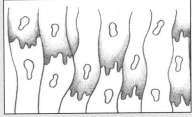

5. Hypoperfusion stimulates baroreceptors, which, in turn, stimulate the adrenal glands to release epinephrine and norepinephrine. This cycle is shown below. These catecholamines increase heart rate and cause peripheral vasoconstriction, further increasing myocardial oxygen demand.

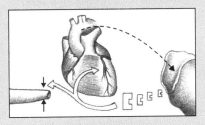

6. Damaged cell membranes in the infarcted area allow intracellular contents into the vascular circulation, as shown below. Ventricular arrhythmias then develop with elevated serum levels of potassium, creatine kinase (CK), CK-MB, aspartate aminotransferase (formerly SGOT), and lactate dehydrogenase.

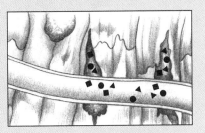

7. All myocardial cells are capable of spontaneous depolarization and repolarization, so the electrical conduction system may be affected by infarct, injury, and ischemia. The illustration below shows an injury site.

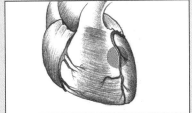

(continued)

Understanding MI *(continued)*

8. Extensive damage to the left ventricle may impair its ability to pump, allowing blood to back up into the left atrium and, eventually, into the pulmonary veins and capillaries, as shown in the illustration below. Crackles may be heard in the lungs on auscultation. Pulmonary artery and capillary wedge pressures are increased.

9. As back pressure rises, fluid crosses the alveolocapillary membrane, impeding diffusion of oxygen (O_2) and carbon dioxide (CO_2). Arterial blood gas measurements may show decreased partial pressure of arterial oxygen and arterial pH and increased partial pressure of arterial carbon dioxide.

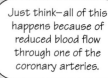

Just think—all of this happens because of reduced blood flow through one of the coronary arteries.

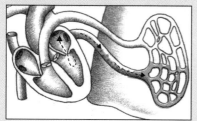

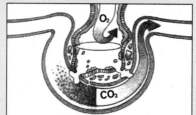

lot of scar tissue forms, contractility may be greatly reduced. The patient may develop heart failure or cardiogenic shock.

Circumflex, left anterior, right coronary

The infarction site depends on the vessels involved. For instance:
• Occlusion of the circumflex coronary artery causes lateral wall infarctions.
• Occlusion of the left anterior coronary artery causes anterior wall infarctions.
• Occlusion of the right coronary artery or one of its branches causes posterior and inferior wall infarctions and right ventricular infarctions. Right ventricular infarctions can also accompany inferior infarctions and may cause right-sided heart failure.

What to look for

The cardinal symptom of MI is persistent, crushing substernal pain that may radiate to the left arm, jaw, neck, or shoulder blades. The pain is often described as heavy, squeezing, or crushing and may persist for 12 hours or more. However, in some patients — particularly the elderly or diabetics — pain may not occur at all. In others, it may be mild and confused with indigestion.

In patients with CAD, angina of increasing frequency, severity, or duration (especially if not provoked by exertion, a heavy meal, or cold and wind) may signal an impending infarction.

Oh no! Fatigue, anxiety, doom

Other clinical effects include:
- a feeling of impending doom
- fatigue
- nausea
- vomiting
- shortness of breath
- cool extremities
- perspiration
- anxiety
- restlessness.

Fever is unusual at the onset of an MI, but a low-grade temperature may develop during the next few days. Blood pressure varies. Hypotension or hypertension may occur. (See *Treating MI,* page 240.)

Don't forget to see our full-color depiction of MI pathophysiology on pages 224 and 225.

What tests tell you

The following tests help diagnose MI:
- Serial 12-lead ECG may be normal or inconclusive during the first few hours after an MI. Abnormalities include serial ST-segment depression in subendocardial MI and ST-segment elevation and Q waves, representing scarring and necrosis, in transmural MI.
- Serum creatine kinase (CK) levels are elevated, especially the CK-MB isoenzyme, the cardiac muscle fraction of CK.
- Echocardiography shows ventricular wall dyskinesia with a transmural MI and is used to evaluate the ejection fraction.
- Radionuclide scans using I.V. technetium TC-99m pentetate can show acutely damaged muscle by picking up accumulations of radionuclide, which appears as a "hot spot" on the film. Myocardial perfusion imaging with thallium-201 reveals a "cold spot" in most patients during the first few hours after a transmural MI.

Battling illness

Treating MI

Arrhythmias, the most common problem during the first 48 hours after myocardial infarction (MI), require antiarrhythmics, a pacemaker (possibly), and cardioversion (rarely). Treatment for MI has three goals:

- to relieve chest pain
- to stabilize heart rhythm
- to reduce cardiac work load.

Drug therapy
Drugs are the mainstay of therapy. Typical drugs include the following:

- thrombolytic agents, such as streptokinase and urokinase to revascularize myocardial tissue
- lidocaine for ventricular arrhythmias or, if it doesn't work, another antiarrhythmic, such as procainamide, quinidine, bretylium, or disopyramide
- I.V. atropine for heart block or bradycardia
- sublingual, topical, transdermal, or I.V. nitroglycerin and calcium channel blockers, such as nifedipine, verapamil, and diltiazem, given by mouth or I.V. to relieve angina
- isosorbide dinitrate, given sublingually, by mouth, or I.V., to relieve pain by redistributing blood to ischemic areas of the myocardium, increasing cardiac output, and reducing myocardial workload
- I.V. morphine (drug of choice), meperidine, or hydromorphone for pain and sedation
- drugs that increase myocardial contractility, such as dobutamine and amrinone
- beta-adrenergic blockers, such as propranolol and timolol, after acute MI to help prevent reinfarction.

Other therapies
Other therapies include the following:

- a temporary pacemaker for heart block or bradycardia
- oxygen administered by face mask or nasal cannula at a modest flow rate for 24 to 48 hours, or at a lower concentration if the patient has chronic obstructive pulmonary disease
- bed rest with bedside commode to decrease cardiac workload
- pulmonary artery catheterization to detect left-sided or right-sided heart failure and to monitor response to treatment (not routinely done)
- intra-aortic balloon pump for cardiogenic shock
- cardiac catheterization, percutaneous transluminal coronary angioplasty, and coronary artery bypass grafting.

Revascularization therapy may be performed on patients under age 70 who don't have a history of cerebrovascular accident, bleeding, G.I. ulcers, marked hypertension, recent surgery, or chest pain lasting longer than 6 hours. It must begin within 6 hours after the onset of symptoms, using I.V. intracoronary or systemic streptokinase or tissue plasminogen activator. The best response occurs when treatment begins within an hour after symptoms first appear.

Pericarditis

The pericardium is the fibroserous sac that envelops, supports, and protects the heart. Inflammation of this sac is called pericarditis.

This condition occurs in both acute and chronic forms. The acute form can be fibrinous or effusive, with serous, purulent, or hemorrhagic exudate. The chronic form, called constrictive pericarditis, is characterized by dense, fibrous pericardial thickening. The prognosis depends on the underlying cause but is typically good in acute pericarditis, unless constriction occurs.

Pathophysiology

Common causes of pericarditis include the following:
• bacterial, fungal, or viral infection (infectious pericarditis)
• neoplasms (primary or metastatic from lungs, breasts, or other organs)
• high-dose radiation to the chest
• uremia
• hypersensitivity or autoimmune disease, such as systemic lupus erythematosus, rheumatoid arthritis, or acute rheumatic fever (most common cause of pericarditis in children)
• drugs, such as hydralazine or procainamide
• idiopathic factors (most common in acute pericarditis)
• postcardiac injury, such as MI (which later causes an autoimmune reaction in the pericardium), trauma, and surgery that leaves the pericardium intact but allows blood to leak into the pericardial cavity
• aortic aneurysm with pericardial leakage (less common)
• myxedema with cholesterol deposits in the pericardium (less common).

Complications

Pericardial effusion is the major complication of acute pericarditis. If fluid accumulates rapidly, cardiac tamponade may occur. This may lead to shock, cardiovascular collapse and, eventually, death.

A scar around the heart

As the pericardium becomes inflamed, it may become thickened and fibrotic. If it doesn't heal completely after an acute episode, it may calcify over a long period of time and form a firm scar around the heart. This scarring interferes with diastolic filling of the ventricles. (See *Understanding pericarditis,* page 243.)

What to look for

In acute pericarditis, a sharp, sudden pain usually starts over the sternum and radiates to the neck, shoulders, back, and arms. However, unlike the pain of MI, this pain is often pleuritic, increasing with deep inspiration and decreasing when the patient sits up and leans forward, pulling the heart away from the diaphragmatic pleurae of the lungs.

Pericardial effusion, the major complication of acute pericarditis, may produce effects of heart failure, such as dyspnea, orthopnea, and tachycardia. It may also produce ill-defined substernal chest pain and a feeling of chest fullness.

If the fluid accumulates rapidly, cardiac tamponade may occur, causing pallor, clammy skin, hypotension, pulsus paradoxus, neck vein distention and, eventually, cardiovascular collapse and death.

Ugh! Fluid retention, ascites, hepatomegaly

Chronic constrictive pericarditis causes a gradual increase in systemic venous pressure and produces symptoms similar to those of chronic right-sided heart failure, including fluid retention, ascites, and hepatomegaly. (See *Treating pericarditis,* page 244.)

What tests tell you

Laboratory test results reflect inflammation and may identify the disorder's cause. They include the following:
• WBC count may be normal or elevated, especially in infectious pericarditis.
• Erythrocyte sedimentation rate (ESR) is elevated.
• Serum CK-MB levels are slightly elevated with associated myocarditis.
• Pericardial fluid culture obtained by open surgical drainage or pericardiocentesis sometimes identifies a causative organism in bacterial or fungal pericarditis.
• BUN levels detect uremia, antistreptolysin-O titers detect rheumatic fever, and purified protein derivative skin test detects tuberculosis.
• ECG shows characteristic changes in acute pericarditis. They include elevated ST segments in the limb leads and most precordial leads. The QRS segments may be diminished when pericardial effusion is present. Rhythm

Now I get it!

Understanding pericarditis

Pericarditis occurs when a pathogen or other substance attacks the pericardium, leading to the events described below.

Inflammation

Pericardial tissue damaged by bacteria or other substances releases chemical mediators of inflammation (such as prostaglandins, histamines, bradykinins, and serotonins) into the surrounding tissue, starting the inflammatory process. Friction occurs as the inflamed pericardial layers rub against each other.

Enhanced phagocytosis

Substances released by the injured tissue stimulate neutrophil production in the bone marrow. Neutrophils then travel to the injury site through the bloodstream and join macrophages in destroying pathogens. Meanwhile, additional macrophages and monocytes migrate to the injured area and continue phagocytosis.

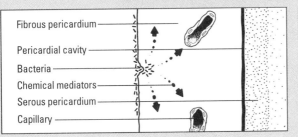

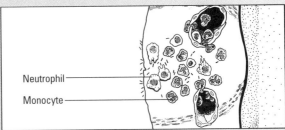

Vasodilation and clotting

Histamines and other chemical mediators cause vasodilation and increased vessel permeability. Local blood flow (hyperemia) increases. Vessel walls leak fluids and proteins (including fibrinogen) into tissues, causing extracellular edema. Clots of fibrinogen and tissue fluid form a wall, blocking tissue spaces and lymph vessels in the injured area. This wall prevents the spread of bacteria and toxins to adjoining healthy tissues.

Exudation

After several days, the infected area fills with an exudate composed of necrotic tissue, dead and dying bacteria, neutrophils, and macrophages. Thinner than pus, this exudate forms until all infection ceases, creating a cavity that remains until tissue destruction stops. The contents of the cavity autolyze and are gradually reabsorbed into healthy tissue.

Initial phagocytosis

Macrophages already present in the tissues begin to phagocytose the invading bacteria but usually fail to stop the infection.

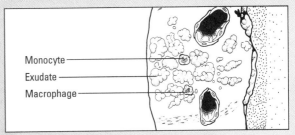

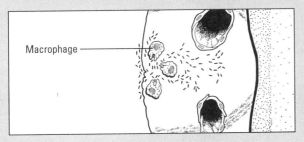

Fibrosis and scarring

As the end products of the infection slowly disappear, fibrosis and scar tissue may form. Scarring, which can be extensive, may ultimately cause heart failure if it restricts movement.

Battling illness

Treating pericarditis

Treatment for pericarditis strives to relieve symptoms, prevent or treat pericardial effusion and cardiac tamponade, and manage the underlying disease.

Drugs and bed rest
In idiopathic pericarditis, post–myocardial infarction pericarditis, and post-thoracotomy pericarditis, treatment is twofold: bed rest as long as fever and pain persist and nonsteroidal anti-inflammatory drugs, such aspirin and indomethacin, to relieve pain and reduce inflammation.

If symptoms continue, the doctor may prescribe corticosteroids. Although they provide rapid and effective relief, corticosteroids must be used cautiously because pericarditis may recur when drug therapy stops.

More involved treatment
When infectious pericarditis results from disease of the left pleural space, mediastinal abscesses, or septicemia, the patient requires antibiotics, surgical drainage, or both. If cardiac tamponade develops, the doctor may perform emergency pericardiocentesis and may inject antibiotics directly into the pericardial sac.

Heavy-duty treatment
Recurrent pericarditis may require partial pericardectomy, which creates a window that allows fluid to drain into the pleural space. In constrictive pericarditis, total pericardectomy may be necessary to permit the heart to fill and contract adequately.

changes may also occur, including atrial ectopic rhythms, such as atrial fibrillation and sinus arrhythmias.
• Echocardiography diagnoses pericardial effusion when it shows an echo-free space between the ventricular wall and the pericardium.

Rheumatic fever and rheumatic heart disease

A systemic inflammatory disease of childhood, acute rheumatic fever develops after infection of the upper respiratory tract with group A beta-hemolytic streptococci. It mainly involves the heart, joints, central nervous system, skin, and subcutaneous tissues and often recurs. If rheumatic fever is not treated, scarring deformity of the cardiac structures results in rheumatic heart disease.

Worldwide, 15 to 20 million new cases of rheumatic fever are reported each year. The disease strikes most often during cool, damp weather in the winter and early spring. In the United States, it's most common in the north.

Rheumatic heart disease refers to the cardiac effects of rheumatic fever.

Rheumatic fever tends to run in families, lending support to the existence of genetic predisposition. Environmental factors also seem to be significant in development of the disorder. For example, in lower socioeconomic groups, the incidence is highest in children between ages 5 and 15, probably due to malnutrition and crowded living conditions.

Pathophysiology

Rheumatic fever appears to be a hypersensitivity reaction. For some reason, antibodies produced to combat streptococci react and produce characteristic lesions at specific tissue sites. Because only about 0.3% of people infected with *Streptococcus* contract rheumatic fever, altered immune response probably is involved in its development or recurrence.

Fewer than 1% of people with strep infections contract rheumatic fever.

Complications

The mitral and aortic valves are often destroyed by rheumatic fever's long-term effects. Their malfunction leads to severe heart inflammation (called carditis) and, occasionally, produces pericardial effusion and fatal heart failure. Of the patients who survive this complication, about 20% die within 10 years.

Carditis develops in up to 50% of patients with rheumatic fever and may affect the endocardium, myocardium, or pericardium during the early acute phase. Later, the heart valves may be damaged, causing chronic valvular disease.

Damage to heart tissue and valves

The extent of heart damage depends on where the infection strikes. For example:
• Myocarditis produces characteristic lesions called Aschoff's bodies in the interstitial tissue of the heart, as well as cellular swelling and fragmentation of interstitial collagen. These lesions lead to formation of progressively fibrotic nodules and interstitial scars.
• Endocarditis causes valve leaflet swelling, erosion along the lines of leaflet closure, and blood, platelet, and fibrin deposits, which form beadlike vegetation. Endocarditis strikes the mitral valve most often in females and the aortic valve in males. It affects the tricuspid valves in both sexes and, rarely, affects the pulmonic valve.

Carditis is the most destructive effect of rheumatic fever.

What to look for

In 95% of patients, rheumatic fever follows a streptococcal infection that appeared a few days to 6 weeks earlier. A temperature of at least 100.4° F (38° C) occurs.

Most patients complain of migratory joint pain or polyarthritis. Swelling, redness, and signs of effusion usually accompany such pain, which most often affects the knees, ankles, elbows, and hips.

Rashes and nodules

About 5% of patients (usually those with carditis) develop a nonpruritic, macular, transient rash called erythema marginatum. This rash gives rise to red lesions with blanched centers. These same patients may also develop firm, movable, nontender subcutaneous nodules about 3 mm to 2 cm in diameter, usually near tendons or bony prominences of joints. These nodules persist for a few days to several weeks. (See *Treating rheumatic fever and rheumatic heart disease.*)

Battling illness

Treating rheumatic fever and rheumatic heart disease

Effective treatment for rheumatic fever and rheumatic heart disease aims to eradicate the streptococcal infection, relieve symptoms, and prevent recurrence, thus reducing the chance of permanent cardiac damage.

Acute phase
During the acute phase, treatment includes penicillin or erythromycin for patients with penicillin hypersensitivity. Salicylates, such as aspirin, relieve fever and minimize joint swelling and pain. If the patient has carditis, or if salicylates fail to relieve pain and inflammation, the doctor may prescribe corticosteroids.

Patients with active carditis require strict bed rest for about 5 weeks during the acute phase, followed by a progressive increase in physical activity. The increase depends on clinical and laboratory findings and the patient's response to treatment.

Long-term treatment
After the acute phase subsides, a monthly intramuscular injection of penicillin G benzathine or daily doses of oral sulfadiazine or penicillin G may be used to prevent recurrence. This treatment usually continues for at least 5 years or until age 25.

Complications
Heart failure requires continued bed rest and diuretics. Severe mitral or aortic valvular dysfunction that causes persistent heart failure will require corrective surgery, such as commissurotomy (separation of the adherent, thickened leaflets of the mitral valve), valvuloplasty (inflation of a balloon within a valve), or valve replacement (with a prosthetic valve). However, this surgery is seldom necessary before late adolescence.

What tests tell you

No specific laboratory tests can determine the presence of rheumatic fever, but the following test results support the diagnosis:

• WBC count and ESR may be elevated during the acute phase; blood studies show slight anemia caused by suppressed erythropoiesis during inflammation.

• C-reactive protein is positive, especially during the acute phase.

• Cardiac enzyme levels may be increased in severe myocarditis.

• Antistreptolysin-O titer is elevated in 95% of patients within 2 months of onset.

• Throat cultures may continue to show group A beta-hemolytic streptococci; however, they usually occur in small numbers and isolating them is difficult.

• ECG reveals no diagnostic changes, but 20% of patients show a prolonged PR interval.

• Chest X-ray shows normal heart size, except with myocarditis, heart failure, and pericardial effusion.

• Echocardiography helps evaluate valvular damage, chamber size, ventricular function, and the presence of a pericardial effusion.

• Cardiac catheterization evaluates valvular damage and left ventricular function in severe cardiac dysfunction.

Quick quiz

1. Major modifiable risk factors for CAD include:
  A. high cholesterol and hypertension.
  B. cigarette smoking and family history.
  C. obesity and age.

Answer: A. Family history and age are risk factors that can be controlled.

2. Half of all cases of hypertrophic cardiomyopathy are caused by:
  A. autoimmune disease.
  B. malnutrition.
  C. a genetic predisposition.

Now I've told you about all my troubles. But let's see if you were paying attention. Take the Incredibly Easy quick quiz.

Answer: C. The other half have unknown causes.

3. The major pathophysiologic effect of cardiac tamponade is:
- A. atelectasis.
- B. distended pericardium.
- C. compressed heart.

Answer: B. The pericardium distends as it fills with fluid.

4. All of the following are complications of MI except:
- A. diabetes mellitus.
- B. cardiogenic shock.
- C. Dressler's syndrome.

Answer: A. Diabetes mellitus is a predisposing factor for MI, not a complication.

Scoring

☆ ☆ ☆   If you answered all four items correctly, hats off! There's only one way to say it — You're all heart!

☆ ☆   If you answered two or three correctly, great job! Your heart is in the right place and so is your nose (that is, your nose is in this book).

☆   If you answered less than two correctly, don't fret! You've shown a lot of heart.

I don't have the heart for any more puns. Let's go on to chapter 9.

Neurologic system

Just the facts

In this chapter you'll learn:

♦ the structures of the neurologic system

♦ how the neurologic system works

♦ causes, pathophysiology, diagnostic tests, and treatments for several common neurologic disorders.

Understanding the neurologic system

The neurologic or nervous system is the body's communication network. It coordinates and organizes the functions of all other body systems. This intricate network has two main divisions:

🖐 the central nervous system (CNS), made up of the brain and spinal cord, which is the body's control center

✌ the peripheral nervous system, which contains cranial and spinal nerves that connect the CNS to remote body parts and which relays and receives messages from those parts.

A functional subdivision

The peripheral nervous system can be divided into the somatic nervous system and the autonomic nervous system. The somatic nervous system regulates voluntary motor control. The autonomic nervous system helps regulate the body's internal environment through involuntary control of organ systems.

The fundamental unit

The neuron, or nerve cell, is the nervous system's fundamental unit. This highly specialized conductor cell receives and transmits electrochemical nerve impulses. Delicate, threadlike nerve fibers called axons and dendrites extend from the central cell body and transmit signals. Axons carry impulses away from the cell body; dendrites carry impulses to the cell body. Most neurons have multiple dendrites but only one axon.

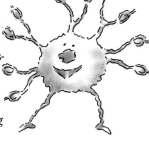

Glad to meet you. I'm the basic unit of the nervous system.

The body's information superhighway

This intricate network of interlocking receptors and transmitters, along with the brain and spinal cord, forms a living computer that controls and regulates every mental and physical function. From birth to death, the nervous system efficiently organizes the body's affairs, controlling the smallest action, thought, or feeling; monitoring communication and the survival instinct; and allowing introspection, wonder, and abstract thought.

Central nervous system

The nervous system is a wonderful thing. Its many parts are well-coordinated and interact constantly...

The CNS consists of the brain and spinal cord. The fragile brain and spinal cord are protected by the bony skull and vertebrae, cerebrospinal fluid (CSF), and three membranes:

the dura mater — a tough, fibrous, leatherlike tissue composed of two layers, the endosteal dura and meningeal dura (The endosteal dura forms the periosteum of the skull and is continuous with the lining of the vertebral canal. The meningeal dura, a thick membrane, covers the brain, dipping between the brain tissue and providing support and protection.)

the pia mater — a continuous layer of connective tissue that covers and contours the spinal tissue and brain

the arachnoid membrane — a thin, fibrous membrane that hugs the brain and spinal cord, though not as precisely as the pia mater.

...this allows humans to think, feel, and act on many different levels at the same time.

The spaces between

Between the dura mater and the arachnoid membrane is the subdural space. Between the pia mater and the arach-

I need
my
space.

noid membrane is the subarachnoid space. Within the subarachnoid space and the brain's four ventricles is CSF, a liquid composed of water and traces of organic materials (especially protein), glucose, and minerals. This fluid protects the brain and spinal tissue from jolts and blows.

Cerebrum

The cerebrum, the largest part of the brain, houses the nerve center that controls sensory and motor activities and intelligence. The outer layer, the cerebral cortex, consists of neuron cell bodies (gray matter). The inner layer consists of axons (white matter) plus basal ganglia, which control motor coordination and steadiness.

The right controls the left and the left controls the right

The cerebrum is divided into the right and left hemispheres. Because motor impulses descending from the brain cross in the medulla, the right hemisphere controls the left side of the body and the left hemisphere controls the right side of the body. Several fissures divide the cerebrum into lobes. Each lobe has a specific function. (See *A close look at lobes and fissures,* page 252.)

Thank your thalamus

The thalamus, a relay center in the cerebrum, further organizes cerebral function by transmitting impulses to and from appropriate areas of the cerebrum. The thalamus is also responsible for primitive emotional responses, such as fear, and for distinguishing pleasant stimuli from unpleasant ones.

Hail to your hypothalamus

The hypothalamus, located beneath the thalamus, is an autonomic center with connections to the brain, spinal cord, autonomic nervous system, and pituitary gland. It regulates temperature control, appetite, blood pressure, breathing, sleep patterns, and peripheral nerve discharges that occur with behavioral and emotional expression. It also partially controls pituitary gland secretion and stress reaction.

Cerebellum and brain stem

Other main parts of the brain are the cerebellum and the brain stem.

I'm rather proud of my cerebrum and its two cerebral hemispheres. It controls sensory and motor activities and intelligence.

My thalamus acts as a relay station. Sensory pathways form synapses in the thalamus on their way to the cerebral cortex.

A close look at lobes and fissures

Several fissures divide the cerebrum into hemispheres and lobes:

• The fissure of Sylvius, or the lateral sulcus, separates the temporal lobe from the frontal and parietal lobes.
• The fissure of Rolando, or the central sulcus, separates the frontal lobes from the parietal lobe.
• The parieto-occipital fissure separates the occipital lobe from the two parietal lobes.

To each lobe, a function

Each lobe has a specific function:

• The frontal lobe controls voluntary muscle movements and contains motor areas such as the one for speech (Broca's motor speech area). It's the center for personality, behavioral functions, intellectual functions (such as judgment, memory, and problem solving), autonomic functions, and cardiac and emotional responses.
• The temporal lobe is the center for taste, hearing, smell, and interpretation of spoken language.
• The parietal lobe coordinates and interprets sensory information from the opposite side of the body.
• The occipital lobe interprets visual stimuli.

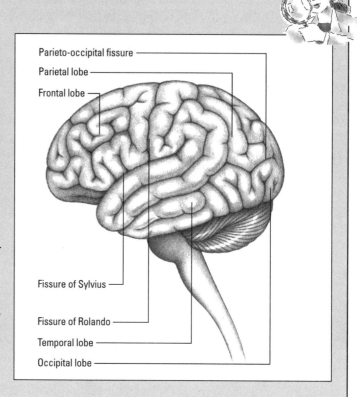

Parieto-occipital fissure
Parietal lobe
Frontal lobe
Fissure of Sylvius
Fissure of Rolando
Temporal lobe
Occipital lobe

Tell 'em about the cerebellum

The cerebellum lies beneath the cerebrum, at the base of the brain. It coordinates muscle movements, controls posture, and maintains equilibrium.

A gem of a stem

The brain stem includes the midbrain, pons, and medulla oblongata. It houses cell bodies for most of the cranial nerves. Along with the thalamus and hypothalamus, it makes up a nerve network called the reticular formation, which acts as an arousal mechanism.

Leave it to the midbrain, pons, and medulla oblongata

The three parts of the brain stem provide two-way conduction between the spinal cord and brain: In addition, they perform the following functions:

• The midbrain is the reflex center for the third and fourth cranial nerves and mediates pupillary reflexes and eye movements.
• The pons helps regulate respirations. It's also the reflex center for the fifth through eighth cranial nerves and mediates chewing, taste, saliva secretion, hearing, and equilibrium.
• The medulla oblongata influences cardiac, respiratory, and vasomotor functions. It's the center for the vomiting, coughing, and hiccuping reflexes.

Spinal cord

The spinal cord extends downward from the brain to the second lumbar vertebra. The spinal cord functions as a two-way conductor pathway between the brain stem and the peripheral nervous system.

A gray area

The spinal cord contains a mass of gray matter divided into horns, which mostly consist of neuron cell bodies. The horns of the spinal cord relay sensations and are needed for voluntary or reflex motor activity.

The outside of the horns is surrounded by white matter consisting of myelinated nerve fibers grouped in vertical columns called tracts. (See *A look inside the spinal cord,* page 254.)

Up and down the tract

The sensory, or ascending, tracts carry sensory impulses up the spinal cord to the brain, while motor, or descending, tracts carry motor impulses down the spinal cord.

From the brain...

The brain's motor impulses reach a descending tract and continue to the peripheral nervous system via upper motor neurons (also called cranial motor neurons).

...to the spinal cord...

Upper motor neurons in the brain conduct impulses from the brain to the spinal cord. Upper motor neurons form two major systems:
• the pyramidal system, or corticospinal tract, which is responsible for fine, skilled movements of skeletal muscle
• the extrapyramidal system, or extracorticospinal tract, which controls gross motor movements.

A look inside the spinal cord

This cross section of the spinal cord shows an H-shaped mass of gray matter divided into horns, which consist primarily of neuron cell bodies.

Cell bodies in the posterior or dorsal horn primarily relay information. Cell bodies in the anterior or ventral horn are needed for voluntary or reflex motor activity.

The illustration below shows the major components of the spinal cord.

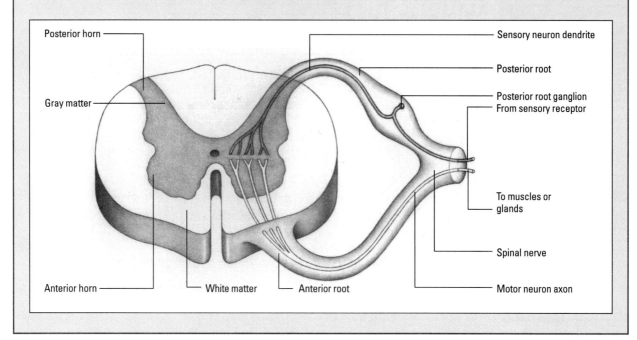

Posterior horn — Sensory neuron dendrite — Posterior root — Posterior root ganglion — From sensory receptor — Gray matter — To muscles or glands — Spinal nerve — Anterior horn — White matter — Anterior root — Motor neuron axon

...to the muscles

Lower motor neurons or spinal motor neurons conduct impulses that originate in upper motor neurons to the muscles.

Peripheral nervous system

Messages transmitted through the spinal cord reach outlying areas through the peripheral nervous system. The peripheral nervous system originates in 31 pairs of spinal nerves arranged in segments and attached to the spinal cord. (See *How spinal nerves are numbered*.)

A route with two roots

Spinal nerves are attached to the spinal cord by two roots:

the anterior or ventral root, which consists of motor fibers that relay impulses from the spinal cord to the glands and muscles

the posterior or dorsal root, which consists of sensory fibers that relay information from receptors to the spinal cord. A swollen area of the posterior root, the posterior root ganglion, is made up of sensory neuron cell bodies.

Ramifications of rami

After leaving the vertebral column, each spinal nerve separates into branches called rami, which distribute peripherally with extensive overlapping. This overlapping reduces the chance of lost sensory or motor function from interruption of a single spinal nerve.

Involuntary yet under control

The autonomic, or involuntary, nervous system is a subdivision of the peripheral nervous system. It controls involuntary body functions, such as digestion, respirations, and cardiovascular function. It's usually divided into two antagonistic systems that balance each other to support homeostasis:

the sympathetic nervous system, which controls energy expenditure, especially in stressful situations, by releasing the adrenergic catecholamine norepinephrine

the parasympathetic nervous system, which helps conserve energy by releasing the cholinergic neurohormone acetylcholine.

> ### How spinal nerves are numbered
>
> Spinal nerves are numbered according to their point of origin in the cord:
>
> • 8 cervical: C1 to C8
> • 12 thoracic: T1 to T12
> • 5 lumbar: L1 to L5
> • 5 sacral: S1 to S5
> • 1 coccygeal.

The autonomic nervous system operates, without conscious control, as the caretaker of the body.

Neurologic disorders

In this section, you will find information about Alzheimer's disease, cerebrovascular accident (CVA), epilepsy, Guillain-Barré syndrome, meningitis, multiple sclerosis (MS), myasthenia gravis, and Parkinson's disease.

Alzheimer's disease

Alzheimer's disease is a progressive degenerative disorder of the cerebral cortex. It accounts for more than half of all cases of dementia. Cortical degeneration is most marked in the frontal lobes, but atrophy occurs in all areas of the cortex. An estimated 5% of people over age 65 have a severe form of this disease, and 12% suffer from mild to moderate dementia.

Because this is a primary progressive dementia, the prognosis is poor. Most patients die 2 to 15 years after the onset of symptoms. The average duration of the illness before death is 8 years.

Like other types of dementia, Alzheimer's disease is marked by mental deterioration.

Pathophysiology

The cause of Alzheimer's disease is unknown. Four factors are thought to contribute:

👆 neurochemical factors, such as deficiencies in the neurotransmitters acetylcholine, somatostatin, substance P, and norepinephrine

✌️ viral factors such as slow-growing CNS viruses

🤟 trauma

🖖 genetic factors.

Researchers believe that up to 70% of Alzheimer's cases may stem from a genetic abnormality located on chromosome 21. Researchers have also isolated a protein substance in brain tissue called amyloid that causes brain damage characteristic of Alzheimer's.

Blame it on brain tissue

The brain tissue of Alzheimer's patients has three distinguishing features:

👆 neurofibrillary tangles formed out of proteins in the neurons

✌️ neuritic plaques

🤟 granulovascular degeneration of neurons.

First microscopic plaques then atrophy

The disease causes degeneration of neuropils (dense complexes of interwoven cytoplasmic processes of nerve cells and neuroglial cells), especially in the frontal, parietal, and occipital lobes. It also causes enlargement of the ventricles (cavities within the brain filled with CSF).

Early cerebral changes include formation of microscopic plaques, consisting of a core surrounded by fibrous tissue. Later on, atrophy of the cerebral cortex becomes strikingly evident.

If a patient has a large number of neuritic plaques, his dementia will be more severe. The plaques contain amyloid, which may exert neurotoxic effects. Evidence suggests that plaques play an important part in bringing about the death of neurons.

Defining characteristics of Alzheimer's are noted on autopsy...

Absence of acetylcholine

Problems with neurotransmitters and the enzymes associated with their metabolism may play a role in the disease. The severity of dementia is directly related to reduction of the amount of the neurotransmitter acetylcholine. On autopsy, the brains of Alzheimer's patients may contain as little as 10% of the normal amount of acetylcholine.

Complications

Complications may include injury resulting from the patient's violent behavior, wandering, or unsupervised activity. Other complications include pneumonia and other infections, especially if the patient doesn't receive enough exercise; malnutrition and dehydration, especially if the patient refuses or forgets to eat; and aspiration.

...they include neurofibrillary tangles in the cytoplasm of neurons, neuritic plaques, and granulovascular degeneration in the neurons.

What to look for

Alzheimer's disease has an insidious onset. At first, changes are barely perceptible, but they gradually lead to serious problems. Patient history is almost always obtained from a family member or caregiver.

Early changes may include forgetfulness, subtle memory loss without loss of social skills or behavior patterns, difficulty learning and retaining new information, inability to concentrate, and deterioration in personal hygiene and appearance. As the disease progresses, signs and symptoms indicate a degenerative disorder of the frontal lobe. They may include:

- difficulty with abstract thinking and activities that require judgment
- progressive difficulty in communicating
- severe deterioration of memory, language, and motor function progressing to coordination loss and an inability to speak or write
- repetitive actions
- restlessness
- irritability, depression, mood swings, paranoia, hostility, and combativeness
- nocturnal awakenings
- disorientation.

Neurologic examination

Neurologic examination confirms many of the problems revealed during the history. In addition, it often reveals an impaired sense of smell (usually an early symptom), an inability to recognize and understand the form and nature of objects by touching them, gait disorders, and tremors. The patient will also have a positive snout reflex; in response to a tap or stroke of the lips or the area just under the nose, the patient grimaces or puckers his lips.

In the final stages, urinary or fecal incontinence, twitching, and seizures commonly occur. (See *Treating Alzheimer's disease.*)

What tests tell you

Alzheimer's disease can't be confirmed until death, when an autopsy reveals pathologic findings.

The following tests help rule out other disorders:
- Positron emission tomography scan measures the metabolic activity of the cerebral cortex and may help confirm early diagnosis.
- Computed tomography (CT) scan may show more brain atrophy than occurs in normal aging.
- Magnetic resonance imaging (MRI) evaluates the condition of the brain and rules out intracranial lesions as the source of dementia.
- Electroencephalogram (EEG) evaluates the brain's electrical activity and may show brain wave slowing late in the disease. It also identifies tumors, abscesses, and other intracranial lesions that might cause symptoms.
- CSF analysis helps determine if signs and symptoms stem from a chronic neurologic infection.

Battling illness

Treating Alzheimer's disease

No cure or definitive treatment for Alzheimer's disease exists. Therapy consists of cerebral vasodilators, such as ergoloid mesylates, isoxsuprine, and cyclandelate to enhance the brain's circulation; hyperbaric oxygen to increase oxygenation to the brain; psychostimulators, such as methylphenidate, to enhance the patient's mood; and antidepressants if depression seems to exacerbate dementia.

Most other drug therapies are experimental. These include choline salts, lecithin, physostigmine, tacrine, enkephalins, and naloxone, which may slow the disease process. Preliminary studies show that long-term estrogen supplements may cut the risk of developing Alzheimer's disease by as much as 50%.

• Cerebral blood flow studies may detect abnormalities in blood flow to the brain.

Oh, no! In CVA, circulation is impaired in the vessels that supply me with blood.

Cerebrovascular accident

Commonly known as stroke, CVA is a sudden impairment of cerebral circulation in one or more of the blood vessels supplying the brain. It interrupts or diminishes oxygen supply, causing serious damage or necrosis in brain tissues.

The sooner, the better

The sooner circulation returns to normal after CVA, the better chances are for complete recovery. About half of those who survive remain permanently disabled and suffer another CVA within weeks, months, or years.

A statement of stats

CVA is the third most common cause of death in the United States and the most common cause of neurologic disability. It strikes about 500,000 people each year, and half of them die as a result. Although it mostly affects older adults, it can strike people of any age. Black men have a higher risk than other population groups.

Transient, progressive, or completed

CVA is classified according to how it progresses:
• Transient ischemic attack (TIA), the least severe type, is caused by a temporary interruption of blood flow, usually in the carotid and vertebrobasilar arteries. (See *Understanding TIA*, page 260.)
• Progressive stroke, also called stroke-in-evolution or thrombus-in-evolution, begins with a slight neurologic deficit and worsens in a day or two.
• Completed stroke, the most severe type, causes maximum neurologic deficits at the onset.

All these factors increase a patient's risk of CVA.

Pathophysiology

Factors that increase the risk of CVA include:
• history of TIA
• atherosclerosis
• hypertension
• arrhythmias
• electrocardiogram changes

Understanding TIA

A transient ischemic attack (TIA) is a recurrent episode of neurologic deficit, lasting from seconds to hours, that clears within 12 to 24 hours. It's usually considered a warning sign of an impending thrombotic cerebrovascular accident. In fact, TIAs have been reported in 50% to 80% of patients who've had a cerebral infarction from thrombosis. The age of onset varies, but incidence rises dramatically after age 50 and is highest among blacks and men.

Interrupting blood flow
In TIA, microemboli released from a thrombus may temporarily interrupt blood flow, especially in the small distal branches of the brain's arterial tree. Small spasms in those arterioles may precede TIA and also impair blood flow.

A transient experience
The most distinctive characteristics of TIAs are the transient duration of neurologic deficits and the complete return of normal function. The signs and symptoms of TIA correlate with the location of the affected artery. They include double vision, unilateral blindness, staggering or uncoordinated gait, unilateral weakness or numbness, falling because of weakness in the legs, dizziness, and speech deficits, such as slurring or thickness.

Preventing a complete stroke
During an active TIA, treatment aims to prevent a completed stroke and consists of aspirin or anticoagulants to minimize the risk of thrombosis. After or between attacks, preventive treatment includes carotid endarterectomy or cerebral microvascular bypass.

- rheumatic heart disease
- diabetes mellitus
- gout
- postural hypotension
- cardiac enlargement
- high serum triglyceride levels
- lack of exercise
- oral contraceptive use
- smoking
- family history of cerebrovascular disease.

Three major causes

Major causes of CVA include:

 thrombosis

 embolism

 hemorrhage.

Number 1 cause

Thrombosis is the most common cause of CVA in middle-aged and elderly people. It usually results from an obstruction in the extracerebral vessels, but sometimes it's intracerebral. The risk increases with obesity, smoking, oral contraceptive use, and surgery.

Number 2 cause

The second most common cause of CVA, embolism is a blood vessel occlusion caused by a fragmented clot, a tumor, fat, bacteria, or air. It can occur at any age, especially in patients with a history of rheumatic heart disease, endocarditis, posttraumatic valvular disease, or myocardial fibrillation or other cardiac arrhythmias. It also occurs after open-heart surgery. Embolism usually develops rapidly — in 10 to 20 seconds — and without warning. The left middle cerebral artery is usually the embolus site.

Number 3 cause

Hemorrhage, the third most common cause of CVA, may also occur suddenly at any age. It arises from chronic hypertension or aneurysms, which cause a sudden rupture of a cerebral artery. (See *Common sites of aneurysm*, page 262.) Increasing cocaine use by younger people has also increased the number of hemorrhagic CVAs because of the severe hypertension caused by this drug.

Cause and effect relationships

Thrombosis, embolus, and hemorrhage affect the body in different ways.

Thrombosis causes congestion and edema in the affected vessel as well as ischemia in the brain tissue supplied by the vessel.

An embolus cuts off circulation in the cerebral vasculature by lodging in a narrow portion of the artery, causing necrosis and edema. If the embolus is septic, and the infection extends beyond the vessel wall, encephalitis may develop. If the infection stays within the vessel wall, an aneurysm may form, which could lead to the sudden rupture of an artery, or cerebral hemorrhage.

In hemorrhage, a brain artery bursts, diminishing blood supply to the area served by the artery. Blood also accumulates deep within the brain, causing even greater damage by further compromising neural tissue.

Complications

Among the many possible complications of CVA are unstable blood pressure from loss of vasomotor control, fluid imbalances, malnutrition, infections such as pneumonia,

Common sites of aneurysm

Cerebral aneurysms usually arise at an arterial junction in the circle of Willis, the circular anasto-mosis forming the major cerebral arteries at the base of the brain. Cerebral aneurysms often rup-ture and cause subarachnoid hemorrhage. The illustration below shows common aneurysm sites.

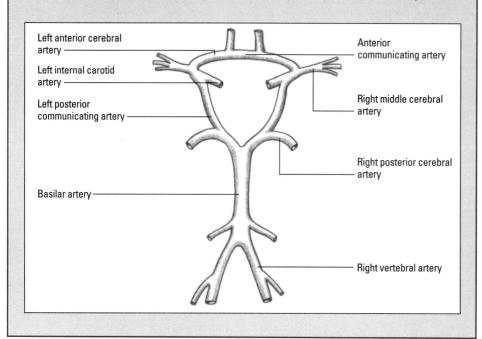

Left anterior cerebral artery

Left internal carotid artery

Left posterior communicating artery

Basilar artery

Anterior communicating artery

Right middle cerebral artery

Right posterior cerebral artery

Right vertebral artery

and sensory impairment, including vision problems. Al-tered level of consciousness, aspiration, contractures, and pulmonary emboli may also occur.

What to look for

When taking the patient's history, you may uncover risk factors for CVA. You may observe loss of consciousness, dizziness, or seizures. Obtain information from a family member or friend if necessary. Neurologic examination provides most of the information about the physical effects of CVA.

Physical findings depend on the following factors:
• the artery affected and the portion of the brain it supplies (See *Neurologic deficits in CVA.*)
• the severity of the damage
• the extent of collateral circulation that develops to help the brain compensate for a decreased blood supply.

Neurologic deficits in CVA

In cerebrovascular accident (CVA), functional loss reflects damage to the brain area normally perfused by the occluded or ruptured artery. While one patient may experience only mild hand weakness, another may develop unilateral paralysis. Hypoxia and ischemia may produce edema that affects distal parts of the brain, causing further neurologic deficits. The signs and symptoms that accompany CVA at different sites are described below.

Middle cerebral artery

- Aphasia
- Dysphasia
- Dyslexia (reading problems)
- Dysgraphia (inability to write)
- Visual field cuts
- Hemiparesis on the affected side, which is more severe in the face and arm than in the leg

Internal carotid artery

- Headaches
- Weakness
- Paralysis
- Numbness
- Sensory changes
- Visual disturbances such as blurring on the affected side
- Altered level of consciousness
- Bruits over the carotid artery
- Aphasia
- Dysphasia
- Ptosis

Anterior cerebral artery

- Confusion
- Weakness
- Numbness on the affected side (especially in the arm)

- Paralysis of the contralateral foot and leg with accompanying footdrop
- Incontinence
- Poor coordination
- Impaired motor and sensory functions
- Personality changes, such as flat affect and distractibility

Vertebral or basilar artery

- Mouth and lip numbness
- Dizziness
- Weakness on the affected side
- Visual deficits, such as color blindness, lack of depth perception, and diplopia
- Poor coordination
- Dysphagia
- Slurred speech
- Amnesia
- Ataxia

Posterior cerebral artery

- Visual field cuts
- Sensory impairment
- Dyslexia
- Coma
- Blindness from ischemia in the occipital area

Reflecting on reflexes

Assessment of motor function and muscle strength often shows a loss of voluntary muscle control and hemiparesis or hemiplegia on one side of the body. In the initial phase, flaccid paralysis with decreased deep tendon reflexes may occur. These reflexes return to normal after the initial phase, accompanied by an increase in muscle tone and, in some cases, muscle spasticity on the affected side.

Sensory impairment: From slight to severe

Vision testing often reveals reduced vision or blindness on the affected side of the body and, in patients with left-sided hemiplegia, problems with visual-spatial relations. Sensory assessment may reveal sensory losses, ranging from slight impairment of touch to the inability to perceive the position and motion of body parts. The patient also may have difficulty interpreting visual, tactile, and auditory stimuli.

Left affects right; right affects left

If the CVA occurs in the brain's left hemisphere, it produces signs and symptoms on the right side of the body. If it occurs in the right hemisphere, signs and symptoms appear on the left side. However, a CVA that damages cranial nerves produces signs on the same side as the hemorrhage. (See *Treating CVA*.)

What tests tell you

The following tests are used to diagnose CVA:
• Cerebral angiography details disruption or displacement of the cerebral circulation by occlusion or hemor-

Remember that my right hemisphere controls the left side of the body and my left hemisphere controls the right side of the body.

Treating CVA

Medical treatment for cerebrovascular accident (CVA) commonly includes physical rehabilitation, dietary and drug regimens to help decrease risk factors, and measures to help the patient adapt to specific deficits, such as speech impairment and paralysis.

Drug therapy
These drugs are commonly used:

• anticonvulsants, such as phenytoin or phenobarbital, to treat or prevent seizures
• stool softeners, such as docusate sodium, to avoid straining, which increases intracranial pressure
• corticosteroids, such as dexamethasone, to minimize cerebral edema
• anticoagulants, such as heparin, warfarin, and ticlopidine, to reduce the risk of thrombotic stroke
• analgesics, such as codeine, to relieve headache that may follow hemorrhagic CVA
• thrombolytic therapy, such as recombinant tissue plasminogen activator, given within the first 3 hours of an ischemic CVA, to restore circulation to the affected brain tissue and limit the extent of brain injury.

Surgery
Depending on the CVA's cause and extent, the patient may also undergo surgery. A craniotomy may be done to remove a hematoma, an endarterectomy to remove atherosclerotic plaque from the inner arterial wall, or extracranial-intracranial bypass to circumvent an artery that's blocked by occlusion or stenosis. Ventricular shunts may be necessary to drain cerebrospinal fluid.

rhage. It's the test of choice for examining the entire cerebral artery.

• Digital subtraction angiography evaluates the patency of the cerebral vessels and identifies their position in the head and neck. It also detects and evaluates lesions and vascular abnormalities.

• CT scan detects structural abnormalities, edema, and lesions, such as nonhemorrhagic infarction and aneurysms. It differentiates CVA from other disorders, such as primary metastatic tumor and subdural, intracerebral, or epidural hematoma. Patients with TIA commonly have a normal CT scan.

• PET scan provides data on cerebral metabolism and cerebral blood flow changes, especially in ischemic stroke.

• Single-photon emission tomography identifies cerebral blood flow and helps diagnose cerebral infarction.

• MRI and magnetic resonance angiography evaluate the lesion's location and size. MRI doesn't distinguish hemorrhage, tumor, or infarction as well as a CT scan, but it provides superior images of the cerebellum and brain stem.

• Transcranial Doppler studies evaluate the velocity of blood flow through major intracranial vessels, which can indicate the vessels' diameter.

• Cerebral blood flow studies measure blood flow to the brain and help detect abnormalities.

• Ophthalmoscopy may show signs of hypertension and atherosclerotic changes in the retinal arteries.

• EEG may detect reduced electrical activity in an area of cortical infarction and is especially useful when CT scan results are inconclusive. It can also differentiate seizure activity from CVA.

• Oculoplethysmography indirectly measures ophthalmic blood flow and carotid artery blood flow.

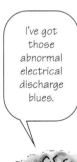

I've got those abnormal electrical discharge blues.

Epilepsy

Also known as seizure disorder, epilepsy is a brain condition characterized by recurrent seizures. Seizures are paroxysmal events associated with abnormal electrical discharges of neurons in the brain. The discharge may trigger a convulsive movement, an interruption of sensation, an alteration in level of consciousness, or a combination of these symptoms. In most patients, epilepsy doesn't affect intelligence.

Strict adherence equals good control

This condition is probably present in 0.5% to 2% of the population and usually occurs in people under age 20. About 80% of patients have good seizure control with strict adherence to prescribed treatment.

Pathophysiology

A group of neurons may lose afferent stimulation (ability to transmit impulses from the periphery toward the central nervous system) and function as an epileptogenic focus. These neurons are hypersensitive and easily activated. In response to changes in the cellular environment, the neurons become hyperactive and fire abnormally.

About half of all cases of epilepsy are idiopathic. No specific cause can be found and the patient has no other neurologic abnormality. In other cases, however, possible causes of epilepsy include:

• genetic abnormalities, such as tuberous sclerosis (tumors and sclerotic patches in the brain) and phenylketonuria (inability to convert phenylalanine into tyrosine)
• perinatal injuries
• metabolic abnormalities, such as hyponatremia, hypocalcemia, hypoglycemia, and pyridoxine deficiency
• brain tumors or other space-occupying lesions of the cortex
• infections, such as meningitis, encephalitis, or brain abscess
• traumatic injury, especially if the dura mater has been penetrated
• ingestion of toxins, such as mercury, lead, or carbon monoxide
• CVA
• hereditary abnormalities (some seizure disorders appear to run in families)
• fever.

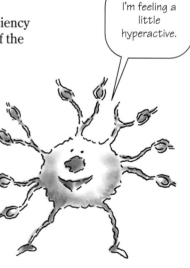

I'm feeling a little hyperactive.

All fired up

This is what happens during a seizure:
• The electronic balance at the neuronal level is altered, causing the neuronal membrane to become susceptible to activation.
• Increased permeability of the cytoplasmic membranes helps hypersensitive neurons fire abnormally. Abnormal

firing may be activated by hyperthermia, hypoglycemia, hyponatremia, hypoxia, or repeated sensory stimulation.
• Once the intensity of a seizure discharge has progressed sufficiently, it spreads to adjacent brain areas. The midbrain, thalamus, and cerebral cortex are most likely to become epileptogenic.
• Excitement feeds back from the primary focus and to other parts of the brain.
• The discharges become less frequent until they stop.

Complications

Depending on the type of seizure, injury may result from a fall at the onset of a seizure or afterward, when the patient is confused. Injury may also result from the rapid, jerking movements that occur during or after a seizure. Anoxia can occur due to airway occlusion by the tongue, aspiration of vomit, or traumatic injury. A continuous seizure state known as status epilepticus can cause respiratory distress and even death. (See *Understanding status epilepticus.*)

What to look for

Signs and symptoms of epilepsy vary depending on the type and cause of the seizure. (See *Understanding types of seizures,* page 268.)

Assessment findings

Physical findings may be normal if the patient doesn't have a seizure during assessment and the cause is idiopathic. If the seizure is caused by an underlying problem, the patient's history and physical examination should uncover related signs and symptoms.

In many cases, the patient's history shows that seizures are unpredictable and unrelated to activities. Occasionally, a patient may report precipitating factors. For example, the seizures may always take place at a certain time, such as during sleep, or after a particular circumstance, such as lack of sleep or emotional stress. He may also report nonspecific symptoms, such as headache, mood changes, lethargy, and myoclonic jerking up to several hours before a seizure. (See *Treating epilepsy,* page 269.)

Understanding status epilepticus

Status epilepticus is a continuous seizure state that must be interrupted by emergency measures. It can occur during all types of seizures. For example, generalized tonic-clonic status epilepticus is a continuous generalized tonic-clonic seizure without an intervening return of consciousness.

Always an emergency
Status epilepticus is accompanied by respiratory distress. It can result from withdrawal of antiepileptic medications, hypoxic or metabolic encephalopathy, acute head trauma, or septicemia secondary to encephalitis or meningitis.

Acting fast
Emergency treatment usually consists of diazepam, phenytoin, or phenobarbital; I.V. dextrose 50% when seizures are secondary to hypoglycemia; and I.V. thiamine in patients with chronic alcoholism or who are undergoing withdrawal.

Understanding types of seizures

Use the guidelines below to understand different seizure types. Keep in mind that some patients may be affected by more than one type.

Partial seizures

Arising from a localized area in the brain, partial seizure activity may spread to the entire brain, causing a generalized seizure. Several types and subtypes of partial seizures exist:

- simple partial, which includes jacksonian and sensory
- complex partial
- secondarily generalized partial seizure (partial onset leading to generalized tonic-clonic seizure).

Jacksonian seizure. A jacksonian seizure begins as a localized motor seizure, characterized by a spread of abnormal activity to adjacent areas of the brain. The patient experiences a stiffening or jerking in one extremity, accompanied by a tingling sensation in the same area. Although the patient in a jacksonian seizure seldom loses consciousness, the seizure may progress to a generalized tonic-clonic seizure.

Sensory seizure. Symptoms of a sensory seizure include hallucinations, flashing lights, tingling sensations, vertigo, déjà vu, and smelling a foul odor.

Complex partial seizure. Signs and symptoms of a complex partial seizure are variable but usually include purposeless behavior, including a glassy stare, picking at clothes, aimless wandering, lip-smacking or chewing motions, and unintelligible speech. An aura may occur first, and seizures may last from a few seconds to 20 minutes. Afterward, mental confusion may last for several minutes, and an observer may mistakenly suspect alcohol or drug intoxication or psychosis. The patient has no memory of his actions during the seizure.

Secondarily generalized seizure. A secondarily-generalized seizure can be simple or complex and can progress to a generalized seizure. An aura may occur first, with loss of consciousness occurring immediately or 1 to 2 minutes later.

Generalized seizures

Generalized seizures cause a generalized electrical abnormality within the brain. Types include:

- absence or petit mal
- myoclonic
- generalized tonic-clonic (grand mal)
- akinetic.

Absence seizure. Absence seizure occurs most often in children. It usually begins with a brief change in the level of consciousness, signaled by blinking or rolling of the eyes, a blank stare, and slight mouth movements. The patient retains his posture and continues preseizure activity without difficulty. Seizures last from 1 to 10 seconds, and impairment is so brief that the patient may be unaware of it. However, if not properly treated, these seizures can recur up to 100 times a day and progress to a generalized tonic-clonic seizure.

Myoclonic seizure. Also called bilateral massive epileptic myoclonus, myoclonic seizure is marked by brief, involuntary muscular jerks of the body or extremities, which may occur in a rhythmic manner, and a brief loss of consciousness.

Generalized tonic-clonic seizure. Typically, a generalized tonic-clonic seizure begins with a loud cry, caused by air rushing from the lungs through the vocal cords. The patient falls to the ground, losing consciousness. The body stiffens (tonic phase) and then alternates between episodes of muscle spasm and relaxation (clonic phase). Tongue biting, incontinence, labored breathing, apnea, and cyanosis may also occur.

The seizure stops in 2 to 5 minutes, when abnormal electrical conduction of the neurons is completed. Afterward, the patient regains consciousness but is somewhat confused. He may have difficulty talking and may have drowsiness, fatigue, headache, muscle soreness, and arm or leg weakness. He may fall into a deep sleep afterwards.

Akinetic seizure. Characterized by a general loss of postural tone and a temporary loss of consciousness, akinetic seizure occurs in young children. Sometimes it's called a "drop attack" because the child falls.

I feel an aura coming on

Some patients report an aura a few seconds or minutes before a generalized seizure. An aura signals the beginning of abnormal electrical discharges within a focal area of the brain. Typical auras include:
- a pungent smell
- nausea or indigestion
- a rising or sinking feeling in the stomach
- a dreamy feeling
- an unusual taste
- a visual disturbance such as a flashing light.

What tests tell you

The following tests are used to diagnose epilepsy:
- EEG showing paroxysmal abnormalities may confirm the diagnosis by providing evidence of continuing seizure tendency. A negative EEG doesn't rule out epilepsy because paroxysmal abnormalities occur intermittently. EEG also helps determine the prognosis and can help classify the disorder.
- CT scan and MRI provide density readings of the brain and may indicate abnormalities in internal structures.

Other tests include serum glucose and calcium studies, skull X-rays, lumbar puncture, brain scan, and cerebral angiography.

Guillain-Barré syndrome

Guillain-Barré syndrome is an acute, rapidly progressive, potentially fatal syndrome. It's associated with segmented demyelination of peripheral nerves. It may also be called acute demyelinating polyneuropathy.

This syndrome occurs equally in both sexes, usually occurring between the ages of 30 and 50. It affects about 2 out of every 100,000 people. As a result of better symptom management, 80% to 90% of patients recover with few or no residual symptoms.

Acute, plateau, recovery

The clinical course of Guillain-Barré syndrome has three phases:

Battling illness

Treating epilepsy

Treatment for epilepsy seeks to reduce the frequency of seizures or prevent their occurrence.

Drug therapy
Drug therapy is specific to the type of seizure. The most commonly prescribed drugs for generalized tonic-clonic and complex partial seizures are phenytoin, carbamazepine, phenobarbital, and primidone, administered individually. Valproic acid, clonazepam, and ethosuximide are commonly prescribed for absence seizures. Lamotrigine is also given as adjunct therapy for partial seizures.

Surgery
If drug therapy fails, treatment may include surgical removal of a demonstrated focal lesion to try to stop seizures. Surgery is also performed when epilepsy results from an underlying problem, such as an intracranial tumor, a brain abscess or cyst, or vascular abnormalities.

The acute phase begins when the first definitive symptom develops and ends 1 to 3 weeks later, when no further deterioration is noted.

The plateau phase lasts for several days to 2 weeks.

The recovery phase, believed to coincide with re-myelination and axonal process regrowth, can last from 4 months to 3 years.

Pathophysiology

The precise cause of Guillain-Barré syndrome is unknown but it's thought to be a cell-mediated, immunologic attack on peripheral nerves in response to a virus. Risk factors include surgery, rabies or swine influenza vaccination, viral illness, Hodgkin's disease or another malignant disease, and lupus erythematosus.

Danger! Demyelination!

An immunologic reaction causes segmental demyelination of the peripheral nerves, which prevents normal transmission of electrical impulses along the sensorimotor nerve roots.

The myelin sheath, which covers the nerve axons and conducts electrical impulses along the nerve pathways, degenerates for unknown reasons. With degeneration comes inflammation, swelling, and patchy demyelination.

As myelin is destroyed, the nodes of Ranvier, located at the junctures of the myelin sheaths, widen. This delays and impairs impulse transmission along the dorsal and ventral nerve roots.

What to look for

Impairment of dorsal nerve roots affects sensory function, so the patient may experience tingling and numbness. Impairment of ventral nerve roots affects motor function, so the patient may experience muscle weakness, immobility, and paralysis.

Other signs and symptoms include muscle stiffness and pain, sensory loss, loss of position sense, and diminished or absent deep tendon reflexes (DTRs).

Symptoms usually follow an ascending pattern, beginning in the legs and progressing to the arms, trunk, and

This syndrome causes inflammation and degenerative changes in both sensory and motor nerve roots....

...that's why signs of sensory and motor impairment occur simultaneously.

face. In mild forms, only cranial nerves may be affected. In some patients, muscle weakness may be absent.

Respiratory risk

The disorder commonly affects respiratory muscles. If the patient dies, it's usually from respiratory complications. Paralysis of the internal and external intercostal muscles leads to a reduction in functional breathing. Vagus nerve paralysis causes a loss of the protective mechanisms that respond to brachial irritation and foreign bodies, as well as a diminished or absent gag reflex. (See *Treating Guillain-Barré syndrome*.)

What tests tell you

• CSF analysis may show a normal white blood cell count, an elevated protein count and, in severe disease, increased CSF pressure. The CSF protein level begins to rise several days after the onset of signs and symptoms, peaking in 4 to 6 weeks, probably because of widespread inflammatory disease of the nerve roots.

Battling illness

Treating Guillain-Barré syndrome

Most patients seek treatment when the disease is in the acute stage. Treatment is primarily supportive and may require endotracheal intubation or tracheotomy if the patient has difficulty clearing secretions and mechanical ventilation if he has respiratory problems.

Continuous electrocardiographic monitoring is necessary to identify the autonomic symptoms, such as cardiac arrhythmias. Propranolol may be given for tachycardia and hypotension, and atropine may be used for bradycardia. Marked hypotension may require volume replacement.

A spontaneous recovery

Most patients recover spontaneously. Intensive physical therapy starts as soon as voluntary movement returns to skeletal muscles, to prevent muscle and joint contractures. However, a small percentage of patients are left with some residual disability.

Alternative approaches

High-dose I.V. immune globulin and plasmapheresis may shorten recovery time. Plasmapheresis temporarily reduces circulating antibodies through removal of the patient's blood, centrifugation of blood to remove plasma, and subsequent reinfusion. It's most effective during the first few weeks of the disease, and patients need less ventilatory support if treatment begins within 2 weeks of onset. The patient may receive three to five plasma exchanges.

• Electromyography may demonstrate repeated firing of the same motor unit instead of widespread sectional stimulation.
• Electrophysiologic testing may reveal marked slowing of nerve conduction velocities.

Meningitis

In meningitis, the brain and spinal cord meninges become inflamed. Inflammation may involve all three meningeal membranes: the dura mater, the arachnoid membrane, and the pia mater.

Early recognition and antibiotics: A winning combination

The prognosis for patients with meningitis is good and complications are rare, especially if the disease is recognized early and the infecting organism responds to antibiotics. However, mortality in untreated meningitis is 70% to 100%. The prognosis is poorer for infants and elderly people.

Pathophysiology

The origin of meningeal inflammation may be:

 bacterial

 viral

 protozoal

 fungal.

The most common causes of meningitis are bacterial and viral. Bacterial infection may be due to *Neisseria meningitidis, Haemophilus influenzae, Streptococcus pneumoniae,* or *Escherichia coli.* Sometimes, no causative organism can be found.

Blame it on bacteria

Bacterial meningitis is one of the most serious infections that may affect infants and children. In most patients, the infection that causes meningitis is secondary to another bacterial infection, such as bacteremia (especially from pneumonia, empyema, osteomyelitis, and endocarditis), sinusitis, otitis media, encephalitis, myelitis, or brain abscess. Respiratory infections increase the risk of bacterial meningitis. Meningitis may also follow a skull fracture, a penetrating head wound, lumbar puncture, or ventricular shunting procedures.

Enter the subarachnoid space

Bacterial meningitis occurs when bacteria enters the subarachnoid space and causes an inflammatory response. Usually, organisms enter the nervous system after having already invaded and infected another region of the body. The organisms gain access to the subarachnoid space and the CSF, where they cause irritation of the tissues bathed by the fluid.

> Bacterial meningitis begins when I enter the subarachnoid space.

The viral version

Meningitis caused by a virus is called aseptic viral meningitis. Aseptic viral meningitis may result from a direct infection or secondary to disease, such as mumps, herpes, measles, or leukemia. Usually, symptoms are mild and the disease is self-limiting. (See *Understanding aseptic viral meningitis.*)

Now I get it!

Understanding aseptic viral meningitis

A benign syndrome, aseptic viral meningitis results from infection with enteroviruses (most common), arboviruses, herpes simplex virus, mumps virus, or lymphocytic choriomeningitis virus.

First a fever

Signs and symptoms of viral meningitis usually begin suddenly with a fever up to 104° F (40° C), drowsiness, confusion, stupor, and slight neck or spine stiffness when the patient bends forward. The patient history may reveal a recent illness.

Other signs and symptoms include headache, nausea, vomiting, abdominal pain, poorly defined chest pain, and sore throat.

What virus is this anyway?

A complete patient history and knowledge of seasonal epidemics are key to differentiating among the many forms of aseptic viral meningitis. Negative bacteriologic cultures and cerebrospinal fluid (CSF) analysis showing pleocytosis (a greater than normal number of cells in the CSF) and increased protein suggest the diagnosis. Isolation of the virus from CSF confirms it.

Bed rest plus

Treatment for viral meningitis includes bed rest, maintenance of fluid and electrolyte balance, analgesics for pain, and exercises to combat residual weakness. Careful handling of excretions and good hand-washing technique prevent the spread of the disease, although isolation isn't necessary.

Risk factors

Infants, children, and elderly people have the highest risk of developing meningitis. Other risk factors include malnourishment, immunosuppression (for example, from radiation therapy), and CNS trauma.

Complications

Potential complications of meningitis include:
- visual impairment
- optic neuritis
- cranial nerve palsies
- deafness
- personality changes
- headache
- paresis or paralysis
- endocarditis
- coma
- vasculitis
- cerebral infarction.

Complications in infants and children may also include:
- sensory hearing loss
- epilepsy
- mental retardation
- hydrocephalus
- subdural effusions.

What to look for

The cardinal signs and symptoms of meningitis are those of infection and increased intracranial pressure:
- headache
- stiff neck and back
- malaise
- photophobia
- chills
- vomiting
- twitching
- seizures
- altered level of consciousness, such as confusion or delirium.

Signs and symptoms in infants and children may also include:
- fretfulness
- refusal to eat

- fever
- vomiting.

Tips for determining type and severity

Findings vary depending on the type and severity of meningitis. In pneumococcal meningitis, the patient history may uncover a recent lung, ear, or sinus infection or endocarditis. It may also reveal alcoholism, sickle cell disease, basal skull fracture, recent splenectomy, or organ transplant. In meningitis caused by *Haemophilus influenzae*, the patient history may reveal a recent respiratory tract or ear infection. In meningococcal meningitis, you may see a petechial, purpuric, or ecchymotic rash on the patient's lower body.

Pinpointing meningeal irritation

Signs of meningeal irritation are nuchal rigidity, exaggerated and symmetrical DTRs, opisthotonos (a spasm in which the back and extremities arch backward so that the body rests on the head and heels), and positive Brudzinski's and Kernig's signs. (See *Treating meningitis,* and *Important signs of meningitis,* page 276.)

Battling illness

Treating meningitis

Treatment for meningitis includes medications, vigorous supportive care, and treatment of coexisting conditions, such as pneumonia and endocarditis.

Antibiotics and more antibiotics

Usually, I.V. antibiotics are given for at least 2 weeks, followed by oral antibiotics. The most common antibiotics are penicillin G, ampicillin, or nafcillin. If the patient is allergic to penicillin, tetracycline, chloramphenicol, or kanamycin are used.

Prophylactic antibiotics may also be used after ventricular shunting procedures, skull fractures, or penetrating head wounds, but this use is controversial.

Other drugs used in treatment include a digitalis glycoside such as digoxin to control arrhythmias, mannitol to decrease cerebral edema, an I.V. anticonvulsant or a sedative to reduce restlessness, and aspirin or acetaminophen to relieve headache and fever.

Bed rest plus

Supportive measures include bed rest, hypothermia, and fluid therapy to prevent dehydration. Isolation is necessary if nasal cultures are positive.

Important signs of meningitis

A positive response to the following tests helps establish a diagnosis of meningitis.

Brudzinski's sign

Place the patient in a dorsal recumbent position, then put your hands behind his neck and bend it forward. Pain and resistance may indicate neck injury, or arthritis. But if the patient also flexes the hips and knees, chances are he has meningeal irritation and inflammation, a sign of meningitis.

Kernig's sign

Place the patient in a supine position. Flex his leg at the hip and knee, and then straighten the knee. Pain or resistance suggests meningitis.

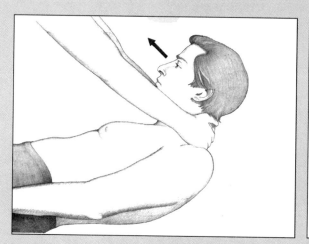

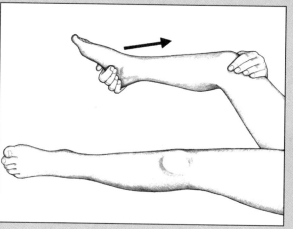

What tests tell you

The following tests are used to diagnose meningitis:
- Vision tests may show diplopia and other visual problems.
- Lumbar puncture shows elevated CSF pressure, cloudy or milky CSF, a high protein level, positive Gram stain and culture that usually identifies the infecting organism (unless it's a virus), and depressed CSF glucose concentration.
- Chest X-rays are important because they may reveal pneumonitis or lung abscess, tubercular lesions, or granulomas secondary to fungal infection.
- Sinus and skull X-rays may help identify the presence of cranial osteomyelitis, paranasal sinusitis, or skull fracture.
- White blood cell (WBC) count usually indicates leukocytosis and abnormal serum electrolyte levels.
- CT scan rules out cerebral hematoma, hemorrhage, or tumor.

Multiple sclerosis

Demyelination is the loss of myelin sheath material essential in nerve impulse transmission.

MS results from progressive demyelination of the white matter of the brain and spinal cord, leading to widespread neurologic dysfunction. The structures most frequently involved are the optic and oculomotor nerves and the spinal nerve tracts. The disorder doesn't affect the peripheral nervous system.

Characterized by exacerbations and remissions, MS is a major cause of chronic disability in people between ages 18 and 40. The incidence is highest in women, in northern urban areas, in higher socioeconomic groups, and in people with a family history of the disease.

The prognosis varies. The disease may progress rapidly, causing death in a few months or disability by early adulthood. However, about 70% of patients lead active, productive lives with prolonged remissions.

Pathophysiology

The exact cause of MS is unknown. It may be caused by a slow-acting viral infection, an autoimmune response of the nervous system, or an allergic response. Other possible causes include trauma, anoxia, toxins, nutritional deficiencies, vascular lesions, and anorexia nervosa, all of which may help destroy axons and the myelin sheath.

In addition, emotional stress, overwork, fatigue, pregnancy, or an acute respiratory tract infection may precede the onset of this illness. Genetic factors may also play a part.

Dots of demyelination

MS affects the white matter of the brain and spinal cord by causing scattered demyelinated lesions that prevent normal neurologic conduction. (See *Understanding myelin breakdown,* page 278.) After the myelin is destroyed, neuroglial tissue in the white matter of the CNS proliferates, forming hard yellow plaques of scar tissue. Proliferation of neuroglial tissue is called gliosis.

Interrupting an impulse

Scar tissue damages the underlying axon fiber so that nerve conduction is disrupted. The symptoms of MS caused by demyelination become irreversible as the dis-

Now I get it!

Understanding myelin breakdown

Myelin plays a key role in speeding electrical impulses to the brain for interpretation. The myelin sheath is a lipoprotein complex formed of glial cells. It protects the neuron's long nerve fiber (axon) much like the insulation on an electrical wire. Because of its high electrical resistance and weak ability to store an electrical charge, the myelin sheath permits conduction of nerve impulses from one node of Ranvier to the next.

Effects of injury

Myelin can be injured by hypoxemia, toxic chemicals, vascular insufficiency, or autoimmune responses. When this occurs, the myelin sheath becomes inflamed and the membrane layers break down into smaller components. These components become well-circumscribed plaques filled with microglial elements, macroglia, and lymphocytes. This process is called demyelination.

The damaged myelin sheath impairs normal conduction, causing partial loss or dispersion of the action potential and consequent neurologic dysfunction.

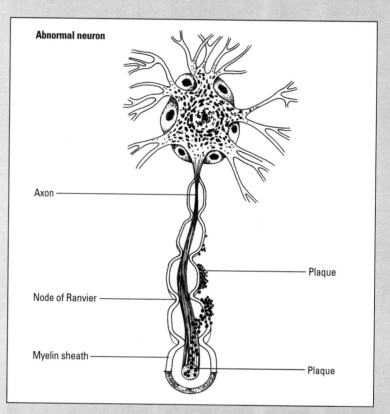

Abnormal neuron

Axon

Plaque

Node of Ranvier

Myelin sheath

Plaque

---

ease progresses. However, remission may result from healing of demyelinated areas by sclerotic tissue.

What to look for

Signs and symptoms of MS depend on the following factors:

 the extent of myelin destruction

 the site of myelin destruction

 the extent of remyelination

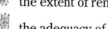 the adequacy of subsequent restored synaptic transmission.

Hard to predict and hard to describe

Symptoms may be unpredictable and difficult for the patient to describe. They may be transient or may last for hours or weeks. Usually, the patient history reveals two initial symptoms: visual problems (caused by an optic neuritis) and sensory impairment such as paresthesia. After the initial episode, findings may vary. They may include blurred vision or diplopia, emotional lability (from involvement of the white matter of the frontal lobes), and dysphagia.

Other signs and symptoms include:
• poorly articulated speech (caused by cerebellar involvement)
• muscle weakness and spasticity (caused by lesions in the corticospinal tracts)
• hyperreflexia
• urinary problems
• intention tremor
• gait ataxia
• paralysis, ranging from monoplegia to quadriplegia
• visual problems, such as scotoma (an area of lost vision in the visual field), optic neuritis, and ophthalmoplegia (paralysis of the eye muscles).

Complications

Complications include injuries from falls, urinary tract infections (UTIs), constipation, joint contractures, pressure ulcers, rectal distention, and pneumonia. (See *Treating multiple sclerosis,* page 280.)

What tests tell you

Diagnosing MS may take years because of remissions. The following tests help diagnose the disease:
• EEG shows abnormalities in one-third of patients.
• CSF analysis reveals elevated immunoglobulin G levels but normal total protein levels. This elevation is significant only when serum gamma globulin levels are normal, and it reflects hyperactivity of the immune system due to chronic demyelination. The WBC count may be slightly increased.
• Evoked potential studies demonstrate slowed conduction of nerve impulses in 80% of patients.
• CT scan may disclose lesions within the brain's white matter.

Treating multiple sclerosis

Treatment for multiple sclerosis (MS) aims to shorten exacerbations and, if possible, relieve neurologic deficits so the patient can resume a near-normal lifestyle.

Drug options
Because MS may have allergic and inflammatory causes, corticotropin, prednisone, or dexamethasone are used to reduce edema of the myelin sheath during exacerbations, relieving symptoms and hastening remissions. However, these drugs don't prevent future exacerbations.

Currently, the preferred treatment during an acute attack is a short course of methylprednisone, with or without a short prednisone taper. Interferon beta-1a or interferon beta-1b may also be given to decrease the frequency of relapses. How these drugs achieve their effect is not clearly understood. Interferon beta-1b, a naturally occurring antiviral and immunoregulatory agent derived from human fibroblasts, is thought to attach to membrane receptors and cause cellular changes, including increased protein synthesis.

Other useful drugs include chlordiazepoxide to mitigate mood swings, baclofen or dantrolene to relieve spasticity, and bethanechol or oxybutynin to relieve urine retention and minimize urinary frequency and urgency.

Bed rest plus
During acute exacerbations, supportive measures include bed rest, massage, prevention of fatigue and pressure ulcers, bowel and bladder training, treatment of bladder infections with antibiotics, physical therapy, and counseling.

• MRI is the most sensitive method of detecting lesions and is also used to evaluate disease progression. More than 90% of patients show lesions when this test is performed.
• Neuropsychological tests may help rule out other disorders.

Myasthenia gravis

Myasthenia gravis produces sporadic, progressive weakness and abnormal fatigue of voluntary skeletal muscles. These effects are exacerbated by exercise and repeated movement.

A menace to muscles

Myasthenia gravis usually affects muscles in the face, lips, tongue, neck, and throat, which are innervated by the cranial nerves, but it can affect any muscle group. Eventually, muscle fibers may degenerate, and weakness, especially of the head, neck, trunk, and limb muscles, may become irreversible. When the disease involves the respiratory system, it may be life-threatening.

Myasthenia gravis affects 2 to 20 people per 100,000 and occurs at any age. However, incidence is highest in women ages 18 to 25 and in men ages 50 to 60. About three times as many women as men develop this disease. About 20% of infants born to mothers with myasthenia gravis have transient or, occasionally, persistent disease.

The disease follows an unpredictable course with periodic exacerbations and remissions. Spontaneous remissions occur in about 25% of patients. No cure exists but, thanks to drug therapy, patients may lead relatively normal lives except during exacerbations.

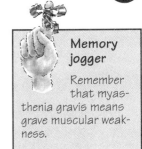

Memory jogger

Remember that myasthenia gravis means grave muscular weakness.

Pathophysiology

The cause of myasthenia gravis is unknown. It commonly accompanies autoimmune and thyroid disorders. In fact, 15% of patients with myasthenia gravis have thymomas.

Interrupting nerve impulses

For some reason, the patient's blood cells and thymus gland produce antibodies that block, destroy, or weaken the neuroreceptors that transmit nerve impulses, causing a failure in transmission of nerve impulses at the neuromuscular junction. (See *What happens in myasthenia gravis,* page 282.)

What to look for

Signs and symptoms of myasthenia gravis vary, depending on the muscles involved and the severity of the disease. However, in all cases, muscle weakness is progressive, and eventually some muscles may lose function entirely.

Common signs and symptoms include:
- extreme muscle weakness
- fatigue
- ptosis
- diplopia
- difficulty chewing and swallowing
- sleepy, masklike expression
- drooping jaw
- bobbing head
- arm or hand muscle weakness.

Now I get it!

What happens in myasthenia gravis

During normal neuromuscular transmission, a motor nerve impulse travels to a motor nerve terminal, stimulating the release of a chemical neurotransmitter called acetylcholine (ACh). When ACh diffuses across the synapse, receptor sites in the motor end plate react and depolarize the muscle fiber. The depolarization spreads through the muscle fiber, causing muscle contraction.

Those darn antibodies

In myasthenia gravis, antibodies attach to the ACh receptor sites. They block, destroy, and weaken these sites, leaving them insensitive to ACh, thereby blocking neuromuscular transmission.

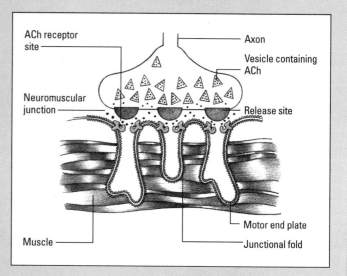

Talking with your patient

The patient may report that she must tilt her head back to see properly. She usually notes that symptoms are milder on awakening and worsen as the day progresses, and that short rest periods temporarily restore muscle function. She may also report that symptoms become more intense during menses and after emotional stress, prolonged exposure to sunlight or cold, or infections.

Respiratory system effects

Auscultation may reveal hypoventilation if the respiratory muscles are involved. This may lead to decreased tidal volume, making breathing difficult and predisposing the patient to pneumonia and other respiratory tract infections.

Progressive weakness of the diaphragm and the intercostal muscles may eventually lead to myasthenic crisis (an acute exacerbation that causes severe respiratory distress). (See *Treating myasthenia gravis*.)

What tests tell you

The following tests are used to diagnose myasthenia gravis:

• Tensilon test confirms diagnosis by temporarily improving muscle function after an I.V. injection of edrophonium or, occasionally, neostigmine. However, long-standing ocular muscle dysfunction may not respond. This test also differentiates a myasthenic crisis from a cholinergic crisis.

• Electromyography measures the electrical potential of muscle cells and helps differentiate nerve disorders from muscle disorders. In myasthenia gravis, the amplitude of motor unit potential falls off with continued use. Muscle contractions decrease with each test, reflecting fatigue.

• Nerve conduction studies measure the speed at which electrical impulses travel along a nerve and also help distinguish nerve disorders from muscle disorders.

• Chest X-ray or CT scan may identify a thymoma.

Battling illness

Treating myasthenia gravis

The main treatment for myasthenia gravis is anticholinesterase drugs, such as neostigmine and pyridostigmine. These drugs counteract fatigue and muscle weakness and restore about 80% of muscle function. However, they become less effective as the disease worsens. Corticosteroids may also help to relieve symptoms.

Options and alternatives

If medications aren't effective, some patients undergo plasmapheresis to remove acetylcholine-receptor antibodies and temporarily lessen the severity of symptoms. Patients with thymomas require thymectomy, which leads to remission in adult-onset disease in about 40% of patients if done within 2 years after diagnosis.

Yikes! It's an emergency!

Myasthenic crisis necessitates immediate hospitalization. Tracheotomy, ventilation with a positive-pressure ventilator, and vigorous suctioning to remove secretions usually bring improvement in a few days. Because anticholinesterase drugs aren't effective in myasthenic crisis, they're discontinued until respiratory function improves.

Parkinson's disease

Parkinson's disease is a slowly progressive movement disorder characterized by muscle rigidity, loss of muscle movement (akinesia), and involuntary tremors. The disease isn't fatal but death may result from aspiration pneumonia or some other infection.

Also called parkinsonism, paralysis agitans, or shaking palsy, Parkinson's disease is one of the most common crippling diseases in the United States. It affects more men than women and usually occurs in middle age or later, striking 1 in every 100 people over age 60.

A deficiency of dopamine is not good news...

Pathophysiology

In most cases, the cause of Parkinson's disease is unknown. However, some cases result from exposure to toxins, such as manganese dust and carbon monoxide, that destroy cells in the substantia nigra of the brain.

A defect in the dopamine pathway...

Parkinson's disease affects the extrapyramidal system, which influences the initiation, modulation, and completion of movement. The extrapyramidal system includes the corpus striatum, globus pallidus, and substantia nigra.

In Parkinson's disease, a dopamine deficiency occurs in the basal ganglia, the dopamine-releasing pathway that connects the substantia nigra to the corpus striatum.

...it prevents my cells from performing their normal function within the central nervous system.

...causes an imbalance of neurotransmitters

Reduction of dopamine in the corpus striatum upsets the normal balance between the inhibitory dopamine and excitatory acetylcholine neurotransmitters. This prevents affected brain cells from performing their normal inhibitory function within the CNS and causes most parkinsonian symptoms.

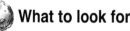

What to look for

Important signs of Parkinson's disease are muscle rigidity, akinesia, and a unilateral "pill-roll" tremor.

Muscle rigidity

Muscle rigidity results in resistance to passive muscle stretching, which may be uniform (lead-pipe rigidity) or jerky (cogwheel rigidity).

Akinesia

Akinesia (absence of movement) causes gait and movement disturbances. The patient walks with his body bent forward, takes a long time initiating movement when performing a purposeful action, pivots with difficulty, and easily loses his balance.

Akinesia may also cause the following signs:
• masklike facial expression
• oculogyric crises, during which the eyes are fixed upward, with involuntary tonic movements
• blepharospasm, in which the eyelids stay closed.

Pill-roll tremor

This insidious tremor begins in the fingers, increases during stress or anxiety, and decreases with purposeful movement and sleep.

More signs and symptoms

Other signs and symptoms of Parkinson's disease include:
• a high-pitched, monotone voice
• drooling
• dysarthria (impaired speech due to a disturbance in muscle control)
• dysphagia (difficulty swallowing)
• fatigue
• muscle cramps in the legs, neck, and trunk
• oily skin
• increased perspiration
• insomnia
• mood changes.

Complications

Common complications of Parkinson's disease include injury from falls, food aspiration due to impaired voluntary movements, UTIs, and skin breakdown due to increased immobility. (See *Treating Parkinson's disease,* page 286.)

Battling illness

Treating Parkinson's disease

Treatment for Parkinson's disease aims to relieve symptoms and keep the patient functional as long as possible. It consists of drugs, physical therapy, and stereotactic neurosurgery in extreme cases.

Looking to levodopa

Drug therapy usually includes levodopa, a dopamine replacement that's most effective in the first few years after it's initiated. It's given in increasing doses until signs and symptoms are relieved or adverse reactions develop. Because adverse effects can be serious, levodopa is frequently given along with carbidopa to halt peripheral dopamine synthesis. Bromocriptine may be given as an additive to reduce the levodopa dose.

When levodopa is ineffective or too toxic, anticholinergics, such as trihexyphenidyl or benztropine, and antihistamines, such as diphenhydramine, are given. Anticholinergics may be used to control tremors and rigidity. They may also be used in combination with levodopa. Antihistamines may help decrease tremors because of their central anticholinergic and sedative effects.

Amantadine, an antiviral agent, is used early in treatment to reduce rigidity, tremors, and akinesia. Patients with mild disease are given deprenyl to slow the progression of the disease and ease symptoms. Tricyclic antidepressants may be given to decrease depression.

Neurosurgery next?

When drug therapy fails, stereotactic neurosurgery is sometimes effective. In this procedure, electrical coagulation, freezing, radioactivity, or ultrasound destroys the ventrolateral nucleus of the thalamus to prevent involuntary movement. Such neurosurgery is most effective in comparatively young, otherwise healthy people with unilateral tremor or muscle rigidity. Like drug therapy, neurosurgery is a palliative measure that can only relieve symptoms.

Physical therapy

Physical therapy helps maintain the patient's normal muscle tone and function. It includes both active and passive range-of-motion exercises, routine daily activities, walking, and baths and massage to help relax muscles.

What tests tell you

Diagnosis of Parkinson's disease is based on the patient's age, history, and signs and symptoms, so laboratory tests are generally of little value. However, urinalysis may reveal decreased dopamine levels, and CT scan or MRI may rule out other disorders such as intracranial tumors.

Quick quiz

1. An epileptic seizure involving loss of consciousness, both tonic and clonic phases, tongue biting, and incontinence is:
   A. a simple partial seizure.
   B. an absence seizure.
   C. a grand mal seizure.

Answer: C. Also called a generalized tonic-clonic seizure, a grand mal seizure stops in 2 to 5 minutes, when abnormal electrical conduction of the neurons is completed.

2. The major causes of CVA are:
   A. thrombosis, embolism, and hemorrhage.
   B. genetic and metabolic abnormalities.
   C. brain and spinal cord tumors.

Answer: A. Thrombosis is the most common cause; then embolism and hemorrhage, in that order.

3. Brudzinski's sign and Kernig's sign are two tests that help diagnose:
   A. CVA.
   B. meningitis.
   C. epilepsy.

Answer: B. A positive response to one or both tests indicates meningeal irritation.

4. A disease characterized by progressive degeneration of the cerebral cortex is:
   A. Alzheimer's disease.
   B. epilepsy.
   C. Guillain-Barré syndrome.

Answer: A. Symptoms of Alzheimer's disease range from recent memory loss to debilitating dementia.

5. Multiple sclerosis is characterized by:
   A. progressive demyelination of the white matter of the CNS.
   B. impairment of cerebral circulation.
   C. deficiency of the neurotransmitter dopamine.

Answer: A. Patches of demyelination cause widespread neurologic dysfunction.

Scoring

☆☆☆   If you answered all five items correctly, unbelievable! Your dedication to dendrites and dopamine is downright unnerving.

☆☆   If you answered three or four correctly, congratulations! We are honored to extend congratulations not only to you, but to both lobes of your cerebrum.

☆   If you answered fewer than three correctly, stay cool. We'll just chalk it up to one of those "lapses in the synapses."

Blood

Just the facts

In this chapter you'll learn:

♦ the structure and function of blood and its components

♦ pathophysiology, signs and symptoms, diagnostic tests, and treatments for common blood disorders.

Understanding blood

Blood is one of the body's major fluid tissues. Pumped by the heart, it continuously circulates through the blood vessels, carrying vital elements to every part of the body. Blood is made of:

• a liquid component — plasma

• formed components — erythrocytes, leukocytes, and thrombocytes suspended in plasma.

Components, do your thing

Each of blood's components perform specific vital functions:

• Plasma carries antibodies and nutrients to tissues and carries wastes away.

• Erythrocytes, also called red blood cells (RBCs), carry oxygen to the tissues and remove carbon dioxide from them.

• Leukocytes, or white blood cells (WBCs), participate in the inflammatory and immune response.

• Thrombocytes, or platelets, along with the coagulation factors in plasma, control clotting. (Coagulation factors are essential components to normal blood clotting.)

Blood is the chief means of transport within the body...

...it circulates through the heart, arteries, capillaries, and veins.

A problem with any of these components may have serious or even deadly consequences.

Plasma

Plasma is a clear, straw-colored fluid that consists mainly of the proteins albumin, globulin, and fibrinogen held in aqueous suspension. Plasma's fluid characteristics, including its osmotic pressure, viscosity, and suspension qualities, depend on its protein content.

Plasma components regulate acid-base balance and immune responses. They also mediate nutrition (by carrying nutrients to tissues) and help to accelerate coagulation.

Other components in plasma include glucose, lipids, amino acids, electrolytes, pigments, hormones, oxygen, and carbon dioxide.

Important products of metabolism that circulate in plasma include urea, uric acid, creatinine, and lactic acid.

Plasma consists mostly of protein in water but also contains nutrients, gases, hormones, products of metabolism, and other substances.

Red blood cells

RBCs in adults are usually produced in the bone marrow. In the fetus, the liver and spleen also participate in RBC production. The RBC production process is called erythropoiesis.

RBC production is regulated by the tissues' demand for oxygen and the blood cells' ability to deliver it. A lack of oxygen in the tissues (anoxia) stimulates the formation and release of erythropoietin, a hormone that activates the bone marrow to produce RBCs. Erythropoiesis is also stimulated by androgens.

The making of an erythrocyte

Erythrocyte formation begins with a precursor, called a stem cell. The stem cell eventually develops into an RBC. Development requires vitamin B_{12}, folic acid, and minerals, such as copper, cobalt and, especially, iron.

When tissues are oxygen-starved, that's when I come out fighting.

Ironclad facts about iron

Iron is a component of hemoglobin and vital to the blood's oxygen-carrying capacity. Iron is found in food and, once consumed, is absorbed in the duodenum and upper jejunum. Once absorbed, iron may be transported to the bone morrow for hemoglobin synthesis. Iron may also be transported to needy tissues, for example, to muscle for myoglobin synthesis.

Unused iron is temporarily stored as ferritin and hemosiderin, most commonly in the liver. It's stored in specialized cells called reticuloendothelial cells until it's released for use in the bone marrow to form new RBCs.

White blood cells

WBCs protect the body against harmful bacteria and infection. They're classified two ways:

Just think, I'm the body's primary defense against infection.

as granular leukocytes (granulocytes), such as basophils, neutrophils, and eosinophils

as nongranular leukocytes, such as lymphocytes, monocytes, and plasma cells.

Most WBCs are produced in bone marrow. Lymphocytes and plasma cells complete their maturation in the lymph nodes.

From hours to years

WBCs have a wide range of life spans; some granulocytes circulate for less than 6 hours, some monocytes may survive for weeks or months, and some lymphocytes last for years. Normally, the number of WBCs range from 5,000 to 10,000/μl.

A white blood cell by any other name

Types of WBCs include:
• neutrophils, the predominant form of granulocyte, which make up about 60% of WBCs; they surround and digest invading organisms and other foreign matter by phagocytosis
• eosinophils, minor granulocytes, which defend against parasites, participate in allergic reactions, and fight lung and skin infections

• basophils, minor granulocytes, which may release heparin and histamine into the blood and participate in delayed allergic reactions

• monocytes, which, along with neutrophils, devour invading organisms by phagocytosis; they also migrate to tissues where they develop into cells called macrophages that participate in immunity

• lymphocytes, which occur mostly in two forms: B cells and T cells; B cells produce antibodies and T cells regulate cell-mediated immunity

• plasma cells, which develop from lymphocytes in tissues, where they produce, store, and release antibodies.

We're lymphocytes, a type of white blood cell.

Platelets

Platelets are small (2 to 4 microns in diameter), colorless, disk-shaped cytoplasmic cells split from cells in bone marrow. They have a life span of 7 to 10 days and perform three vital functions:

☝ shrinking damaged blood vessels to minimize blood loss

✌ forming hemostatic plugs in injured blood vessels

🖐 with plasma, providing materials that accelerate blood coagulation, especially factor III.

Mesh, clump, and prevent hemorrhage

In a complex process called hemostasis, platelets, plasma, and coagulation factors interact to control bleeding. When tissue injury occurs, blood vessels at the injury site constrict and platelets mesh or clump to help prevent hemorrhage. (See *Understanding clotting*.)

Clotting begins within minutes when a blood vessel is injured or severed.

Now I get it!

Understanding clotting

When a blood vessel is severed or injured, clotting begins within minutes to stop loss of blood. Coagulation factors are essential to normal blood clotting. Absent, decreased, or excess coagulation factors may lead to a clotting abnormality. Coagulation factors are commonly designated by Roman numerals.

 Clotting may be initiated through two different pathways, the intrinsic pathway or the extrinsic pathway. The intrinsic pathway is activated when plasma comes in contact with damaged vessel surfaces. The extrinsic pathway is activated when tissue thrombo-plastin, a substance released by damaged endothelial cells, comes in contact with one of the clotting factors.

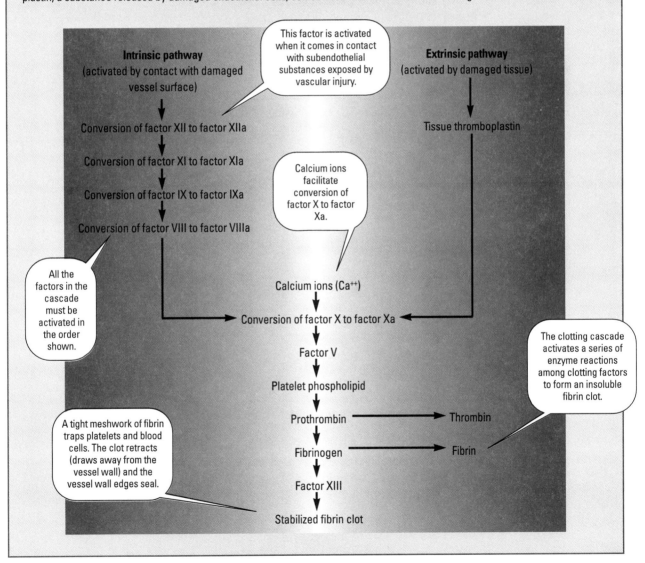

Blood dyscrasias

Blood elements are prone to various types of dysfunction. An abnormal or pathologic condition of the blood is called a dyscrasia.

Erythrocytes, leukocytes, and platelets are manufactured in the bone marrow. Bone marrow cells and their precursors are especially vulnerable to physiologic changes that can affect cell production. Why? Because bone marrow cells reproduce rapidly and have a short life span. In addition, storage of circulating cells in the bone marrow is minimal.

Blood disorders may be primary or secondary and quantitative or qualititative. They may involve some or all blood components.

Qualitative or quantitative

Qualitative blood disorders stem from intrinsic cell abnormalities or plasma component dysfunction. Quantitative blood disorders result from increased or decreased cell production or cell destruction.

It all leads to bad blood

Blood disorders may be caused by:
- trauma
- chronic disease such as cirrhosis
- surgery
- malnutrition
- drugs
- toxins
- radiation
- genetic and congenital defects
- sepsis.

When healthy blood cells are in short supply, the result is pancytopenia.

Oh my! Depression and destruction

Depressed bone marrow production or increased destruction of mature blood cells can result in:
- decreased RBCs (anemia)
- decreased platelets (thrombocytopenia)
- decreased granulocytes (granulocytopenia).

These three conditions are collectively called pancytopenia.

Marrow multiplication

The production of bone marrow components may increase. This results in myeloproliferative disorders such as leukemia.

RBC disorders

RBC disorders follow any condition that destroys or inhibits the formation of these cells. Examples include anemia, hemorrhage, and hypervolemia.

WBC disorders

A temporary increase in the production and release of mature WBCs is a normal response to infection.

Those blasted blasts!

Perils of a proliferation of precursors

An increase in WBC precursors and their accumulation in bone marrow or lymphoid tissue signals leukemia. These nonfunctioning WBCs (called blasts) leave little to love:
• They don't protect against infection.
• They crowd out other vital components, such as RBCs, platelets, and mature WBCs.
• They spill into the bloodstream, sometimes infiltrating organs and impairing their function.

Where'd that deficiency come from?

WBC deficiencies may result from:
• inadequate cell production
• drug reactions
• ionizing radiation
• infiltrated bone marrow (cancer)
• congenital defects
• aplastic anemias
• folic acid deficiency
• hypersplenism.
    The most common types of WBC deficiencies are granulocytopenia and lymphocytopenia; monocytopenia occurs less frequently.

Platelet disorders

Platelet disorders include:
• platelet decrease (thrombocytopenia)

• platelet excess (thrombocytosis)
• platelet dysfunction (thrombocytopathy).

Decrease

Thrombocytopenia may result from a congenital deficiency, or it may be acquired. Causes include:
• exposure to drugs and ionizing radiation
• cancerous infiltration of bone marrow
• abnormal sequestration (blood accumulation and pooling) in the spleen
• infection.

Excess

Thrombocytosis occurs as a result of certain diseases such as cancer.

Dysfunction

Thrombocytopathy usually results from disease, such as uremia or liver failure, or adverse effects of medications, such as salicylates and nonsteroidal anti-inflammatory drugs.

> Decrease, excess, or dysfunction. It all leads to trouble.

Blood disorders

The disorders discussed below include disseminated intravascular coagulation (DIC), idiopathic thrombocytopenia purpura (ITP), and thrombocytopenia.

Disseminated intravascular coagulation

In DIC, clotting and hemorrhage occur in the vascular system at the same time. It may also be called consumption coagulopathy or defibrination syndrome.

DIC causes small blood vessel blockage, organ tissue damage (necrosis), depletion of circulating clotting factors and platelets, and activation of a clot-dissolving process called fibrinolysis. This, in turn, can lead to severe hemorrhage.

Overzealous clotting is characteristic of DIC.

Pathophysiology

With DIC, clotting in the microcirculation usually affects the kidneys and extremities but may also occur in the brain, lungs, pituitary and adrenal glands, and GI mucosa.

DIC may result from:

Here's a list of things that can lead to DIC.

• infection — gram-negative or gram-positive septicemia; viral, fungal, rickettsial, or protozoal infection
• obstetric complications — abruptio placentae, amniotic fluid embolism, retained dead fetus, eclampsia, septic abortion, postpartum hemorrhage
• neoplastic disease — acute leukemia, metastatic carcinoma, lymphomas
• disorders that produce necrosis — extensive burns and trauma, brain tissue destruction, transplant rejection, liver necrosis, anorexia
• other disorders and conditions — heatstroke, shock, poisonous snakebite, cirrhosis, fat embolism, incompatible blood transfusion, drug reactions, cardiac arrest, surgery necessitating cardiopulmonary bypass, giant hemangioma, severe venous thrombosis, purpura fulminans, adrenal disease, adult respiratory distress syndrome, diabetic ketoacidosis, pulmonary embolism, and sickle cell anemia.

We just don't know

No one is certain why these conditions and disorders lead to DIC. Furthermore, whether they lead to it through a common mechanism is also uncertain. In many patients, DIC may be triggered by the entrance of foreign protein into the circulation or by vascular endothelial injury.

Three (count 'em) pathologic processes

DIC usually develops in association with three pathologic processes:

 damage to the endothelium

 release of tissue thromboplastin

 activation of factor X.

In DIC, blood goes from one extreme to the other...

DIC arises from the series of events described below:
• One of DIC's many causes triggers the coagulation system.
• Excess fibrin is formed (triggered by the action of thrombin, an enzyme) and becomes trapped in the microvasculature along with platelets, causing clots.
• Blood flow to the tissues decreases, causing acidemia, blood stasis, and tissue hypoxia; organ failure may result.
• Both fibrinolysis and antithrombotic mechanisms lead to anticoagulation.
• Platelets and coagulation factors are consumed and massive hemorrhage may ensue. (See *Understanding DIC*.)

...first it clots; then it dissolves, leading to hemorrhage.

Now I get it!

Understanding DIC

The simplified illustration below shows the pathophysiology of DIC. Circulating thrombin activates both coagulation and fibrinolysis, leading to paradoxical bleeding and clotting.

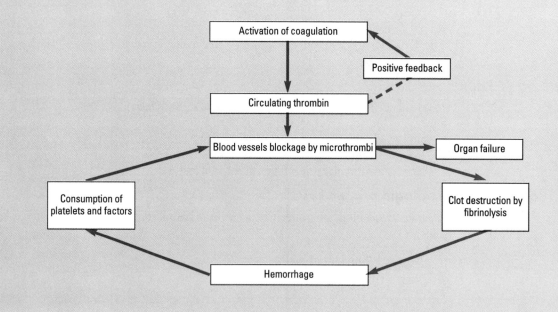

What to look for

The most significant clinical feature of DIC is abnormal bleeding without a history of serious hemorrhagic disorder. Other signs and symptoms include:
- cutaneous oozing
- petechiae or blood blisters
- bleeding from surgical or I.V. sites
- bleeding from the GI tract
- cyanosis of the extremities
- severe muscle, back, abdominal, and chest pain
- nausea and vomiting
- seizures
- oliguria
- shock
- confusion.

   DIC is usually acute, although it may be chronic in cancer patients. The prognosis depends on timeliness of detection, the severity and site of the hemorrhage, and treatment of the underlying disease or condition. (See *Treating DIC*.)

What tests tell you

Abnormal bleeding with no other blood disorder suggests DIC. The following test results support the diagnosis:
- Platelet count is decreased, usually to less than 100,000/µl, because platelets are consumed during thrombosis.
- Fibrinogen levels are decreased to less than 150 mg/dl because fibrinogen is consumed in clot formation. Levels may be normal if elevated by hepatitis or pregnancy.
- Prothrombin time is prolonged to more than 15 seconds.
- Partial thromboplastin time is prolonged to more than 60 to 80 seconds.
- Fibrin degradation products are increased, often to greater than 45 mcg/ml. Increases are produced by excess fibrin clots broken down by plasmin.
- D-dimer test is positive at less than 1:8 dilution.
- Other blood test results include positive fibrin monomers, diminished levels of factors V and VIII, fragmentation of RBCs, and hemoglobin levels decreased to less than 10 g/dl.
- Renal status test results demonstrate reduced urine output (less than 30 ml/hour), elevated blood urea nitrogen

Battling illness

Treating DIC

Successful management of disseminated intravascular coagulation (DIC) requires prompt treatment of the underlying disorder.

Support may be sufficient
Treatment may be highly specific or it may be supportive. Supportive care is appropriate if the underlying disorder is self-limiting or if the patient isn't actively bleeding.

In case of bleeding
Active bleeding may require administration of fresh-frozen plasma, platelets, or packed red blood cells to support hemostasis.

Heparin therapy
Heparin therapy is controversial. It may be used early in the disease to prevent microclotting or as a last resort in the patient who is actively bleeding. If thrombosis does occur, heparin therapy is usually mandatory. In most cases, it's administered along with transfusion therapy.

levels (greater than 25 mg/dl), and elevated serum creatinine levels (greater than 1.3 mg/dl).

Final confirmation of the diagnosis may be difficult because similar test results also occur in other disorders such as primary fibrinolysis. However, fibrin degradation products and D-dimer tests are considered specific and diagnostic of DIC. Additional tests may determine the underlying cause.

Idiopathic thrombocytopenic purpura

Thrombocytopenia that results from immunologic platelet destruction is known as idiopathic thrombocytopenic purpura (ITP). It occurs in two forms:

☝ acute (also called postviral thrombocytopenia), which usually affects children between ages 2 and 6

✌ chronic (also called Werlhof's disease, purpura hemorrhagica, essential thrombocytopenia, or autoimmune thrombocytopenia), which affects adults under age 50, especially women between ages 20 and 40.

Pathophysiology

ITP is an autoimmune disorder. Antibodies that reduce the life span of platelets appear in nearly all patients.

One follows infection, the other doesn't

Acute ITP usually follows a viral infection, such as rubella or chickenpox, and can result from immunization with a live vaccine.

Chronic ITP seldom follows infection and is often linked with immunologic disorders, such as systemic lupus erythematosus and human immunodeficiency virus (HIV) infection. Chronic ITP affects women more commonly than men.

A plague on platelets

ITP occurs when circulating immunoglobulin G (IgG) molecules react with host platelets, which are then destroyed by phagocytosis in the spleen and, to a lesser degree, in the liver. Normally, the life span of platelets in circulation is 7 to 10 days. In ITP, platelets survive 1 to 3 days or less.

In ITP, platelets are destroyed by the immune system.

Memory jogger

To help you remember how ITP progresses, use the acronym ATP:

Antibodies develop to the body's own platelets.

Trapped platelets appear in the spleen and liver.

Phagocytosis causes thrombocytopenia.

Yikes! Hemorrhage, purpuric lesions, lymphoma

Hemorrhage can severely complicate ITP. Cerebral hemorrhage is a major complication during the initial phase of the disease and is most likely to occur if the patient's platelet count falls below 500/µl. Potentially fatal purpuric lesions (caused by hemorrhage into tissues) may also occur in vital organs, such as the brain and kidneys. ITP is also often a precursor to lymphoma.

What to look for

The following signs and symptoms indicate decreased platelets:
- nosebleed
- oral bleeding
- purpura
- petechiae
- excessive menstruation.

In the acute form, sudden bleeding usually follows a recent viral illness, although it may not occur until 21 days after the virus strikes. In the chronic form, the onset of bleeding is insidious.

The prognosis for acute ITP is excellent; nearly four out of five patients recover completely without specific treatment. The prognosis for chronic ITP is good; transient remissions lasting weeks or even years are common, especially in women. (See *Treating ITP*, page 302.)

What tests tell you

The following tests help diagnose ITP:
- Platelet count less than 20,000/µl and prolonged bleeding time suggest ITP. Platelet size and appearance may be

Battling illness

Treating ITP

Acute idiopathic thrombocytopenic purpura (ITP) may be allowed to run its course without intervention. Alternatively, it may be treated with glucocorticoids or immune globulin. Treatment with plasmapheresis or platelet pheresis with transfusion has met with limited success.

For chronic ITP, corticosteroids may be used to suppress phagocytic activity, promote capillary integrity, and enhance platelet production.

Splenectomy

Patients who don't respond spontaneously to treatment within 1 to 4 months or who require high doses of corticosteroids to maintain platelet counts require splenectomy. This procedure is up to 85% successful in adults when splenomegaly accompanies the initial thrombocytopenia.

Looking at alternatives

Alternative treatments include immunosuppressants, such as cyclophosphamide or vincristine sulfate, and high-dose I.V. immune globulin, which is effective in 85% of adults.

Immunosuppressant use requires weighing the risks against the benefits. Immune globulin has a rapid effect, raising platelet counts within 1 to 5 days, but this effect lasts only about 1 to 2 weeks. Immune globulin is usually administered to prepare severely thrombocytic patients for emergency surgery.

abnormal, and anemia may be present if bleeding has occurred.
• Bone marrow studies show an abundance of megakaryocytes (platelet precursors) and a circulating platelet survival time of only several hours to a few days.
• Humoral tests that measure platelet-associated IgG may help establish the diagnosis. However, they're nonspecific, so their usefulness is limited. Half of all patients with thrombocytopenia show an increased IgG level.

Thrombocytopenia

Thrombocytopenia is characterized by a deficient number of circulating platelets. It's the most common cause of hemorrhagic disorders.

Pathophysiology

Thrombocytopenia may be congenital or acquired, but the acquired form is more common. In either case, it usually results from:

• decreased or defective platelet production in the bone marrow
• increased destruction outside the marrow caused by an underlying disorder
• sequestration (increase in the amount of blood in a limited vascular area)
• blood loss. (See *Factors that decrease platelet count*.)

Platelets play a vital role in coagulation ...that's why this disorder threatens the body's ability to control bleeding.

Four (count 'em) mechanisms

In thrombocytopenia, lack of platelets can cause inadequate hemostasis. Four mechanisms are responsible:

 decreased platelet production

 decreased platelet survival

 pooling of blood in the spleen

Factors that decrease platelet count

Decreased platelet count may result from diminished or defective platelet production, increased peripheral destruction of platelets, sequestration (separation of a portion of the circulating blood in a specific body part), or blood loss. More specific causes are listed below.

Diminished or defective production

Congenital

• Wiskott-Aldrich syndrome
• Maternal ingestion of thiazides
• Neonatal rubella

Acquired

• Aplastic anemia
• Marrow infiltration (acute and chronic leukemia, tumor)
• Nutritional deficiency (B_{12}, folic acid)
• Myelosuppressive agents
• Drugs that directly influence platelet production (thiazides, alcohol, hormones)
• Radiation
• Viral infections (measles, dengue)

Increased peripheral destruction

Congenital

• Nonimmune (prematurity, erythroblastosis fetalis, infection)
• Immune (drug-induced, especially with quinine and quinidine; posttransfusion purpura; acute and chronic idiopathic thrombocytopenic purpura; sepsis; alcohol)

Acquired

• Invasive lines or devices
• Intra-aortic balloon pump
• Prosthetic cardiac valves

Sequestration
• Hypersplenism
• Hypothermia

Loss
• Hemorrhage
• Bleeding

intravascular dilution of circulating platelets.

Platelets are produced by giant cells in bone marrow called megakaryocytes. Platelet production falls when the number of megakaryocytes is reduced or when platelet production becomes dysfunctional.

What to look for

Thrombocytopenia typically produces a sudden onset of petechiae or blood blisters in the skin and bleeding into any mucous membrane. Nearly all patients are otherwise symptom-free, although some may complain of malaise, fatigue, and general weakness.

In adults, large blood-filled blisters usually appear in the mouth. In severe disease, hemorrhage may lead to tachycardia, shortness of breath, loss of consciousness, and death.

Prognosis is excellent in drug-induced thrombocytopenia if the offending drug is withdrawn. Recovery may be immediate. In other cases, the prognosis depends on the patient's response to treatment of the underlying cause. (See *Treating thrombocytopenia*.)

What tests tell you

The following tests help establish a diagnosis of thrombocytopenia:
- Platelet count is decreased, usually to less than 100,000/µl in adults.
- Bleeding time is prolonged.
- Prothrombin and partial thromboplastin times are normal.
- Platelet antibody studies can help determine why the platelet count is low and can also be used to select treatment.
- Platelet survival studies help differentiate between ineffective platelet production and platelet destruction as causes of thrombocytopenia.
- Bone marrow studies determine the number, size, and cytoplasmic maturity of megakaryocytes in severe disease. This helps identify ineffective platelet production as the cause and rules out malignant disease at the same time.

Battling illness

Treating thrombocytopenia

Withdrawing the offending drug or treating the underlying cause, if possible, is essential. Other treatments include:

- administration of corticosteroids to increase platelet production

- administration of lithium carbonate or folate to stimulate bone marrow production of platelets

- I.V. administration of gamma globulin for severe or refractory thrombocytopenia (still experimental)

- platelet transfusion to stop episodic abnormal bleeding caused by a low platelet count (only minimally effective if platelet destruction results from an immune disorder; may be reserved for life-threatening bleeding)

- splenectomy to correct disease caused by platelet destruction because the spleen acts as the primary site of platelet removal and antibody production.

Quick quiz

1. A vital function of platelets is to:
   A. form hemostatic plugs in injured blood vessels.
   B. regulate acid-base balance and immune responses.
   C. protect the body against harmful bacteria and infection.

Answer: A. Platelets also minimize blood loss by causing damaged blood vessels to contract and provide materials that accelerate blood coagulation.

2. The normal life span of an RBC is:
   A. 90 days.
   B. 120 days.
   C. 240 days.

Answer: B. A normal RBC is viable for approximately 120 days.

3. Thrombocytopenia is characterized by:
   A. not enough circulating platelets.
   B. too many circulating platelets.
   C. decreased RBC production.

Answer: A. Platelet deficiency may be due to decreased or defective production of platelets or increased destruction of platelets.

4. The initial response to tissue injury in the extrinsic pathway is the release of:
   A. prothrombin.
   B. thrombin.
   C. tissue thromboplastin.

Answer: C. Tissue thromboplastin is also called factor III.

5. DIC is marked by:
   A. clotting and hemorrhage.
   B. clotting deficiency and immune dysfunction.
   C. hemorrhagic and fibrinolytic coagulopathy.

Answer: A. In DIC, clotting and hemorrhage occur simultaneously in the vascular system.

Scoring

☆☆☆   If you answered all five items correctly, you're golden! Indeed, your knowledge of red and white blood cells has us green with envy!

☆☆   If you answered three or four correctly, congrats! You're a connoisseur of liquid and formed components.

☆   If you answered fewer than three correctly, don't stop circulating. Never give up. Need inspiration? Think of the stem cell precursor that eventually emerges to become a mighty RBC.

Digestive system

Just the facts

In this chapter you'll learn:

♦ the structures and functions of the digestive system and related organs

♦ pathophysiology, signs and symptoms, diagnostic tests, and treatments for common digestive disorders.

Understanding the digestive system

The digestive system is the body's food processing complex. It performs the critical task of supplying essential nutrients to fuel the other organs and body systems. The digestive system has two major components:

 the GI tract, or alimentary canal

 the accessory glands and organs.

A malfunction along the GI tract or in one of the accessory glands or organs can produce far-reaching metabolic effects, which may become life-threatening. (See *A close look at the digestive system,* page 308.)

GI tract

The GI tract is basically a hollow, muscular tube that begins in the mouth and ends at the anus. It includes the following structures:
• mouth
• pharynx
• esophagus

A close look at the digestive system

This illustration shows the organs of the GI tract and several accessory organs.

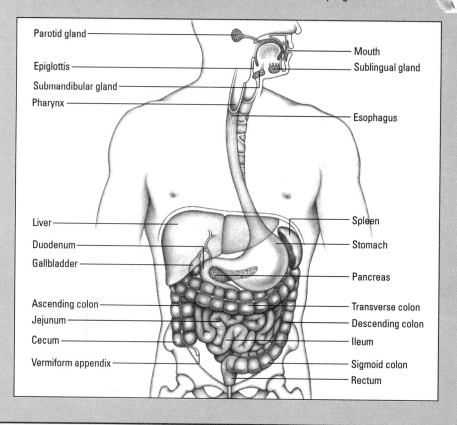

Parotid gland
Epiglottis
Submandibular gland
Pharynx
Mouth
Sublingual gland
Esophagus
Liver
Duodenum
Gallbladder
Ascending colon
Jejunum
Cecum
Vermiform appendix
Spleen
Stomach
Pancreas
Transverse colon
Descending colon
Ileum
Sigmoid colon
Rectum

- stomach
- small intestine
- large intestine.

Mouth and esophagus

When a person smells, tastes, chews, or merely thinks of food, the digestive system gets ready to go to work.

Before you even swallow

The digestive process begins in the mouth. Chewing and salivation soften food, making it easy to swallow. An enzyme in saliva, called ptyalin, begins to convert starches to sugars even before food is swallowed.

> Digestion begins long before food reaches the stomach.

Riding the peristaltic wave

When a person swallows, the upper esophageal sphincter relaxes, allowing food to enter the esophagus. In the esophagus, peristaltic waves activated by the glossopharyngeal nerve propel food toward the stomach.

Stomach

Digestion occurs in two phases:

 the cephalic phase

 the gastric phase.

By the time food is traveling through the esophagus on its way to the stomach, the cephalic phase of digestion has begun. In this phase, the stomach secretes hydrochloric acid and pepsin, digestive juices that help break down food.

Exit the esophagus, enter the stomach

The gastric phase of digestion begins when food passes the cardiac sphincter, a circle of muscle at the end of the esophagus. The food exits the esophagus and enters the stomach, causing the stomach wall to distend. This stimulates the mucosal lining of the stomach to release the hormone gastrin. Gastrin serves two purposes:
- stimulation of gastric juice secretion
- stimulation of the stomach's motor functions.

> The hormone gastrin sets the stomach in motion.

Juice it up

Gastric juice secretions are highly acidic, with a pH of 0.9 to 1.5. In addition to hydrochloric acid and pepsin, the gastric juices contain intrinsic factor (which helps the body use vitamin B_{12}) and proteolytic enzymes (which help the body use protein). The gastric juices mix with the food, which becomes a thick, gruel-like material called chyme.

Hold it, mix it, and parcel it out

The stomach has three major motor functions:
- holding food
- mixing food by peristaltic contractions with gastric juices

• slowly parceling chyme into the small intestine for further digestion and absorption.

Small intestine

Nearly all digestion takes place in the small intestine, which is about 20′ (6 m) long. Chyme passes through the small intestine propelled by peristaltic contractions. The small intestine has three major sections:
• the duodenum
• the jejunum
• the ileum.

Dedicated to the duodenum

The duodenum is a 10″ (25-cm) long, C-shaped curve of the small intestine that extends from the stomach. Food passes from the stomach to the duodenum through a narrow opening called the pylorus.

The duodenum also has an opening through which bile and pancreatic enzymes enter the intestine to neutralize the acidic chyme and aid digestion. This opening is called Oddi's sphincter.

It's just a jejunum

The jejunum extends from the duodenum and forms the longest portion of the small intestine. The jejunum leads to the ileum, the narrowest portion of the small intestine.

Digestive secretions convert carbohydrates, proteins, and fats into sugars, amino acids, fatty acids, and glycerin. Along with water and electrolytes, these nutrients are absorbed through the intestinal mucosa into the bloodstream for use by the body. Nonnutrients, such as vegetable fibers, are carried through the intestine.

The small intestine ends at the ileocecal valve, located in the lower right part of the abdomen. The ileocecal valve is a sphincter that empties nutrient-depleted chyme into the large intestine.

Large intestine

After chyme passes through the small intestine, it enters the ascending colon at the cecum, the pouchlike beginning of the large intestine. By this time, the chyme consists of mostly indigestible material. From the ascending colon, chyme passes through the transverse colon, and then down through the descending colon to the rectum and anal canal, where it's finally expelled.

Harboring bacteria

The large intestine produces no hormones or digestive enzymes. Rather, it's where absorption takes place. The large intestine absorbs nearly all of the water remaining in the chyme plus large amounts of sodium and chloride.

The large intestine also harbors the bacteria *Escherichia coli, Enterobacter aerogenes, Clostridium welchii,* and *Lactobacillus bifidus,* which help produce vitamin K and break down cellulose into usable carbohydrates.

In the lower part of the descending colon, long, relatively sluggish contractions cause propulsive waves known as mass movements. These movements propel the intestinal contents into the rectum and produce the urge to defecate.

Yeah, it's true.
I hang out in the large intestine.

Accessory glands and organs

The liver, gallbladder, and pancreas contribute several substances — enzymes, bile, and hormones — that are vital to digestion. These structures deliver their secretions to the duodenum through the ampulla of Vater.

Liver

A large, highly vascular organ, the liver is enclosed in a fibrous capsule in the upper right area of the abdomen. The liver performs many complex and important functions, many of which are related to digestion and nutrition:
• It plays an important role in carbohydrate metabolism.
• It filters and detoxifies blood, removing foreign substances, such as drugs, alcohol, and other toxins.
• It removes naturally occurring ammonia from body fluids, converting it to urea for excretion in urine.
• It produces plasma proteins, nonessential amino acids, and vitamin A.
• It stores essential nutrients, such as iron and vitamins K, D, and B_{12}.
• It produces bile to aid digestion.

• It converts glucose to glycogen and stores it as fuel for the muscles.
• It stores fats and converts excess sugars to fats for storage in other parts of the body.

Gallbladder

The gallbladder is a small, pear-shaped organ nestled under the liver and joined to the larger organ by the cystic duct. The gallbladder's job is to store and concentrate bile produced by the liver. Bile is a clear yellowish liquid that helps break down fats and neutralize gastric secretions in the chyme.

Delivery to the duodenum

Secretion of the hormone cholecystokinin causes the gallbladder to contract and the ampulla of Vater to relax. This allows the release of bile into the common bile duct for delivery to the duodenum. When the ampulla of Vater closes, bile shunts to the gallbladder for storage.

Pancreas

The pancreas lies behind the stomach, with its head and neck extending into the curve of the duodenum and its tail lying against the spleen. The pancreas is made up of two types of tissue:
• exocrine tissue, from which enzymes are secreted through ducts to the digestive system
• endocrine tissue, from which hormones are secreted into the blood.

Exocrine action

The pancreas's exocrine function involves small, scattered glands — called acini — that secrete more than 1,000 ml

Exocrine tissue releases digestive enzymes...

...and endocrine tissue releases hormones, such as insulin and glucagon.

of digestive enzymes daily. These lobules of enzyme-producing cells release their secretions into small ducts that merge to form the pancreatic duct. The pancreatic duct runs the length of the pancreas and joins the bile duct from the gallbladder before entering the duodenum. Vagus nerve stimulation and the release of two hormones (secretin and cholecystokinin)control the rate and amount of pancreatic secretion.

Endocrine action

The endocrine function involves the islets of Langerhans, microscopic structures scattered throughout the pancreas. Over 1 million islets house two major cell types:
• alpha cells, which secrete glucagon, a hormone that stimulates glucose formation in the liver
• beta cells, which secrete insulin to promote carbohydrate metabolism.
   Both hormones flow directly into the blood. Release of both hormones is stimulated by blood glucose levels.

Digestive disorders

The disorders discussed below include appendicitis, cholecystitis, cirrhosis, Crohn's disease, pancreatitis, peptic ulcer, ulcerative colitis, and viral hepatitis.

Appendicitis

Appendicitis occurs when the appendix becomes inflamed. It's the most common major surgical emergency.

Let's get specific

More precisely, this disorder is an inflammation of the vermiform appendix, a small, fingerlike projection attached to the cecum just below the ileocecal valve.

What am I here for?

Although the appendix has no known function, it does regularly fill and empty itself of food.
   Appendicitis may occur at any age and affects both sexes equally; however, between puberty and age 25, it's more prevalent in men.

In appendicitis, the appendix is inflamed and may be blocked.

Pathophysiology

Until recently, appendicitis was thought to result from an obstruction by a fecal mass, a stricture, barium ingestion, or viral infection. Obstruction does sometimes occur, but mucosal ulceration usually happens first. Although the exact cause of the ulceration is unknown, it may be viral in nature.

Appendicitis, one step at a time

After ulceration occurs, appendicitis progresses this way:
• Inflammation accompanies the ulceration and temporarily obstructs the appendix.
• Obstruction, if present, is usually caused by stool accumulation around vegetable fibers ("fecalith" is a fancy name for this).
• Mucus outflow is blocked, which distends the organ.
• Pressure within the appendix increases, and the appendix contracts.
• Bacteria multiply and inflammation and pressure continue to increase, affecting blood flow to the organ and causing severe abdominal pain.

Ulceration and inflammation...

...may lead to blockage,...

...which may cause swelling and rupture.

Yikes! Infection, clotting, tissue decay

Inflammation can lead to infection, clotting, tissue decay, and perforation of the appendix. If the appendix ruptures or perforates, the infected contents spill into the abdominal cavity, causing peritonitis, the most common and dangerous complication.

What to look for

Distention and contractions of the appendix cause pain, usually in the upper right abdominal area. As inflammation spreads, the pain becomes more severe and localized in the lower right part of the abdomen, which becomes tender. If the appendix is in back of the cecum or in the pelvis, the patient may have flank tenderness instead of abdominal tenderness.

Other signs and symptoms include:
- anorexia
- nausea or vomiting
- low-grade fever.

Signs and symptoms of rupture include:
- pain
- tenderness
- spasm, followed by a brief cessation of abdominal pain.

Untreated appendicitis is invariably fatal. In recent times, the use of antibiotics has reduced the incidence of death from appendicitis. (See *Treating appendicitis.*)

What tests tell you

The following tests are used to diagnose appendicitis:
- White blood cell (WBC) count is moderately high with an increased number of immature cells.
- X-ray with a radiographic contrast agent aids diagnosis. Failure of the organ to fill with contrast agent indicates appendicitis.

Differential diagnosis rules out illnesses with similar symptoms, such as bladder infection, diverticulitis, gastritis, ovarian cyst, pancreatitis, renal colic, and uterine disease.

Battling illness

Treating appendicitis

Appendectomy is the only effective treatment for appendicitis. If peritonitis develops, treatment involves gastric intubation, parenteral replacement of fluids and electrolytes, and antibiotic administration.

Cholecystitis

In cholecystitis, the gallbladder becomes inflamed. Usually, a gallstone becomes lodged in the cystic duct, causing painful gallbladder distention. Cholecystitis may be acute or chronic. The acute form is most common during middle age; the chronic form, among elderly people.

Pathophysiology

Cholecystitis is caused by the formation of calculi called gallstones. The exact cause of gallstone formation is unknown. Abnormal metabolism of cholesterol and bile salts play an important role. Acute cholecystitis may also result from poor or absent blood flow to the gallbladder.

Risk factors

The following risk factors predispose a person to gallstones:
• obesity and a high-calorie, high-cholesterol diet
• increased estrogen levels from oral contraceptives, hormone replacement therapy, or pregnancy
• use of clofibrate, an antilipemic drug
• diabetes mellitus, ileal disease, blood disorders, liver disease, or pancreatitis.

First stones and then...

In acute cholecystitis, inflammation of the gallbladder wall usually develops after a gallstone lodges in the cystic duct. (See *Understanding gallstone formation.*) When bile flow is blocked, the gallbladder becomes inflamed and distended. Bacterial growth, usually *Escherichia coli*, may contribute to the inflammation. Then the following sequence of events occurs:
• Edema of the gallbladder and sometimes the cystic duct occurs.
• Edema obstructs bile flow, which chemically irritates the gallbladder.
• Cells in the gallbladder wall may become oxygen starved and die as the distended organ presses on vessels and impairs blood flow.
• Dead cells slough off.
• An exudate covers ulcerated areas, causing the gallbladder to adhere to surrounding structures.

Ugh! Pus, fluid, gangrene

Cholecystitis may lead to complications:
• Pus may accumulate in the gallbladder (empyema).
• Fluid may accumulate in the gallbladder (hydrops).
• The gallbladder may become distended with mucous secretions (mucocele).
• Gangrene may occur, leading to perforation, peritonitis, abnormal passages in the tissues (fistulas), and pancreatitis.
• Chronic cholecystitis may develop.

What causes a gallstone to form?

No one can say for sure but excess of cholesterol in the bile appears to be a factor.

Now I get it!

Understanding gallstone formation

Abnormal metabolism of cholesterol and bile salts plays an important role in gallstone formation. Bile is made continuously by the liver and is concentrated and stored in the gallbladder until the duodenum needs it to help digest fat. Changes in the composition of bile may allow gallstones to form. Changes to the absorptive ability of the gallbladder lining may also contribute to gallstone formation.

Too much cholesterol
Certain conditions, such as age, obesity, and estrogen imbalance, cause the liver to secrete bile that's abnormally high in cholesterol or lacking the proper concentration of bile salts.

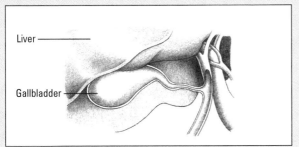

Inside the gallbladder
When the gallbladder concentrates this bile, inflammation may or may not occur. Excessive water and bile salts are reabsorbed, making the bile less soluble. Cholesterol, calcium, and bilirubin precipitate into gallstones.

Fat entering the duodenum causes the intestinal mucosa to secrete the hormone cholecystokinin, which stimulates the gallbladder to contract and empty. If a stone lodges in the cystic duct, the gallbladder contracts but can't empty.

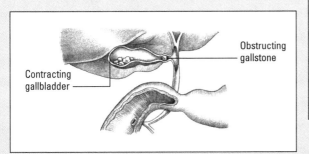

Jaundice, irritation, inflammation
If a stone lodges in the common bile duct, the bile flow into the duodenum becomes obstructed. Bilirubin is absorbed into the blood, causing jaundice.

Biliary narrowing and swelling of the tissue around the stone can also cause irritation and inflammation of the common bile duct.

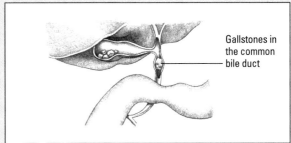

Up the biliary tree
Inflammation can progress up the biliary tree and cause infection of any of the bile ducts. This causes scar tissue, fluid accumulation, cirrhosis, portal hypertension, and bleeding.

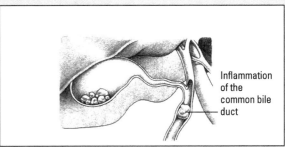

• Cholangitis (bile duct infection) may develop.

What to look for

Signs and symptoms of acute cholecystitis often strike after meals that are rich in fats. They may occur at night, suddenly awakening the patient. They include:
• acute abdominal pain in the right upper quadrant that may radiate to the back, between the shoulders, or to the front of the chest
• colic due to passage of gallstones along the bile duct (biliary colic)
• belching
• flatulence
• indigestion
• light-headedness
• nausea
• vomiting
• chills
• low-grade fever
• jaundice (caused by bile in the blood) and clay-colored stools, if a stone obstructs the common bile duct.

Cholecystitis accounts for 10% to 25% of all patients requiring gallbladder surgery. The prognosis is good with treatment. (See *Treating cholecystitis.*)

What tests tell you

The following tests are used to diagnose cholecystitis:
• X-rays reveal gallstones if they contain enough calcium to be radiopaque and also help disclose porcelain gallbladder, limy bile, and gallstone ileus.
• Ultrasonography confirms gallstones as small as 2 mm and distinguishes between obstructive and nonobstructive jaundice.
• Oral cholecystography confirms the presence of gallstones, although this test is gradually being replaced by ultrasonography.
• Technetium-labeled scan indicates cystic duct obstruction and acute or chronic cholecystitis if the gallbladder can't be seen.
• Percutaneous transhepatic cholangiography, performed with fluoroscopy, supports the diagnosis of obstructive jaundice and reveals calculi in the ducts.

Battling illness

Treating cholecystitis

Surgery is the most common treatment for gallbladder and bile duct disease. Procedures include:

• gallbladder removal (cholecystectomy), with or without X-ray of the bile ducts (operative cholangiography)
• creation of an opening into the common bile duct for drainage (choledochostomy)
• exploration of the common bile duct.

Alternatives to surgery
Other invasive procedures include:

• insertion of a flexible catheter through a sinus tract into the common bile duct and removal of stones using a basket-shaped tool guided by fluoroscopy
• endoscopic retrograde cholangiopancreatography, which removes stones with a balloon or basket-shaped tool passed through an endoscope
• lithotripsy, which breaks up gallstones with ultrasonic waves (contraindicated in patients with pacemakers or implantable defibrillators)
• stone dissolution therapy with oral chenodeoxycholic acid or ursodeoxycholic acid (of limited use).

Diet, drugs, and more
Other treatments include:

• a low-fat diet with replacement of vitamins A, D, E, and K and administration of bile salts to facilitate digestion and vitamin absorption
• narcotics, especially meperidine, to relieve pain during an acute attack
• antispasmodics and anticholinergics to relax smooth muscles and decrease ductal tone and spasm
• antiemetics to reduce nausea and vomiting
• a nasogastric tube connected to intermittent low-pressure suction to relieve vomiting
• cholestyramine if the patient has obstructive jaundice with severe itching from accumulation of bile salts in the skin
• aromatherapy, using a few drops of rosemary oil in a warm bath or inhaled directly, to improve the gallbladder's function.

• Blood studies may reveal high levels of serum alkaline phosphatase, lactate dehydrogenase, aspartate aminotransferase, and total bilirubin. The icteric index, a measure of bilirubin in the blood, is elevated.
• WBC count is slightly elevated during a cholecystitis attack.

Cirrhosis

Cirrhosis is a chronic liver disease. It's characterized by widespread destruction of hepatic cells, which are replaced by fibrous cells. This process is called fibrotic regeneration. Cirrhosis is a common cause of death in the United States and, among people ages 35 to 55, the fourth leading cause of death. It can occur at any age.

In Laënnec's cirrhosis, malnutrition, lack of protein, and heavy drinking cause liver damage.

Pathophysiology

There are many types of cirrhosis, each with a different cause. The most common are:

• *Laënnec's cirrhosis.* Stemming from chronic alcoholism and malnutrition, Laënnec's cirrhosis is also called portal, nutritional, or alcoholic cirrhosis. It's most prevalent among malnourished alcoholic men and accounts for more than half of all cirrhosis cases in the United States. Many alcoholics never develop the disease, however, while others develop it even with adequate nutrition.

• *postnecrotic cirrhosis.* Usually a complication of viral hepatitis (inflammation of the liver), postnecrotic cirrhosis may also occur after exposure to liver toxins, such as arsenic, carbon tetrachloride, or phosphorus. This form is more common in women and is the most common type of cirrhosis worldwide.

• *biliary cirrhosis.* Prolonged bile duct obstruction or inflammation is the cause of this disorder.

• *cardiac cirrhosis.* This type of cirrhosis is caused by prolonged venous congestion in the liver from right-sided heart failure.

In addition, some patients develop idiopathic cirrhosis (with no known cause).

Other types of cirrhosis are caused by toxins, bile obstruction, and poor blood flow.

Detail about liver damage

Cirrhosis is characterized by irreversible chronic injury of the liver, extensive fibrosis, and nodular tissue growth. These changes result from:

• liver cell death (hepatocyte necrosis)
• collapse of the liver's supporting structure (the reticulin network)
• distortion of the vascular bed (blood vessels throughout the liver)
• nodular regeneration of remaining liver tissue.

Oh my! Malfunction, jaundice, and portal hypertension

When the liver begins to malfunction, blood clotting disorders (coagulopathies), jaundice, edema, and a variety of metabolic problems develop.

Fibrosis and the distortion of blood vessels may impede blood flow in the capillary branches of the portal vein and hepatic artery, leading to portal hypertension (elevated pressure in the portal vein). Increased pressure may lead to the development of esophageal varices, enlarged, tortuous veins in the lower part of the esophagus, the area where it meets the stomach. Esophageal varices may easily rupture and leak large amounts of blood into the upper GI tract.

What to look for

Early signs and symptoms of cirrhosis are vague but usually include loss of appetite, indigestion, nausea, vomiting, constipation, diarrhea, and dull abdominal ache.

What happens later

Late-stage symptoms affect several body systems and include:
• respiratory effects — fluid in the lungs and weak chest expansion, leading to hypoxia
• central nervous system effects — lethargy, mental changes, slurred speech, asterixis (a motor disturbance marked by intermittent lapses in posture), peripheral nerve damage
• hematologic effects — nosebleeds, easy bruising, bleeding gums, anemia
• endocrine effects — testicular atrophy, menstrual irregularities, gynecomastia, loss of chest and axillary hair
• skin effects — severe itching and dryness, poor tissue turgor, abnormal pigmentation, spider veins
• hepatic effects — jaundice, enlarged liver (hepatomegaly), fluid in the abdomen (ascites), and edema
• renal effects — insufficiency that may progress to failure
• miscellaneous effects — musty breath, enlarged superficial abdominal veins, muscle atrophy, pain in the upper right abdominal quadrant that worsens when the patient sits up or leans forward, palpable liver or spleen, temperature of 101° to 103° F (38.3° to 39.4° C), bleeding from esophageal varices (See *Treating cirrhosis,* page 322.)

What tests tell you

The following tests help confirm cirrhosis:
• Liver biopsy, the definitive test, reveals tissue destruction and fibrosis.
• Abdominal X-ray shows liver size, cysts or gas within the biliary tract or liver, liver calcification, and massive fluid accumulation (ascites).

Battling illness

Treating cirrhosis

Therapy aims to remove or alleviate the underlying cause of cirrhosis, prevent further liver damage, and prevent or treat complications.

Drug therapy

Drug therapy requires special caution because the cirrhotic liver can't detoxify harmful substances efficiently. These drugs are used:

• vitamins and nutritional supplements to help heal damaged liver cells and improve nutritional status
• antacids to reduce gastric distress and decrease the potential for GI bleeding
• potassium-sparing diuretics such as furosemide to reduce fluid accumulation
• vasopressin to treat esophageal varices.

Noninvasive procedures

To control bleeding from esophageal varices or other GI hemorrhage, two measures are attempted first:

gastric intubation, in which the stomach is lavaged until the contents are clear and antacids and histamine antagonists are administered if the bleeding is caused by a gastric ulcer

esophageal balloon tamponade, in which bleeding vessels are compressed to stop blood loss from esophageal varices

Surgery

In patients with ascites, paracentesis may be used to relieve abdominal pressure. A shunt may be inserted to divert ascites into venous circulation. This treatment causes weight loss, decreased abdominal girth, increased sodium excretion from the kidneys, and improved urine output.

Sclerotherapy

If conservative treatment fails to stop hemorrhaging, a sclerosing agent is injected into the oozing vessels to cause clotting and sclerosis. If bleeding from the varices doesn't stop in 2 to 5 minutes, a second injection is given below the bleeding site. Sclerotherapy also may be performed on nonbleeding varices to prevent hemorrhaging.

Last resort

As a last resort, portal-systemic shunts may be inserted to control bleeding from esophageal varices and decrease portal hypertension. These shunts divert a portion of the portal vein blood flow away from the liver. This procedure is seldom performed because it can cause bleeding, infection, and shunt thrombosis. Massive hemorrhage requires blood transfusions to maintain blood pressure.

• Computed tomography (CT) and liver scans show liver size, abnormal masses, and hepatic blood flow and obstruction.
• Esophagogastroduodenoscopy reveals bleeding esophageal varices, stomach irritation or ulceration, or duodenal bleeding and irritation.
• Blood studies show elevated liver enzymes, total serum bilirubin, and indirect bilirubin levels. Total serum albumin and protein levels decrease; prothrombin time is prolonged; hemoglobin, hematocrit, and serum electrolyte levels decrease; and vitamins A, C, and K are deficient.
• Urine studies show increased levels of bilirubin and urobilinogen.

• Fecal studies show decreased fecal urobilinogen levels.

Crohn's disease

Crohn's is an inflammatory bowel disease. It may affect any part of the GI tract but usually involves the end of the ileum. It extends through all layers of the intestinal wall and may involve lymph nodes and supporting membranes in the area.

Crohn's disease is most prevalent in adults ages 20 to 40. It tends to run in families. Up to 20% of patients have a positive family history.

In Crohn's disease, severe inflammation may occur in any part of the GI tract...

What's in a name?

...inflammation extends through the intestinal wall...

When Crohn's disease affects only the small bowel, it's known as regional enteritis. When it involves the colon or only affects the colon, it's known as Crohn's disease of the colon. Crohn's disease of the colon is also called granulomatous colitis; however, not all patients develop granulomas (tumorlike masses of granulation tissue).

Pathophysiology

Although researchers are still studying Crohn's disease, possible causes include the following:
• lymphatic obstruction
• infection
• allergies
• immune disorders, such as altered immunoglobulin A production and increased suppressor T-cell activity.

Genetic factors may also play a role; Crohn's disease sometimes occurs in identical twins, and 10% to 20% of patients with the disease have one or more affected relatives. However, no simple pattern of inheritance has been identified.

...and the bowel becomes a patchwork of healthy and diseased segments as inflammation slowly spreads.

Crohn's disease: Step-by-step

In Crohn's disease, inflammation spreads slowly and progressively. Here's what happens:
• Lymph nodes enlarge and lymph flow in the submucosa is blocked.
• Lymphatic obstruction causes edema, mucosal ulceration, fissures, abscesses and, sometimes, granulomas. Mucosal ulcerations are called skipping lesions because they are not continuous as in ulcerative colitis.

• Oval, elevated patches of closely packed lymph follicles — called Peyer's patches — develop on the lining of the small intestine.

• Fibrosis occurs, thickening the bowel wall and causing stenosis, or narrowing of the lumen. (See *Changes to the bowel in Crohn's disease.*)

• Inflammation of the serous membrane (serositis) develops, inflamed bowel loops adhere to other diseased or normal loops, and diseased bowel segments become interspersed with healthy ones.

• Eventually, diseased parts of the bowel become thicker, narrower, and shorter.

Complications

Severe diarrhea and corrosion of the perineal area by enzymes can cause anal fistula, the most common complica-

---

Changes to the bowel in Crohn's disease

As Crohn's disease progresses, fibrosis thickens the bowel wall and narrows the lumen. Narrowing — or stenosis — can occur in any part of the intestine and cause varying degrees of intestinal obstruction. At first, the mucosa may appear normal, but as the disease progresses it takes on a "cobblestone" appearance as shown below.

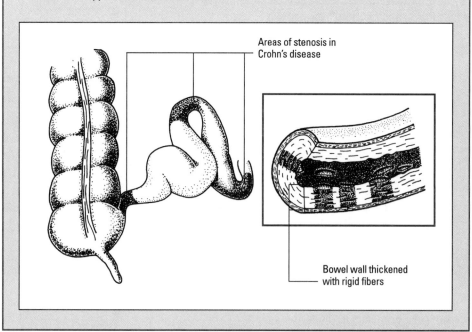

Areas of stenosis in Crohn's disease

Bowel wall thickened with rigid fibers

Battling illness

Treating Crohn's disease

Treatment for Crohn's disease requires drug therapy, lifestyle changes and, sometimes, surgery. During acute attacks, maintaining fluid and electrolyte balance is the key. Debilitated patients need total parenteral nutrition to provide adequate calories and nutrition while resting the bowel.

Drug therapy
The following drugs combat inflammation and relieve symptoms:

- Corticosteroids, such as prednisone, reduce diarrhea, pain, and bleeding by decreasing inflammation.
- Immunosuppressants, such as azathioprine, suppress the body's response to antigens.
- Sulfasalazine reduces inflammation.
- Metronidazole treats perianal complications.
- Antidiarrheals, such as diphenoxylate and atropine, combat diarrhea but aren't used in patients with significant bowel obstruction.
- Narcotics control pain and diarrhea.

Lifestyle changes
Stress reduction and reduced physical activity rest the bowel and allow it to heal. Vitamin supplements compensate for the bowel's inability to absorb vitamins. Dietary changes decrease bowel activity while still providing adequate nutrition. The following foods are usually eliminated:

- fruits, vegetables, and other high-fiber foods
- dairy products, spicy and fatty foods, and other foods and liquids that irritate the mucosa
- carbonated or caffeinated beverages and other foods or liquids that stimulate excessive intestinal activity.

Surgery
Surgery is necessary if bowel perforation, massive hemorrhage, fistulas, or acute intestinal obstruction develop. Colectomy with ileostomy is often performed in patients with extensive disease of the large intestine and rectum.

tion. A perineal abscess may also develop during the active inflammatory state. Fistulas may develop to the bladder or vagina or even to the skin in an old scar area.

Other complications include:
- intestinal obstruction
- nutrient deficiencies caused by malabsorption of bile salts and vitamin B_{12} and poor digestion
- fluid imbalances
- rarely, inflammation of abdominal linings (peritonitis).
(See *Treating Crohn's disease*.)

What to look for

Initially, the patient experiences malaise and diarrhea, often with pain in the right lower quadrant or generalized abdominal pain and fever.

Chronic diarrhea results from bile salt malabsorption, loss of healthy intestinal surface area, and bacterial growth. Weight loss, nausea, and vomiting also occur. Stools may be bloody.

What tests tell you

The following tests and results support a diagnosis of Crohn's disease:
• Fecal occult test shows minute amounts of blood in stools.
• Small-bowel X-ray shows irregular mucosa, ulceration, and stiffening.
• Barium enema reveals the string sign (segments of stricture separated by normal bowel) and may also show fissures and narrowing of the bowel.
• Sigmoidoscopy and colonoscopy show patchy areas of inflammation, which helps rule out ulcerative colitis. The mucosal surface has a cobblestone appearance. When the colon is involved, ulcers may be seen.
• Biopsy performed during sigmoidoscopy or colonoscopy reveals granulomas in up to half of all specimens.

Laboratory tests indicate increased WBC count and erythrocyte sedimentation rate. Other findings include decreased potassium, calcium, magnesium, and hemoglobin levels in the blood.

Pancreatitis

Pancreatitis is inflammation of the pancreas. It occurs in acute and chronic forms. In men, the disorder is commonly linked to alcoholism, trauma, or peptic ulcer; in women, to biliary tract disease.

The prognosis is good when pancreatitis follows biliary tract disease but poor when it is a complication of alcoholism. Mortality reaches 60% when pancreatitis causes tissue destruction (necrosis) or hemorrhage.

Pancreatitis may start with a stomachache and progress to swelling, tissue damage, or bleeding.

In pancreatitis, the pancreas harms itself...

Pathophysiology

The most common causes of pancreatitis are biliary tract disease and alcoholism, but it can also result from the following:

...enzymes normally excreted by the pancreas digest pancreatic tissue.

- abnormal organ structure
- metabolic or endocrine disorders, such as high cholesterol levels or overactive thyroid
- pancreatic cysts or tumors
- penetrating peptic ulcers
- blunt trauma or surgical trauma
- drugs, such as glucocorticoids, sulfonamides, thiazides, and oral contraceptives
- kidney failure or transplantation
- endoscopic examination of the bile ducts and pancreas.

In addition, heredity may predispose a patient to pancreatitis. In some patients, emotional or neurogenic factors play a part.

Chronic pancreatitis

Chronic pancreatitis is a persistent inflammation that produces irreversible changes in the structure and function of the pancreas. It sometimes follows an episode of acute pancreatitis. Here's what probably happens:

- Protein precipitates block the pancreatic duct and eventually harden or calcify.
- Structural changes lead to fibrosis and atrophy of the glands.
- Growths called pseudocysts, containing pancreatic enzymes and tissue debris, form.
- An abscess results if these growths become infected.

Acute pancreatitis

Acute pancreatitis occurs in two forms:

edematous (interstitial), causing fluid accumulation and swelling

necrotizing, causing cell death and tissue damage.

The inflammation that occurs with both types is caused by premature activation of enzymes, which causes tissue damage.

Two theories to explain active enzymes

Normally, the acini in the pancreas secrete enzymes in an inactive form. Two theories explain why enzymes become prematurely activated:

☝ A toxic agent, such as alcohol, alters the way the pancreas secretes enzymes. Alcohol probably increases pancreatic secretion, alters the metabolism of the acinar cells, and encourages duct obstruction by causing pancreatic secretory proteins to precipitate.

✌ A reflux of duodenal contents containing activated enzymes enters the pancreatic duct, activating other enzymes and setting up a cycle of more pancreatic damage.

Why does it hurt so bad?

Pain can be caused by several factors, including:

☝ escape of inflammatory exudate and enzymes into the back of the peritoneum

✌ edema and distention of the pancreatic capsule

🖐 obstruction of the biliary tract.

Oh no! Diabetes mellitus, hemorrhage, diabetic acidosis

If pancreatitis damages the islets of Langerhans, diabetes mellitus may result. Sudden, severe pancreatitis causes massive hemorrhage and total destruction of the pancreas. This may lead to diabetic acidosis, shock, or coma.

What to look for

In many patients, the only symptom of mild pancreatitis is steady epigastric pain centered close to the navel and unrelieved by vomiting.

Acute pancreatitis causes severe, persistent, piercing abdominal pain, usually in the midepigastric region, although it may be generalized or occur in the left upper quadrant radiating to the back. The pain usually begins suddenly after eating a large meal or drinking alcohol. It increases when the patient lies on his back and is relieved when he rests on his knees and upper chest. (See *Treating pancreatitis.*)

Endocrine damage in pancreatitis can cause diabetes mellitus.

Battling illness

Treating pancreatitis

The goals of treatment are to maintain circulation and fluid volume, relieve pain, and decrease pancreatic secretions.

Acute pancreatitis

• Shock is the most common cause of death in the early stages, so I.V. replacement of electrolytes and proteins is necessary.
• Metabolic acidosis requires fluid volume replacement.
• Blood transfusions may be needed.
• Food and fluids are withheld to allow the pancreas to rest and to reduce pancreatic enzyme secretion.
• Nasogastric tube suctioning decreases stomach distention and suppresses pancreatic secretions.

Drug therapy

• Meperidine relieves abdominal pain.
• Antacids neutralize gastric secretions.
• Histamine antagonists, such as cimetidine, famotidine, or ranitidine, decrease hydrochloric acid production.
• Antibiotics, such as clindamycin or gentamicin, fight bacterial infections.
• Anticholinergics reduce vagal stimulation, decrease GI motility, and inhibit pancreatic enzyme secretion.
• Insulin corrects hyperglycemia.

Surgery

Surgical drainage is necessary for a pancreatic abscess or a pseudocyst. A laparotomy may be needed if biliary tract obstruction causes acute pancreatitis.

Chronic pancreatitis

Pain control measures are similar to those for acute pancreatitis. Meperidine is the drug of choice; however, pentazocine is also effective. Other treatment depends on the cause.

Surgery relieves abdominal pain, restores pancreatic drainage, and reduces the frequency of attacks. Patients with an abscess or pseudocyst, biliary tract disease, or a fibrotic pancreatic sphincter may undergo surgery. Surgery may also help relieve obstruction and allow drainage of pancreatic secretions.

What tests tell you

The following tests are used to diagnose pancreatitis:
• Dramatically elevated serum amylase and lipase levels confirm acute pancreatitis. Dramatically elevated amylase levels are also found in urine, ascites, and pleural fluid.
• Blood and urine glucose tests may reveal transient glucose in urine (glucosuria) and hyperglycemia. In chronic pancreatitis, serum glucose levels may be transiently elevated.
• WBC count is elevated.
• Serum bilirubin levels are elevated in both acute and chronic pancreatitis.
• Blood calcium levels may be decreased.
• Stool analysis shows elevated lipid and trypsin levels in chronic pancreatitis.

• Abdominal and chest X-rays detect pleural effusions and differentiate pancreatitis from diseases that cause similar symptoms.
• CT scan and ultrasonography show an enlarged pancreas and pancreatic cysts and pseudocysts.
• Endoscopic retrograde cholangiopancreatography shows the anatomy of the pancreas; identifies ductal system abnormalities, such as calcification or strictures; and differentiates pancreatitis from other disorders such as pancreatic cancer.

Peptic ulcer

A peptic ulcer is a circumscribed lesion in the mucosal membrane. Peptic ulcers can develop in the lower esophagus, stomach, duodenum, or jejunum. (See *A close look at peptic ulcers*.)

Pathophysiology

The major forms of peptic ulcer are duodenal ulcer and gastric ulcer. Both are chronic conditions.

Duodenal ulcers

Duodenal ulcers affect the upper part of the small intestine. This type of ulcer accounts for about 80% of peptic ulcers, occurs mostly in men between ages 20 and 50, and follows a chronic course of remissions and exacerbations. About 5% to 10% of patients develop complications that make surgery necessary.

Gastric ulcers

Gastric ulcers affect the stomach lining (mucosa). They're most common in middle-aged and elderly men, especially poor and undernourished men. They commonly occur in chronic users of aspirin or alcohol.

There are three major causes of peptic ulcers:

bacterial infection with *Helicobacter pylori*

use of nonsteroidal anti-inflammatory drugs

hypersecretory states such as Zollinger-Ellison syndrome.

Researchers are still discovering the exact mechanisms of ulcer formation. Predisposing factors include:

Chronic use of aspirin or alcohol may cause gastric ulcers.

Bacteria are another major cause of gastric ulcers.

A close look at peptic ulcers

This illustration shows different degrees of peptic ulceration. Lesions that don't extend below the mucosal lining (epithelium) are called erosions. Lesions of both acute and chronic ulcers can extend through the epithelium and may perforate the stomach wall. Chronic ulcers also have scar tissue at the base.

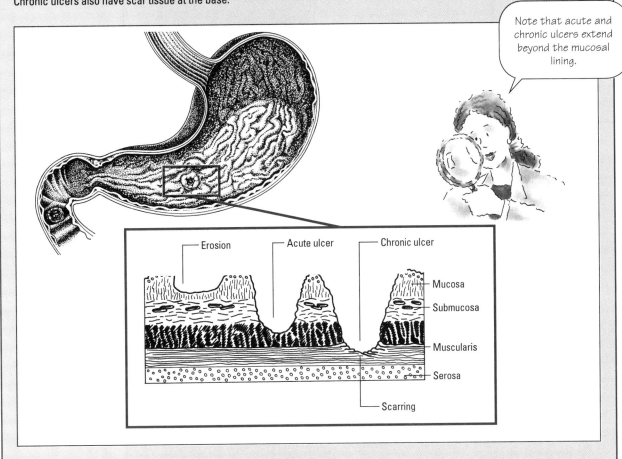

Note that acute and chronic ulcers extend beyond the mucosal lining.

- blood type A (common in people with gastric ulcers) and blood type O (common in those with duodenal ulcers)
- genetic factors
- exposure to irritants
- trauma
- stress and anxiety
- normal aging.

There's a hole in my mucous coat

In a peptic ulcer due to *H. pylori,* acid adds to the effects of the bacterial infection. *H. pylori* releases a toxin that destroys the stomach's mucous coat, reducing the epithelium's resistance to acid digestion and causing gastritis and ulcer disease.

Uh oh! GI hemorrhage, hypovolemic shock, obstruction

A possible complication of severe ulceration is erosion of the mucosa. This can cause GI hemorrhage, which can progress to hypovolemic shock, perforation, and obstruction. Obstruction of the pylorus may cause the stomach to distend with food and fluid, block blood flow, and cause tissue damage.

Darn! A hole in the duodenal wall!

The ulcer crater may extend beyond the duodenal wall into nearby structures, such as the pancreas or liver. This phenomenon is called penetration and is a fairly common complication of duodenal ulcer.

What to look for

The patient with a gastric ulcer may report:
• recent loss of weight or appetite
• pain, heartburn, or indigestion
• a feeling of abdominal fullness or distention
• pain triggered or aggravated by eating.

Sharp, gnawing, burning, or hard to define

The patient with a duodenal ulcer may describe the pain as sharp, gnawing, burning, boring, aching, or hard to define. He may liken it to hunger, abdominal pressure, or fullness. Typically, pain occurs 90 minutes to 3 hours after eating. However, eating often reduces the pain, so the patient may report a recent weight gain. The patient may also have pale skin from anemia caused by blood loss. (See *Treating peptic ulcer.*)

> Bacteria can destroy the stomach's protective coat and expose it to acid.

Battling illness

Treating peptic ulcer

Treatment for peptic ulcer aims to relieve symptoms.

Drug therapy
Bismuth and other antimicrobial agents may help to fight *Helicobacter pylori.* Antacids and histamine-receptor antagonists may also be prescribed.

For duodenal ulcers, coating agents and anticholinergics may be prescribed. For gastric ulcers, sedatives and tranquilizers may be used.

Diet and activity
Rest, decreased activity, and eating six small meals daily or small hourly meals may be recommended. A bland diet is controversial.

More drastic measures
GI bleeding may require iced saline lavage through a nasogastric tube, possibly containing norepinephrine. Gastroscopy allows visualization of the bleeding site and coagulation by laser or cautery to control bleeding.

Surgery is indicated if the patient doesn't respond to other treatment or if he has a perforation, suspected cancer, or other complications.

What tests tell you

The following tests are used to diagnose peptic ulcer:
- Barium swallow and upper GI or small-bowel series may pinpoint the ulcer in a patient whose symptoms aren't severe.
- Upper GI endoscopy or esophagogastroduodenoscopy confirms an ulcer and permits cytologic studies and biopsy to rule out *H. pylori* or cancer. Endoscopy is the major diagnostic test for peptic ulcers.
- Upper GI tract X-ray reveals mucosal abnormalities.
- Stool analysis may detect occult blood in stools.
- WBC count is elevated; other blood tests may also disclose clinical signs of infection.
- Gastric secretory studies show excess hydrochloric acid (hyperchlorhydria).
- Carbon-13 urea breath test reflects activity of *H. pylori*.

Ulcerative colitis

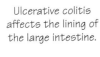

Ulcerative colitis affects the lining of the large intestine.

This inflammatory disease causes ulcerations of the mucosa in the colon. It commonly occurs as a chronic condition.

As many as 1 out of 1,000 people may have ulcerative colitis. Peak occurrence is between ages 15 and 20 and between ages 55 and 60. It's more prevalent among women, people of Jewish ancestry, and whites.

Pathophysiology

Although the cause of ulcerative colitis is unknown, it may be related to an abnormal immune response in the GI tract, possibly associated with genetic factors. Lymphocytes (T cells) in people with ulcerative colitis may have cytotoxic effects on the epithelial cells of the colon.

Stress doesn't cause the disorder, but it can increase the severity of an attack. Although no specific organism has been linked to ulcerative colitis, infection hasn't been ruled out.

Ulcerative colitis: Step-by-step

Ulcerative colitis damages the large intestine's mucosal and submucosal layers. Here's how it progresses:
- Usually, the disease originates in the rectum and lower colon. Then it spreads to the entire colon.

• The mucosa develops diffuse ulceration, with hemorrhage, congestion, edema, and exudative inflammation Unlike Crohn's disease, ulcerations are continuous.
• Abscesses formed in the mucosa drain purulent pus, become necrotic, and ulcerate.
• Sloughing occurs, causing bloody, mucus-filled stools.

Looking closer at the colon

As ulcerative colitis progresses, the colon undergoes changes described below:
• Initially, the colon's mucosal surface becomes dark, red, and velvety.
• Abscesses form and coalesce into ulcers.
• Necrosis of the mucosa occurs.
• As abscesses heal, scarring and thickening may appear in the bowel's inner muscle layer.
• As granulation tissue replaces the muscle layer, the colon narrows, shortens, and loses its characteristic pouches (haustral folds).

Yikes! Obstruction, dehydration, imbalances

Progression of ulcerative colitis may lead to intestinal obstruction, dehydration, and major fluid and electrolyte imbalances. Malabsorption is common, and chronic anemia may result from loss of blood in the stools.

What to look for

The hallmark of ulcerative colitis is recurrent bloody diarrhea — often containing pus and mucus — alternating with symptom-free remissions. Accumulation of blood and

First, ulceration strikes the bowel.

Next, abscesses form and cause tissue damage.

Finally, dead tissue is sloughed off and expelled with stool.

Battling illness

Treating ulcerative colitis

The goals of treatment are to control inflammation, replace lost nutrients and blood, and prevent complications. Supportive measures include bed rest, I.V. fluid replacement, and blood transfusions.

Drug therapy
Medications include:

• corticotropin and adrenal corticosteroids, such as prednisone, prednisolone, and hydrocortisone, to control inflammation
• sulfasalazine, which has anti-inflammatory and antimicrobial properties
• antispasmodics, such as tincture of belladonna
• antidiarrheals, such as diphenoxylate and atropine, for patients with frequent, troublesome diarrhea whose ulcerative colitis is otherwise under control
• iron supplements to correct anemia.

Diet therapy
Patients with severe disease usually need total parenteral nutrition (TPN) and are allowed nothing by mouth. TPN is also used for patients awaiting surgery or those dehydrated or debilitated from excessive diarrhea. This treatment rests the intestinal tract, decreases stool volume, and restores nitrogen balance.

Patients with moderate signs and symptoms may receive supplemental drinks and elemental feedings. A low-residue diet may be ordered for the patient with mild disease.

Surgery
Surgery is performed if the patient has massive dilation of the colon (toxic megacolon), if he doesn't respond to drugs and supportive measures, or if he finds the symptoms unbearable.

The most common surgical technique is proctocolectomy with ileostomy, although pouch ileostomy and ileoanal reservoir are also done.

mucus in the bowel causes cramping abdominal pain, rectal urgency, and diarrhea. Other symptoms include irritability, weight loss, weakness, anorexia, nausea, and vomiting. (See *Treating ulcerative colitis.*)

What tests tell you

The following tests are used to diagnose ulcerative colitis:
• Sigmoidoscopy confirms rectal involvement by showing mucosal friability (vulnerability to breakdown) and flattening and thick, inflammatory exudate.
• Colonoscopy shows the extent of the disease, strictured areas, and pseudopolyps. It isn't performed when the patient has active signs and symptoms.

• Biopsy during colonoscopy can help confirm the diagnosis.

• Barium enema is used to show the extent of the disease, detect complications, and identify cancer. It's not performed in a patient with active signs and symptoms.

• Stool specimen analysis reveals blood, pus, and mucus but no disease-causing organisms.

• Other laboratory tests show decreased serum potassium, magnesium, and albumin levels; decreased WBC count; decrease hemoglobin; and prolonged prothrombin time. Increase of the erythrocyte sedimentation rate correlates with the severity of the attack.

Viral hepatitis

Usually, liver cells grow back with little residual damage...

Viral hepatitis is a common infection of the liver. In most patients, damaged liver cells eventually regenerate with little or no permanent damage. However, old age and serious underlying disorders make complications more likely. More than 70,000 cases are reported annually in the United States.

Pathophysiology

Viral hepatitis is marked by liver cell destruction, tissue death (necrosis), and self-destruction of cells (autolysis). It leads to anorexia, jaundice, and hepatomegaly. Five types of viral hepatitis are recognized:

• type A — increasing among homosexuals and in people with human immunodeficiency virus (HIV) infection

• type B — also increasing among HIV-positive people; accounts for 5% to 10% of posttransfusion hepatitis cases in the United States

• type C — accounts for about 20% of all viral hepatitis and most cases that follow transfusion

• type D — in the United States, confined to people frequently exposed to blood and blood products, such as I.V. drug users and hemophiliacs

• type E — formerly grouped with type C under the name non-A, non-B hepatitis; in the United States, mainly occurs in people who have visited an endemic area, such as India, Africa, Asia, or Central America.

The five major forms of viral hepatitis result from infection with the causative viruses A, B, C, D, or E. (See *Viral hepatitis from A to E,* page 338.)

...but it can take months of rest and a proper diet.

Injury and necrosis in varying degrees

Despite the different causes, changes to the liver are usually similar in each type of viral hepatitis. Varying degrees of liver cell injury and necrosis occur. These changes in the liver are completely reversible when the acute phase of the disease subsides.

> Injury and necrosis, oh my!

More trouble

A fairly common complication is chronic persistent hepatitis, which prolongs recovery up to 8 months. Some patients also suffer relapses. A few may develop chronic active hepatitis, which destroys part of the liver and causes cirrhosis. In rare cases, severe and sudden (fulminant) hepatic failure and death may result from massive tissue loss. Primary hepatocellular carcinoma is a late complication that can cause death within 5 years, but it's rare in the United States.

What to look for

Signs and symptoms of viral hepatitis progress in three stages:

1. prodromal

2. clinical

3. recovery.

Prodromal stage

In the prodromal stage, the following signs and symptoms may be caused by circulating immune complexes: fatigue, anorexia, mild weight loss, generalized malaise, depression, headache, weakness, joint pain (arthralgia), muscle pain (myalgia), intolerance to light (photophobia), nausea and vomiting, changes in the senses of taste and smell, fever of 100° to 102° F (37.8° to 38.9° C), right upper quadrant tenderness, and dark-colored urine and clay-colored stools (1 to 5 days before the onset of the clinical jaundice stage). During this phase, the infection is highly transmissible.

Viral hepatitis from A to E

The following chart compares the features of each type of viral hepatitis.

Feature	Hepatitis A	Hepatitis B	Hepatitis C	Hepatitis D	Hepatitis E
Incubation	15 to 45 days	30 to 180 days	15 to 160 days	14 to 64 days	14 to 60 days
Onset	Acute	Insidious	Insidious	Acute and chronic	Acute
Age-group most affected	Children, young adults	Any age	More common in adults	Any age	Ages 20 to 40
Transmission	Fecal-oral, sexual (especially oral-anal contact), non-percutaneous (sexual, maternal-neonatal), percutaneous (rare)	Blood-borne; parenteral route, sexual, maternal-neonatal; virus is shed in all body fluids	Blood-borne; parenteral route	Parenteral route; most people infected with hepatitis D are also infected with hepatitis B	Primarily fecal-oral
Severity	Mild	Often severe	Moderate	Can be severe and lead to fulminant hepatitis	Highly virulent with common progression to fulminant hepatitis and hepatic failure, especially in pregnant patients
Prognosis	Generally good	Worsens with age and debility	Moderate	Fair; worsens in chronic cases; can lead to chronic hepatitis D and chronic liver disease	Good unless pregnant
Progression to chronicity	None	Occasional	10% to 50% of cases	Occasional	None

Clinical stage

Also called the icteric stage, the clinical stage begins 1 to 2 weeks after the prodromal stage. It's the phase of actual illness.

If the patient progresses to this stage, he may have these signs and symptoms: itching, abdominal pain or tenderness, indigestion, appetite loss (in early clinical stage), and jaundice.

Jaundice lasts for 1 to 2 weeks and indicates that the damaged liver can't remove bilirubin from the blood. However, jaundice doesn't indicate disease severity and, occasionally, hepatitis occurs without jaundice.

Recovery stage

Recovery begins with the resolution of jaundice and lasts 2 to 6 weeks in uncomplicated cases. The prognosis is poor if edema and hepatic encephalopathy develop. (See *Treating viral hepatitis*.)

What tests tell you

The following tests are used to diagnose viral hepatitis:
- Hepatitis profile establishes the type of hepatitis.
- Liver function studies show disease stage.
- Prothrombin time is prolonged.
- WBC count is elevated.
- Liver biopsy may be performed if chronic hepatitis is suspected.

Quick quiz

1. Bleeding from esophageal varices usually stems from:
    A. esophageal perforation.
    B. pulmonary hypertension.
    C. portal hypertension.

Answer: C. Increased pressure within the portal veins causes them to bulge, leading to rupture and bleeding.

2. The two phases of digestion are:
    A. cephalic and gastric.
    B. salivation and secretion.
    C. esophageal and abdominal.

Answer: A. The cephalic phase begins when the food is on its way to the stomach. Food entering the stomach initiates the gastric phase.

Battling illness

Treating viral hepatitis

Hepatitis C has been treated somewhat successfully with interferon alfa. No specific drug therapy has been developed for other types of viral hepatitis. Instead, the patient is advised to rest in the early stages of the illness and combat anorexia by eating small, high-calorie, high-protein meals.

Protein intake should be reduced if signs of precoma — lethargy, confusion, or mental changes — develop. Large meals are usually better tolerated in the morning because many patients have nausea late in the day.

Acute cases

In acute viral hepatitis, hospitalization is usually required only if severe symptoms or complications occur. Parenteral nutrition may be needed if the patient can't eat because of persistent vomiting.

3. The initial event in appendicitis is usually:
    A. lymph node enlargement.
    B. obstruction of the appendiceal lumen.
    C. ulceration of the mucosa.

Answer: C. Although an obstruction can be identified in some cases, ulceration of the mucosa usually occurs first.

4. In women, pancreatitis is usually associated with:
    A. biliary tract disease.
    B. alcoholism.
    C. allergies.

Answer: A. Trauma and peptic ulcer are leading causes in men.

Scoring

☆☆☆   If you answered all four items correctly, excellent! When it comes to understanding the GI system, you're on the right tract.

☆☆   If you answered two or three correctly, relax and enjoy! Your performance on this test is not difficult to digest.

☆   If you answered one or none correctly, don't worry. Take your chyme and you'll absorb all the important facts.

Stop! I don't think I can stomach any more puns.

Renal system

Just the facts

In this chapter you'll learn:

♦ how the renal system functions

♦ the pathophysiology, signs and symptoms, diagnostic tests, and treatments for several common renal system disorders.

Understanding the renal system

Together with the urinary system, the renal system serves as the body's water treatment plant. These systems work together to collect the body's waste products and expel them as urine.

The kidneys are located on each side of the abdomen near the lower back. These compact organs contain an amazingly efficient filtration system. The byproduct of this filtration is urine, which contains water and waste products. (See *A close look at the kidney*, page 342).

WOW!

I filter about 45 gallons of fluid a day!

It's downhill from here

Once produced by the kidneys, urine passes through the urinary system and is expelled from the body. The other structures of the system, extending downward from the kidneys, include the:

• ureters — 16″ to 18″ (41- to 46-cm) muscular tubes that contract rhythmically to transport urine from each kidney to the bladder

A close look at the kidney

Illustrated below is a kidney and next to it an enlargement of a nephron, the kidney's functional unit. Major structures of the kidney include the following:

• medulla—inner portion of the kidney, composed of renal pyramids and tubular structures
• renal artery—supplies blood to the kidney
• renal pyramid—channels output to renal pelvis for excretion
• renal calyx—channels formed urine from the renal pyramids to the renal pelvis
• renal vein—about 99% of filtered blood is circulated through the renal vein back to the general circulation; the remaining 1%, which contains waste products, undergoes further processing in the kidney
• renal pelvis—once blood that contains waste products is processed in the kidney, formed urine is channeled to the renal pelvis
• ureter—tube that terminates in the urethra; urine enters the urethra for excretion
• cortex—outer layer of the kidney.

Note the nephron

The nephron is the functional and structural unit of the kidney. It's two main activities are selective reabsorption and secretion of ions and mechanical filtration of fluids, wastes, electrolytes, and acids and bases.

Components of the nephron include the following:

• glomerulus—a network of twisted capillaries which acts as a filter for the passage of protein-free and red blood cell–free filtrate to the proximal convoluted tubules
• Bowman's capsule—contains the glomerulus and acts as a filter for urine
• proximal convoluted tubule—site of reabsorption of glucose, amino acids, metabolites, and electrolytes from filtrate; reabsorbed substances return to circulation
• loop of Henle—a U-shaped nephron tubule located in the medulla and extending from the proximal convoluted tubule to the distal convoluted tubule; site for further concentration of filtrate through reabsorption
• distal convoluted tubule—site from which filtrate enters the collecting tubule
• collecting tubule—releases urine.

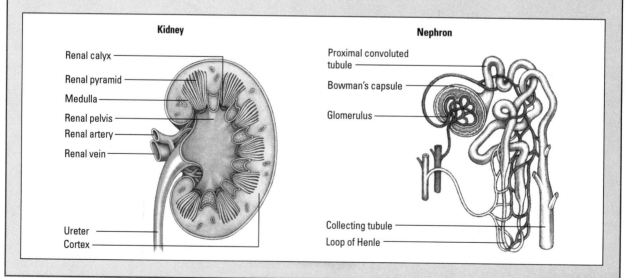

Kidney

Renal calyx
Renal pyramid
Medulla
Renal pelvis
Renal artery
Renal vein
Ureter
Cortex

Nephron

Proximal convoluted tubule
Bowman's capsule
Glomerulus
Collecting tubule
Loop of Henle

• urinary bladder — a sac with muscular walls that collects and holds urine that is expelled from the ureters every few seconds
• urethra — a narrow tube leading out of the body through which urine is expelled from the bladder.

Regulating, balancing, detoxifying

The kidneys perform the following vital functions:
• regulating the electrolyte concentration, acid-base balance, and amount of body fluids
• detoxifying the blood and eliminating wastes
• regulating blood pressure
• aiding red blood cell (RBC) formation (erythropoiesis).

Regulating, balancing, detoxifying — yep, I do it all.

Maintaining fluid and acid-base balance

Two important kidney functions are maintaining fluid and acid-base balance.

Fluid balance

The kidneys maintain fluid balance in the body by regulating the amount and makeup of the fluids inside and around the cells. Kidneys maintain the volume and composition of extracellular fluid and, to a lesser extent, intracellular fluid. They do this by continuously exchanging water and solutes, such as hydrogen, sodium, potassium, chloride, bicarbonate, sulfate, and phosphate ions, across their cell membranes.

Acid-base balance

To regulate acid-base balance, the kidneys:

 secrete hydrogen ions

 reabsorb sodium and bicarbonate ions

 acidify phosphate salts

 produce ammonia.

All of these regulating activities keep the blood at its normal pH of 7.37 to 7.43. Acidosis occurs when the pH falls below 7.35, and alkalosis occurs when the pH is greater than 7.45.

Thanks to two hormones

Hormones partially control the kidneys' role in fluid balance. This control depends on the response of specialized sensory nerve endings (osmoreceptors) to changes in osmolality (the ionic concentration of a solution).

The two hormones involved are:

 antidiuretic hormone (ADH), produced by the pituitary gland

aldosterone, produced by the adrenal cortex.

All about ADH

ADH alters the collecting tubules' permeability to water. When ADH concentration in plasma is high, the tubules are most permeable to water. This creates a highly concentrated but small volume of urine. If ADH concentration is low, the tubules are less permeable to water. This creates a larger volume of less concentrated urine.

All that's known about aldosterone

Aldosterone regulates water reabsorption by the distal tubules and changes urine concentration by increasing sodium reabsorption. A high plasma aldosterone concentration increases sodium and water reabsorption by the tubules and decreases sodium and water excretion in the urine. A low plasma aldosterone concentration promotes sodium and water excretion.

Aldosterone helps control the secretion of potassium by the distal tubules. A high aldosterone concentration increases the excretion of potassium. Other factors that affect potassium secretion include:
• the amount of potassium ingested
• the number of hydrogen ions secreted
• potassium levels in the cells
• the amount of sodium in the distal tubule
• the glomerular filtration rate (GFR), the rate at which plasma is filtered as it flows through the glomerular capillary filtration membrane.

Problems in hormone concentration may cause fluctuations in sodium and potassium concentrations that, in turn, may lead to hypertension.

ADH controls the concentration of body fluids by altering the permeability of the kidneys, thereby conserving water.

Aldosterone regulates the reabsorption of sodium and the excretion of potassium by the kidneys.

Going against the current

The kidneys concentrate urine through the countercurrent exchange system. In this system, fluid flows in opposite directions through parallel tubes, up and down parallel sides of the loops of Henle. A concentration gradient causes fluid exchange. The longer the loop the greater the concentration gradient.

Filtering, reabsorbing, and secreting — that's how I do my job.

Waste collection

The kidneys collect and eliminate wastes from the body in a three-step process:

🖐 by glomerular filtration — filtering the blood that flows through the kidney's blood vessels, or glomeruli

✌ by tubular reabsorption — reabsorbing filtered fluid through the minute canals (tubules) that make up the kidney

🖖 by tubular secretion — release of the filtered substance by the tubules.

Clear the way

Clearance is the complete removal of a substance from the blood. Clearance is commonly described as the amount of blood that can be cleared in a specific amount of time. For example, creatinine clearance is the volume of blood in milliliters that the kidneys can clear of creatinine in 1 minute.

Some substances are filtered out of the blood by the glomeruli. Dissolved substances that remain in the fluid may or may not be reabsorbed by the renal tubular cells. (See *Understanding the glomerular filtration rate*, page 346.)

Nephrons can only do so much

In patients whose kidneys have shrunk from disease, healthy nephrons (the filtering units of the kidney) enlarge to compensate. But as nephron damage progresses, the enlargement can no longer adequately compensate, and the GFR slows.

Healthy nephrons are great but they can't compensate for damage forever.

Now I get it!

Understanding the glomerular filtration rate

The glomerular filtration rate (GFR) is the rate at which the glomeruli filter blood. The normal GFR is about 120 ml/minute. GFR depends on:
• permeability of capillary walls
• vascular pressure
• filtration pressure.

GFR and clearance
Clearance is the complete removal of a substance from the blood. The most accurate measure of glomerular filtration is creatinine clearance. That's because creatinine is filtered by the glomeruli but not reabsorbed by the tubules.

Equal to, greater than, or less than
Here's more about how the GFR affects clearance measurements for a substance in the blood:
• If the tubules neither reabsorb nor secrete the substance — as happens with creatinine — clearance equals the GFR.
• If the tubules reabsorb the substance, clearance is less than the GFR.
• If the tubules secrete the substance, clearance exceeds the GFR.
• If the tubules reabsorb and secrete the substance, clearance may be less than, equal to, or greater than the GFR.

Don't saturate the system

The amount of a substance that's reabsorbed or secreted depends on the substance's maximum tubular transport capacity — the maximum amount of a substance that can be reabsorbed or secreted in a minute without saturating the renal system.

For example, in diabetes mellitus, excess glucose in the blood overwhelms the renal tubules and causes glucose to appear in the urine (glycosuria). In other cases, when glomeruli are damaged, protein appears in the urine (proteinuria) because the large protein molecules pass into the urine instead of being reabsorbed.

Blood pressure regulation

The kidneys help regulate blood pressure by producing and secreting the enzyme renin in response to an actual or perceived decline in extracellular fluid volume. Renin, in turn, forms angiotensin I, which is converted to the more potent angiotensin II.

By secreting renin, I help to regulate blood pressure.

Angiotensin II raises low arterial blood pressure levels by:
- increasing peripheral vasoconstriction
- stimulating aldosterone secretion.

The increase in aldosterone promotes the reabsorption of sodium and water to correct the fluid deficit and inadequate blood flow (renal ischemia).

Hypertension: A serious threat

Hypertension can stem from a fluid and electrolyte imbalance as well as renin-angiotensin hyperactivity. High blood pressure can damage blood vessels. High blood pressure may also cause hardening of the kidneys (nephrosclerosis), one of the leading causes of chronic renal failure.

RBC production

Erythropoietin is a hormone that prompts the bone marrow to increase RBC production. The kidneys secrete erythropoietin when the oxygen supply in tissue blood drops. They also produce active vitamin D and help regulate calcium balance and bone metabolism. Loss of renal function results in chronic anemia and insufficient calcium levels (hypocalcemia) because of a decrease in erythropoietin.

Renal disorders

The disorders discussed below include acute renal failure, acute tubular necrosis, benign prostatic hyperplasia, chronic renal failure, prostatitis, and renal calculi (kidney stones).

Acute renal failure

Acute renal failure is a sudden interruption of renal function. It can be caused by obstruction, poor circulation, or kidney disease.

In acute renal failure, my ability to function comes to a sudden halt.

Pathophysiology

Acute renal failure may be classified as prerenal, intrarenal, or postrenal. Each type has separate causes. (See *Causes of acute renal failure*.)

Prerenal failure results from conditions that diminish blood flow to the kidneys (hypoperfusion). Examples include hypovolemia, hypotension, vasoconstriction, or inadequate cardiac output. One condition, prerenal azotemia (excess nitrogenous waste products in the blood), accounts for 40% to 80% of all cases of acute renal failure. Azotemia occurs as a response to renal hypoperfusion. Usually, it can be rapidly reversed by restoring renal blood flow and glomerular filtration.

Intrarenal failure, also called intrinsic or parenchymal renal failure, results from damage to the filtering structures of the kidneys, usually from acute tubular necrosis, a disorder that causes cell death, or from nephrotoxic substances such as certain antibiotics.

Postrenal failure results from bilateral obstruction of urine outflow, as in prostatic hyperplasia or bladder outlet obstruction.

With treatment, each type of acute renal failure passes through three distinct phases:

- oliguric (decreased urine secretion)

- diuretic (increased urine secretion)

- recovery.

Prerenal failure is caused by any condition that reduces blood flow to the kidneys.

Intrarenal failure is caused by damage to the filtering structure of the kidneys.

Oliguric phase

Oliguria is a decreased urine output (less than 400 ml/24 hours). Prerenal oliguria results from decreased blood flow to the kidney. Before damage occurs, the kidney responds to decreased blood flow by conserving sodium and water. Once damage occurs, the kidney's ability to conserve sodium is impaired. Untreated prerenal oliguria may lead to acute tubular necrosis.

Diuretic and recovery phases

The diuretic phase is marked by increased urine secretion. High urine volume (more than 400 ml/24 hours) has two causes:

Postrenal failure is caused by any condition that blocks urine flow from both kidneys.

Causes of acute renal failure

Acute renal failure is classified as prerenal, intrarenal, or postrenal. All conditions that lead to prerenal failure impair blood flow to the kidneys (renal perfusion), resulting in decreased GFR and increased tubular reabsorption of sodium and water. Intrarenal failure results from damage to the kidneys themselves; postrenal failure, from obstructed urine flow. The table shows the causes of each type of acute renal failure.

Prerenal failure	Intrarenal failure	Postrenal failure

Cardiovascular disorders

- Arrhythmias
- Cardiac tamponade
- Cardiogenic shock
- Heart failure
- Myocardial infarction

Hypovolemia

- Burns
- Dehydration
- Diuretic overuse
- Hemorrhage
- Hypovolemic shock
- Trauma

Peripheral vasodilation

- Antihypertensive drugs
- Sepsis

Renovascular obstruction

- Arterial embolism
- Arterial or venous thrombosis
- Tumor

Severe vasoconstriction

- Disseminated intravascular coagulation
- Eclampsia
- Malignant hypertension
- Vasculitis

Acute tubular necrosis

- Ischemic damage to renal parenchyma from unrecognized or poorly treated prerenal failure
- Nephrotoxins, including analgesics such as phenacetin, anesthetics such as methoxyflurane, antibiotics such as gentamicin, heavy metals such as lead, radiographic contrast media, and organic solvents
- Obstetric complications, such as eclampsia, postpartum renal failure, septic abortion, and uterine hemorrhage
- Pigment release, such as crush injury, myopathy, sepsis and transfusion reaction

Other parenchymal disorders

- Acute glomerulonephritis
- Acute interstitial nephritis
- Acute pyelonephritis
- Bilateral renal vein thrombosis
- Malignant nephrosclerosis
- Papillary necrosis
- Periarteritis nodosa (inflammatory disease of the arteries)
- Renal myeloma
- Sickle cell disease
- Systemic lupus erythematosus
- Vasculitis

Bladder obstruction

- Anticholinergic drugs
- Autonomic nerve dysfunction
- Infection
- Tumor

Ureteral obstruction

- Blood clots
- Calculi
- Edema or inflammation
- Necrotic renal papillae
- Retroperitoneal fibrosis or hemorrhage
- Surgery (accidental ligation)
- Tumor
- Uric acid crystals

Urethral obstruction

- Prostatic hyperplasia or tumor
- Strictures

the kidney's inability to conserve sodium and water

osmotic diuresis produced by high blood urea nitrogen (BUN) levels.

These conditions can lead to deficits of potassium, sodium, and water that can be deadly if left untreated. But if the cause of the diuresis is corrected, azotemia gradually disappears and the patient improves greatly — leading to the recovery stage.

Uh Oh! Complications

Primary damage to the renal tubules or blood vessels results in kidney failure (intrarenal failure). The causes of intrarenal failure are classified as nephrotoxic, inflammatory, or ischemic.

When the damage is caused by nephrotoxicity or inflammation, the delicate layer under the epithelium (basement membrane) becomes irreparably damaged, often proceeding to chronic renal failure. Severe or prolonged lack of blood flow (ischemia) may lead to renal damage (ischemic parenchymal injury) and excess nitrogen in the blood (intrinsic renal azotemia).

What to look for

The signs and symptoms of prerenal failure depend on the cause. If the underlying problem is a change in blood pressure and volume, the patient may have:
- oliguria
- tachycardia
- hypotension
- dry mucous membranes
- flat neck veins
- lethargy progressing to coma
- decreased cardiac output and cool, clammy skin in patients with heart failure.

As renal failure progresses, the patient has the following signs and symptoms of uremia:
- confusion
- GI complaints
- fluid in the lungs
- infection.

About 5% of all hospitalized patients develop acute renal failure. The condition is usually reversible with treat-

Acute renal failure is usually reversible with medical treatment but can be fatal without it.

Battling illness

Treating acute renal failure

Supportive measures for acute renal failure include:
- a high-calorie diet that is low in protein, sodium, and potassium
- fluid and electrolyte balance
- monitoring for signs and symptoms of uremia
- fluid restriction
- diuretic therapy during the oliguric phase
- prevention of infection.

Halting hyperkalemia
Meticulous electrolyte monitoring is needed to detect excess potassium in the blood (hyperkalemia). Symptoms include malaise, loss of appetite, numbness and tingling, muscle weakness, and ECG changes. If these symptoms occur, hypertonic glucose, insulin, and sodium bicarbonate are given I.V., and sodium polystyrene sulfonate (Kayexalate) is given by mouth or enema.

If the above measures fail to control uremia, the patient may need hemodialysis or peritoneal dialysis.

ment, but if it's not treated, it may progress to end-stage renal disease, excess urea in the blood (prerenal azotemia or uremia), and death. (See *Treating acute renal failure*.)

What tests tell you

The following tests are used to diagnose acute renal failure:

• Blood studies reveal elevated BUN, serum creatinine, and potassium levels and decreased blood pH, bicarbonate, hematocrit, and hemoglobin levels.

• Urine specimens show casts, cellular debris, decreased specific gravity and, in glomerular diseases, proteinuria and urine osmolality close to serum osmolality. Urine sodium level is less than 20 mEq/L if oliguria results from decreased perfusion and more than 40 mEq/L if it results from an intrarenal problem.

• Creatinine clearance test measures the GFR and is used to estimate the number of remaining functioning nephrons.

• Electrocardiogram (ECG) shows tall, peaked T waves, a widening QRS complex, and disappearing P waves if increased blood potassium (hyperkalemia) is present.

• Other studies that help determine the cause of renal failure include kidney ultrasonography, plain films of the abdomen, kidney-ureter-bladder (KUB) radiography, excretory urography, renal scan, retrograde pyelography, computed tomography scans, and nephrotomography.

Acute tubular necrosis

Acute tubular necrosis causes about 75% of all cases of acute renal failure. Also called acute tubulointerstitial nephritis, this disorder destroys the tubular segment of the nephron, causing renal failure and uremia (excess by-products of protein metabolism in the blood).

Pathophysiology

Acute tubular necrosis may follow two types of injury to the kidney:

☝ ischemic injury, the most common cause

✌ nephrotoxic injury, usually in debilitated patients, such as the critically ill or those who've undergone extensive surgery.

Oy vey! Damage to my tubules can cause renal failure.

Ischemic injury

In ischemic injury, blood flow to the kidneys is disrupted. This may be caused by:
- circulatory collapse
- severe hypotension
- trauma
- hemorrhage
- dehydration
- cardiogenic or septic shock
- surgery
- anesthetics
- transfusion reactions.

The longer blood flow is interrupted, the worse the kidney damage.

Nephrotoxic injury

Nephrotoxic injury can result from:
- ingesting or inhaling toxic chemicals, such as carbon tetrachloride, heavy metals, and methoxyflurane anesthetics
- a hypersensitivity reaction of the kidneys to such substances as antibiotics and radiographic contrast agents.

Let's get specific

Some specific causes of acute tubular necrosis and their effects include:
- diseased tubular epithelium that allows glomerular filtrate to leak through the membranes and be reabsorbed into the blood
- obstructed urine flow from the collection of damaged cells, casts, RBCs, and other cellular debris within the tubular walls
- ischemic injury to glomerular epithelial cells, causing cellular collapse and poor glomerular capillary permeability
- ischemic injury to the vascular endothelium, eventually causing cellular swelling, and tubular obstruction.

Here's a rundown of some more specific causes of acute tubular necrosis and their effects.

Deep or shallow lesions

Deep or shallow lesions may occur in acute tubular necrosis:

With ischemic injury, necrosis creates deep lesions, destroying both tubular epithelium and the basement membrane (the delicate layer underlying the epithelium). Ischemic injury causes patches of necrosis in the tubules.

Nephrotoxic damage may be reversible.

A close look at acute tubular necrosis

In acute tubular necrosis caused by ischemia, patches of necrosis occur, usually in the straight portion of the proximal tubules. In areas without lesions, tubules are usually dilated. In acute tubular necrosis caused by nephrotoxicity, the tubules have a more uniform appearance.

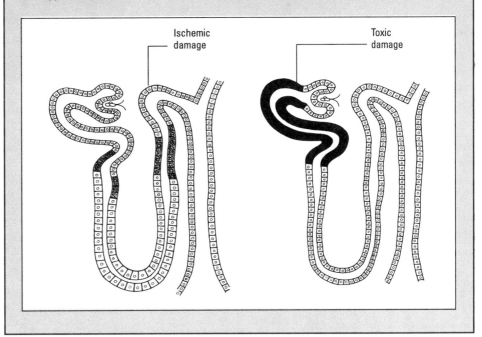

Ischemia can also cause lesions in the connective tissue of the kidney.

 With nephrotoxic injury, necrosis occurs only in the epithelium of the tubules, leaving the basement membrane of the nephrons intact. This type of damage may be reversible. (See *A close look at acute tubular necrosis*.)

Four ways to damage cells

Toxicity takes a toll. Nephrotoxic agents can injure tubular cells four ways:

 by direct cellular toxic effects

by coagulation and destruction (lysis) of RBCs

Ouch!

by oxygen deprivation (hypoxia)

 by crystal formation of solutes.

Oh no! Infection, hemorrhage, other complications

There are several common complications of acute tubular necrosis :
• Infections (frequently septicemia) complicate up to 70% of all cases and are the leading cause of death.
• GI hemorrhage, fluid and electrolyte imbalance, and cardiovascular dysfunction may occur during the acute phase or the recovery phase.
• Neurologic complications are common in elderly patients and occur occasionally in younger patients.
• Excess blood calcium (hypercalcemia) may occur during the recovery phase.

What to look for

Early-stage acute tubular necrosis may be hard to spot because the patient's primary disease may obscure the signs and symptoms. The first recognizable effect may be decreased urine output, usually less than 400 ml/24 hours.

Other signs and symptoms depend on the severity of systemic involvement and may include:
• bleeding abnormalities
• vomiting of blood
• dry skin and mucous membranes
• lethargy
• confusion
• agitation
• edema
• fluid and electrolyte imbalances
• muscle weakness with hyperkalemia
• tachycardia and an irregular rhythm.

Mortality can be as high as 70%, depending on complications from underlying diseases. Nonoliguric forms of acute tubular necrosis have a better prognosis. (See *Treating acute tubular necrosis*.)

What tests tell you

Acute tubular necrosis is hard to diagnose except in advanced stages. The following tests are commonly done:

Battling illness

Treating acute tubular necrosis

Acute tubular necrosis requires vigorous supportive measures during the acute phase until normal kidney function is restored. Treatment includes the following:

• Initially, diuretics are given and fluids are infused to flush tubules of cellular casts and debris and to replace lost fluids.

• Projected and calculated fluid losses require daily replacement.

• Transfusion of packed RBCs is administered for anemia.

• Non-nephrotoxic antibiotics are given for infection.

• Hyperkalemia requires an emergency I.V. infusion of 50% glucose, regular insulin, and sodium bicarbonate.

• Sodium polystyrene sulfonate is given by mouth or by enema to reduce potassium levels.

• Hemodialysis or peritoneal dialysis is used to prevent severe fluid and electrolyte imbalance and uremia.

• Urinalysis shows dilute urine, low osmolality, high sodium levels, and urine sediment containing RBCs and casts.
• Blood studies reveal high BUN and serum creatinine levels, low serum protein levels, anemia, platelet adherence defects, metabolic acidosis, and hyperkalemia.
• ECG may show arrhythmias from electrolyte imbalances and, with hyperkalemia, a widening QRS complex, disappearing P waves, and tall, peaked T waves.

Recent evidence suggests a link between benign prostatic hyperplasia and hormonal activity in men.

Benign prostatic hyperplasia

Although most men over age 50 have some prostate enlargement, in benign prostatic hyperplasia, the prostate gland enlarges enough to compress the urethra and cause urinary obstruction.

Pathophysiology

Recent evidence suggests a link between benign prostatic hyperplasia and hormonal activity. As men age, production of hormones that stimulate male characteristics (androgens) decreases and estrogen production increases. This causes an androgen-estrogen imbalance and high levels of dihydrotestosterone, the main prostatic intracellular androgen.

Other possible causes of prostate enlargement include:
• tumor
• arteriosclerosis
• inflammation
• metabolic or nutritional disturbances.

How hormonal havoc happens

In benign prostatic hyperplasia, here's what happens:
• Increased estrogen levels prompt androgen receptors in the prostate gland to increase.
• This causes an overgrowth of normal cells (hyperplasia) that begins around the urethra.
• Growth eventually causes areas of poor blood flow and tissue damage (necrosis) in adjacent prostatic tissue. The center and side lobes of the prostate usually grow, but not the posterior lobe.
• As the prostate enlarges, it may extend into the bladder and decrease urine flow by compressing or distorting the urethra.

An enlarged prostate may compress the urethra.

Beyond benign

Urinary obstruction can lead to complications. Enlargement that blocks the urethra and pushes up the bladder can stop urine flow and cause urinary tract infection (UTI) or calculi. Bladder muscles may thicken and a pouch (diverticulum) may form in the bladder that retains urine when the rest of the bladder empties.

Other complications of this disorder include:
• formation of a fibrous cord of connective tissue in the bladder wall (bladder wall trabeculation)
• detrusor muscle enlargement
• narrowing of the urethra
• incontinence
• acute or chronic renal failure
• distention of the innermost area of the kidney — the renal pelvis and calices — with urine (hydronephrosis).

What to look for

A patient's signs and symptoms depend on the extent of the prostate's enlargement and the lobes affected. Usually, the patient complains of the following group of symptoms, known as prostatism:
• decreased urine stream size and force
• interrupted urine stream
• urinary hesitancy causing straining and a feeling of incomplete voiding.

As the obstruction increases, the patient may report:
• frequent urination with nocturia
• dribbling
• urine retention
• incontinence
• blood in the urine.

An incompletely emptied and distended bladder is visible as a midline bulge, and an enlarged prostate is palpable with rectal digital exam.

Depending on the size of the prostate, the patient's age and health, and the extent of obstruction, this disorder may be treated surgically or symptomatically. (See *Treating benign prostatic hyperplasia*.)

What tests tell you

The following tests are used to diagnose benign prostatic hyperplasia:

Battling illness

Treating benign prostatic hyperplasia

Conservative treatments for relieving symptoms of an enlarged prostate include:

- prostate massage
- sitz baths
- short-term fluid restriction to prevent bladder distention
- antimicrobials if infection occurs
- regular sexual intercourse to relieve prostatic congestion
- terazosin to improve urine flow rates
- finasteride to reduce prostate size.

Surgery

Surgery is the only effective therapy for acute urine retention, kidney distention (hydronephrosis), severe hematuria, recurrent urinary tract infections, or other intolerable symptoms. A transurethral resection — in which tissue is removed with a wire loop and an electric current — may be performed if the prostate weighs less than 2 oz (57 g).

Other transurethral procedures include vaporization of the prostate or a prostate incision with a scalpel or laser. Open surgical removal of the prostate is usually reserved for cancer of the prostate.

- Excretory urography may indicate urinary tract obstruction, hydronephrosis, calculi or tumors, and filling and emptying defects in the bladder.
- Elevated BUN and serum creatinine levels suggest impaired renal function.
- Urinalysis and urine culture show hematuria, pyuria, and UTI.
- Cystourethroscopy is performed when symptoms are severe to determine the best surgical procedure. It can show prostate enlargement, bladder wall changes, calculi, and a raised bladder.
- Prostate-specific antigen test rules out prostatic cancer.

Chronic renal failure

Chronic renal failure is usually the end result of gradual tissue destruction and loss of kidney function. Occasionally, however, it results from a rapidly progressing disease of sudden onset that destroys the nephrons and causes irreversible kidney damage.

Pathophysiology

Chronic renal failure often progresses through four stages. (See *Stages of chronic renal failure.*)

Chronic renal failure may result from:
• chronic glomerular disease such as glomerulonephritis, which affects the capillaries in the glomeruli
• chronic infections, such as chronic pyelonephritis and tuberculosis
• congenital anomalies such as polycystic kidney disease
• vascular diseases, such as hypertension and nephrosclerosis, which causes hardening of the kidneys
• obstructions such as kidney stones
• collagen diseases such as lupus erythematosus
• nephrotoxic agents such as long-term aminoglycoside therapy
• endocrine diseases such as diabetic neuropathy.

Nephrons can't compensate forever

Nephron damage is progressive. Once damaged, nephrons are no longer able to function. Healthy nephrons compensate for destroyed nephrons by enlarging and increasing their clearance capacity. The kidneys can maintain relatively normal function until about 75% of the nephrons are nonfunctional. Eventually, the healthy glomeruli are so overburdened they become sclerotic and stiff, leading to their destruction as well. If this condition continues unchecked, toxins accumulate and produce potentially fatal changes in all major organ systems.

Uh oh! Anemia, neuropathy, other problems

Even if the patient can tolerate life-sustaining maintenance dialysis or a kidney transplant, he may still have anemia, nervous system effects (peripheral neuropathy), cardiopulmonary and GI complications, sexual dysfunction, and skeletal defects.

What to look for

Few symptoms develop until more than 75% of glomerular filtration is lost. Then the remaining normal tissue deteriorates progressively. Symptoms worsen as kidney function decreases. Profound changes affect all body systems. The major findings include:
• hypervolemia (abnormal increase in plasma volume)
• hyperkalemia

Stages of chronic renal failure

Chronic renal failure may progress through the following stages:

reduced renal reserve. GFR is 35% to 50% of the normal rate.

renal insufficiency. GFR is 20% to 35% of the normal rate.

renal failure. GFR is 20% to 25% of the normal rate.

end-stage renal disease. GFR is less than 20% of the normal rate.

Healthy nephrons work hard to keep me going but, eventually, they're overwhelmed.

- hypocalcemia
- azotemia
- metabolic acidosis
- anemia
- peripheral neuropathy.

Other signs and symptoms, by body system, include:
- renal — dry mouth, fatigue, nausea, hypotension, loss of skin turgor, listlessness that may progress to somnolence and confusion, decreased or dilute urine, irregular pulses, edema
- cardiovascular — hypertension, irregular pulse, life-threatening arrhythmias, heart failure
- respiratory — infection, crackles, pleuritic pain
- GI — gum sores and bleeding, hiccups, metallic taste, anorexia, nausea, vomiting, ammonia smell to the breath, abdominal pain on palpation
- skin — pallid, yellowish-bronze color; dry, scaly skin; thin, brittle nails; dry, brittle hair that may change color and fall out easily; severe itching; white, flaky urea deposits (uremic frost) in critically ill patients
- neurologic — altered level of consciousness, muscle cramps and twitching, pain, burning, and itching in legs and feet (restless leg syndrome)
- endocrine — growth retardation in children, infertility, decreased libido, amenorrhea, impotence
- hematologic — GI bleeding and hemorrhage from body orifices, easy bruising
- musculoskeletal — fractures, bone and muscle pain, abnormal gait, impaired bone growth and bowed legs in children.

The progression of chronic renal failure can sometimes be slowed, but it's ultimately irreversible, culminating in end-stage renal disease. Although it's fatal without treatment, dialysis or a kidney transplant can sustain life. (See *Treating chronic renal failure,* page 360.)

What tests tell you

The following tests are used to diagnose chronic renal failure:
- Blood studies show decreased arterial pH and bicarbonate levels, low hemoglobin and hematocrit levels, and elevated BUN, serum creatinine, sodium, and potassium levels.
- Arterial blood gas analysis reveals metabolic acidosis.

Battling illness

Treating chronic renal failure

Treatment for chronic renal failure may consist of conservative measures, such as diet and fluid restrictions, drug therapy, or emergency interventions.

Conservative measures

Conservative treatment consists of:

- a low-protein diet to reduce end products of protein metabolism that the kidneys can't excrete
- a high-protein diet for patients on continuous peritoneal dialysis
- a high-calorie diet to prevent ketoacidosis (the accumulation of ketones, such as acetone, in the blood) and tissue atrophy
- sodium and potassium restrictions
- fluid restrictions to maintain fluid balance.

Drug therapy

The following drugs are used to treat chronic renal failure:

- loop diuretics such furosemide (if some renal function remains) to maintain fluid balance
- digitalis glycosides to mobilize fluids that cause edema
- antihypertensives to control blood pressure and edema
- antiemetics to relieve nausea and vomiting
- cimetidine or ranitidine to decrease gastric irritation
- methylcellulose or docusate to prevent constipation
- iron and folate supplements or red blood cell (RBC) infusion for anemia
- synthetic erythropoietin to stimulate the bone marrow to produce RBCs
- antipruritics, such as trimeprazine or diphenhydramine, to relieve itching
- aluminum hydroxide gel to lower serum phosphate levels
- supplementary vitamins, particularly B and D, and essential amino acids.

Yikes! It's an emergency

Potassium levels in the blood must be monitored to detect hyperkalemia. Emergency treatment includes dialysis therapy, oral or rectal administration of cation exchange resins such as sodium polystyrene sulfonate, and I.V. administration of calcium gluconate, sodium bicarbonate, 50% hypertonic glucose, and regular insulin.

Cardiac tamponade resulting from pericardial effusion may require emergency pericardiocentesis or surgery. Intensive dialysis and thoracentesis can relieve pulmonary edema and pleural effusion. Hemodialysis or peritoneal dialysis can help control end-stage renal disease.

- Urine specific gravity becomes fixed at 1.010; urinalysis may show proteinuria, glycosuria, RBCs, leukocytes, casts, or crystals, depending on the cause.
- X-ray studies, including KUB radiography, excretory urography, nephrotomography, renal scan, and renal arteriography, show reduced kidney size.
- Renal biopsy is used to identify underlying disease.
- EEG shows changes that indicate brain disease (metabolic encephalopathy).

Prostatitis

Prostatitis is an acute or chronic inflammation of the prostate gland. Acute prostatitis is usually due to bacterial invasion. Chronic prostate inflammation is the most common cause of repeated UTI in men.

Chronic prostatitis affects up to 35% of men over age 50.

Pathophysiology

About 80% of bacterial prostatitis cases result from *Escherichia coli* infection. The remaining 20% results from infection by *Klebsiella, Enterobacter, Proteus, Pseudomonas, Serratia, Streptococcus, Staphylococcus,* and diphtheroids, which are contaminants from normal flora of the urethra.

These organisms probably spread to the prostate gland by one of these four methods:

Bacteria may spread to the prostate in the bloodstream, by moving up from the urethra, from the rectum, or in a backup of infected bladder urine in the prostatic ducts.

☝ through the bloodstream

✌ from ascending urethral infection

🤟 from invasion by rectal bacteria through the lymphatic vessels

🖐 from reflux of infected urine from the bladder into prostatic ducts.

Chronic prostatitis is usually caused by bacterial invasion from the urethra. Less common means of acute or chronic infection are urethral procedures performed with instruments, such as cystoscopy and catheterization, and infrequent or excessive sexual intercourse.

Acute, chronic, or sans infection

Types of prostatitis include:
• acute bacterial prostatitis, an ascending infection of the urinary tract
• chronic bacterial prostatitis, marked by recurrent UTI and persistent pathogenic bacteria
• prostatic inflammation without infection, the most common type of prostatitis.

UTI and beyond

UTI is the most common complication of prostatitis. An untreated infection can progress to these other problems:
• prostatic abscess
• acute urine retention from prostatic edema

• inflammation of the kidney (pyelonephritis)
• inflammation of the epididymis, where spermatozoa are stored (epididymitis).

What to look for

Signs and symptoms vary according to the type of prostatitis.

Acute bacterial prostatitis

In acute bacterial prostatitis, the patient has sudden onset of fever, chills, low back pain, muscle pain (myalgia), pelvic area (perineal) fullness, joint pain (arthralgia), urinary urgency and frequency, cloudy urine, painful urination (dysuria), nocturia, and transient erectile dysfunction.

Some degree of urinary obstruction also may occur. The bladder may feel distended when palpated. When palpated rectally, the prostate is tender, abnormally hard, swollen, and warm. (See *Treating prostatitis*.)

Chronic bacterial prostatitis

Chronic bacterial prostatitis may develop from acute prostatitis that doesn't clear up with antibiotics. Some patients are symptom-free, but usually the same signs and symptoms as in the acute form are present, although to a lesser degree.

Other signs and symptoms may include urethral discharge and painful ejaculation leading to sexual dysfunction. The prostate may feel soft, and a dry, crackling sound or sensation (crepitation) may be evident on palpation if prostatic calculi are present.

Nonbacterial prostatitis

The patient with nonbacterial prostatitis usually complains of dysuria, mild perineal or low back pain, and nocturia. The patient may experience pain on ejaculation. The prostate gland usually feels normal upon palpation.

What tests tell you

The following tests confirm the diagnosis of prostatitis:
• X-ray of the pelvis may show prostatic calculi.
• Smears of prostatic secretions reveal inflammatory cells but often no causative organism in nonbacterial prostatitis.
• Urodynamic evaluation may reveal detrusor muscle hyperreflexia and pelvic floor myalgia from chronic spasms.

Battling illness

Treating prostatitis

Treatment for prostatitis may consist of drug therapy, surgery, or supportive measures.

Drug therapy
The following drugs are used to treat prostatitis:
• systemic antibiotics for acute prostatitis
• aminoglycosides in combination with penicillins or cephalosporins in severe cases
• co-trimoxazole to prevent chronic prostatitis
• co-trimoxazole, carbenicillin, nitrofurantoin, erythromycin, or tetracycline for *Escherichia coli*
• anticholinergics, analgesics, and minocycline, doxycycline, or erythromycin for nonbacterial prostatitis

Supportive therapy
Supportive therapy includes bed rest, plenty of fluids, sitz baths, and stool softeners. Ejaculation or regular sexual intercourse using a condom to promote drainage of prostatic secretions is pre-scribed for chronic prostatitis. Regular prostatic massage for several weeks or months may also be effective.

Surgery
If drug therapy is unsuccessful, transurethral resection of the prostate may be done. This proce-dure may lead to retrograde ejaculation and sterility, so it's usually not done on young men. Total prostatectomy is curative but may cause impotence and incontinence.

A firm diagnosis also depends on urine cultures, rectal examinations, and cultures for bacterial growth.

Renal calculi

Renal calculi, or kidney stones, may form anywhere in the urinary tract, but they usually develop in the renal pelvis or calices. Calculi form when substances that normally dissolve in the urine precipitate. They vary in size, shape, and number. (See *A close look at renal calculi,* page 364.)

About 1 in 1,000 Americans require hospitalization at some time for renal calculi. They're more common in men than women and rare in blacks and children.

Stones are formed from calcium, phosphate, or urate compounds that are normally dissolved in the urine.

Pathophysiology

Renal calculi are particularly prevalent in specific geographic areas such as the southeastern United States. Although their exact cause is unknown, predisposing factors include:

• *dehydration.* Decreased water and urine excretion concentrates calculus-forming substances.

• *infection.* Infected, scarred tissue provides a site for calculus development. Calculi may become infected if bacteria are the nucleus in calculi formation. Calculi that result from *Proteus* infections may lead to destruction of kidney tissue.

• *changes in urine pH.* Consistently acidic or alkaline urine provides a favorable medium for calculus formation.

• *obstruction.* Urinary stasis allows calculus constituents to collect and adhere, forming calculi. Obstruction also encourages infection, which compounds the obstruction.

This list of causes leaves no stone unturned!

A close look at renal calculi

Renal calculi vary in size and type. Small calculi may remain in the renal pelvis or pass down the ureter. A staghorn calculus is a cast of the innermost part of the kidney — the calyx and renal pelvis. A staghorn calculus may develop from a calculus that stays in the kidney.

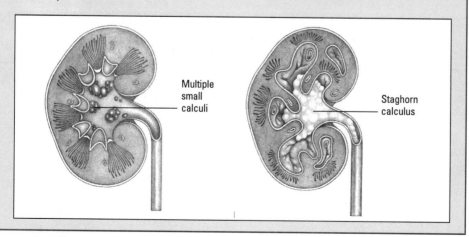

Multiple small calculi

Staghorn calculus

• *immobilization.* Immobility from spinal cord injury or other disorders allows calcium to be released into the circulation and, eventually, to be filtered by the kidneys.

• *diet.* Increased intake of calcium or oxalate-rich foods encourages calculi formation.

• *metabolic factors.* Hyperparathyroidism, renal tubular acidosis, elevated uric acid (usually with gout), defective oxalate metabolism, a genetic defect in cystine metabolism, and excessive intake of vitamin D or dietary calcium may predispose a person to renal calculi.

A delicate balance breaks down

Renal calculi usually arise because the delicate excretory balance breaks down. Here's how it happens:

• Urine becomes concentrated with insoluble materials.

• Crystals form from these materials and then consolidate, forming calculi. These calculi contain an organic mucoprotein framework and crystalloids, such as calcium, oxalate, phosphate, urate, uric acid, struvite, cystine, and xanthine.

They're called kidney stones...

...but they can form anywhere in the urinary tract.

• Mucoprotein is reabsorbed by the tubules, establishing a site for calculi formation.

• Calculi remain in the renal pelvis and damage or destroy kidney tissue or they enter the ureter.

• Large calculi in the kidneys may cause tissue damage (pressure necrosis).

• In certain locations, calculi obstruct urine, which collects in the renal pelvis (hydronephrosis). These calculi also tend to recur. Intractable pain and serious bleeding can result.

• Initially, hydrostatic pressure increases in the collection system near the obstruction, forcing nearby renal struc-

tures to dilate as well. The farther the obstruction is from the kidney, the less serious the dilation because the pressure is diffused over a larger surface area.
• With a complete obstruction, pressure in the renal pelvis and tubules increases, the GFR falls, and a disruption occurs in the junctional complexes between tubular cells. If left untreated, tubular atrophy and destruction of the medulla leave connective tissue in place of the glomeruli, causing irreversible damage.

> Ouch! The pain from stones can be excruciating!

What to look for

The key symptom of renal calculi is severe pain, which usually occurs when large calculi obstruct the opening of the ureter and increase the frequency and force of peristaltic contractions. Pain may travel from the lower back to the sides and then to the pubic region and external genitalia. Pain intensity fluctuates and may be excruciating at its peak.

The patient with calculi in the renal pelvis and calices may complain of more constant, dull pain. He also may report back pain from obstruction within a kidney and severe abdominal pain from calculi traveling down a ureter. Severe pain is typically accompanied by nausea, vomiting and, possibly, fever and chills.

Other signs and symptoms include:
• hematuria when stones abrade a ureter
• abdominal distention
• oliguria from an obstruction in urine flow.

Most small calculi can be flushed out of a person's system by drinking lots of fluids. (See *Treating renal calculi*.)

What tests tell you

Diagnosis is based on clinical features and the following tests:
• KUB radiography reveals most renal calculi.
• Excretory urography helps confirm the diagnosis and shows the size and location of calculi.
• Kidney ultrasonography is easy to perform, noninvasive, and nontoxic and detects obstructions not seen on the KUB radiography.

Battling illness

Treating renal calculi

Ninety percent of renal calculi are smaller than 5 mm in diameter and may pass naturally with vigorous hydration (more than 3 L/day). Other treatments may include drug therapy for infection or other effects of illness and measures to prevent recurrence of calculi. If calculi are too large for natural passage, they may be removed by surgery or other means.

Drug therapy
Drug therapy may include:
• antimicrobial agents for infection (varying with the cultured organism)
• analgesics, such as meperidine and morphine, for pain
• diuretics to prevent urinary stasis and further calculi formation
• thiazides to decrease calcium excretion into the urine
• methenamine mandelate to suppress calculi formation when infection is present.

Preventive measures
Measures to prevent recurrence of renal calculi include:

• a low-calcium ion oxalate diet
• oxalate-binding cholestyramine for absorptive hypercalciuria
• parathyroidectomy for hyperparathyroidism
• allopurinol for uric acid calculi
• daily oral doses of ascorbic acid to acidify the urine.

Removing calculi
Calculi lodged in the ureter may be removed by inserting a cystoscope through the urethra and then manipulating the calculi with catheters or retrieval instruments. A flank or lower abdominal approach may be needed to extract calculi from other areas, such as the kidney calyx or renal pelvis. Percutaneous ultrasonic lithotripsy and extracorporeal shock wave lithotripsy shatter the calculi into fragments for removal by suction or natural passage.

• Urinalysis may indicate pus in the urine (pyuria), a sign of UTI.
• 24-hour urine collection reveals calcium oxalate, phosphorus, and uric acid levels. Three separate collections, along with blood samples, are needed for accurate testing.
• Calculus analysis shows mineral content.

These tests may identify the cause of calculus formation:
• Blood calcium and phosphorus levels detect hyperparathyroidism and show an increased calcium level in proportion to normal serum protein levels.
• Blood protein levels determine the level of free calcium unbound to protein.

• Differential diagnosis rules out appendicitis, cholecystitis, peptic ulcer, and pancreatitis as sources of pain.

Quick quiz

1. The most accurate measurement of glomerular filtration is:

    A. blood pressure.
    B. intake and output.
    C. creatinine clearance.

Answer: C. This is because creatinine is only filtered by the glomeruli and not reabsorbed by the tubules.

2. Prerenal failure results from:

    A. bilateral obstruction of urine outflow.
    B. conditions that diminish blood flow to the kidneys.
    C. damage to the kidneys themselves.

Answer: B. One such condition, prerenal azotemia, accounts for between 40% and 80% of all cases of acute renal failure.

3. The main complication of benign prostatic hyperplasia is:

    A. urinary obstruction.
    B. pancreatitis.
    C. prostatitis.

Answer: A. Depending on the size of the enlarged prostate and resulting complications, the obstruction may be treated surgically or symptomatically.

Scoring

 If you answered all three items correctly, wow! Your ability to concentrate, absorb, and secrete data about the kidneys is amazing!

    If you answered one or two correctly, fantastic! Your knowledge of things renal is not venal.

    If you answered one or none correctly, don't freak out. We recommend reviewing the chapter once more. After all, that's what reabsorption is all about.

I don't know which is more painful—a stone in my calices or listening to these puns!

Musculoskeletal system

Just the facts

In this chapter you'll learn:

♦ about the function of the musculoskeletal system

♦ about bone formation, growth, and renewal

♦ pathophysiology, diagnostic tests, and treatment for two musculoskeletal disorders.

Understanding the musculoskeletal system

The structures of the musculoskeletal work together to provide support and produce movement. This system consists of:
- muscles
- bones
- cartilage
- joints
- bursae
- tendons
- ligaments.

These structures give the human body its shape and ability to move.

Muscles

There are three major muscle types:

  skeletal (voluntary, striated muscles)

  visceral (involuntary, smooth muscles)

  cardiac.

A close look at skeletal muscles

The human body has about 600 skeletal muscles. Each muscle is classified by the kind of movement for which it's responsible. For example, flexors permit the bending of joints, or flexion; abductors permit shortening so that joints are straightened or abducted. The illustrations below show some of the major muscles. The illustration on the next page shows interior structures of the muscles.

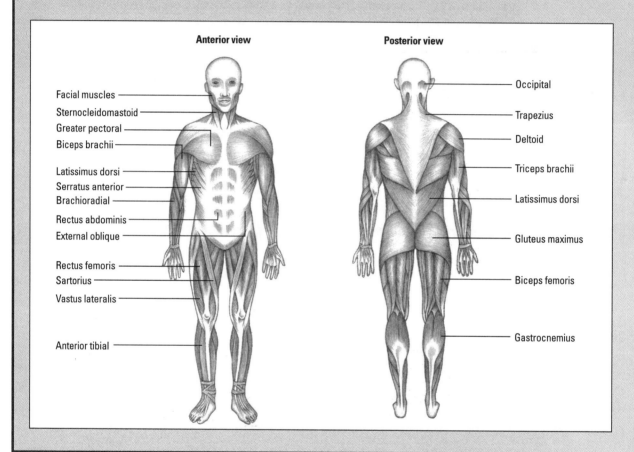

Anterior view

- Facial muscles
- Sternocleidomastoid
- Greater pectoral
- Biceps brachii
- Latissimus dorsi
- Serratus anterior
- Brachioradial
- Rectus abdominis
- External oblique
- Rectus femoris
- Sartorius
- Vastus lateralis
- Anterior tibial

Posterior view

- Occipital
- Trapezius
- Deltoid
- Triceps brachii
- Latissimus dorsi
- Gluteus maximus
- Biceps femoris
- Gastrocnemius

This chapter focuses on skeletal muscle, which is attached to bone. Skeletal muscle cells are arranged in long bands or strips, called striations. Skeletal muscle is voluntary, meaning it can be contracted at will. (See *A close look at skeletal muscles*.)

Muscle develops when existing muscle fibers grow. Exercise, nutrition, sex, and genetic constitution account for variations in muscle strength and size among individuals.

Muscle structure

Each muscle contains cell groups called muscle fibers that extend the length of the muscle. A sheath of connective tissues—called the perimysium—binds the fibers into a bundle, or fasciculus. A stronger sheath, the epimysium, binds fasciculi together to form the fleshy part of the muscle. Extending beyond the muscle, the epimysium becomes a tendon.

Each muscle fiber is surrounded by a plasma membrane, the sarcolemma. Within the sarcoplasm (or cytoplasm) of the muscle fiber lie tiny myofibrils. Arranged lengthwise, myofibrils contain still finer fibers, called thick fibers and thin fibers.

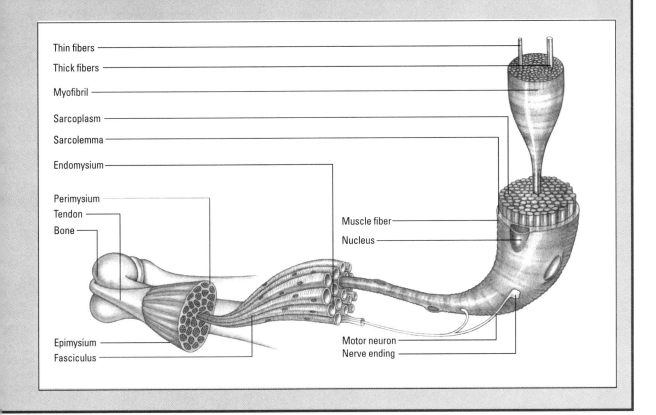

- Thin fibers
- Thick fibers
- Myofibril
- Sarcoplasm
- Sarcolemma
- Endomysium
- Perimysium
- Tendon
- Bone
- Muscle fiber
- Nucleus
- Epimysium
- Fasciculus
- Motor neuron
- Nerve ending

Bones

There are 206 separate bones in the human body. (See *A close look at the bones,* page 372.)

Bones are classified by shape and location:
- long, such as arm and leg bones (the humerus, radius, femur, tibia, ulna, and fibula)
- short, such as wrist and ankle bones (the carpals and tarsals)

A close look at the bones

The human skeleton contains 206 bones; 80 form the axial skeleton and 126 form the appendicular skeleton. The illustration below shows some of the major bones and bone groups. The illustration on the next page depicts the interior structure of a bone.

Bone consists of layers of calcified matrix containing spaces occupied by osteocytes (bone cells). Bone layers (lamellae) are arranged concentrically about central canals (haversian canals). Small cavities (lacunae) lying between the lamellae contain osteocytes. Tiny canals (canaliculi) connect the lacunae. These canals form the structural units of bone and provide nutrients to bone tissue.

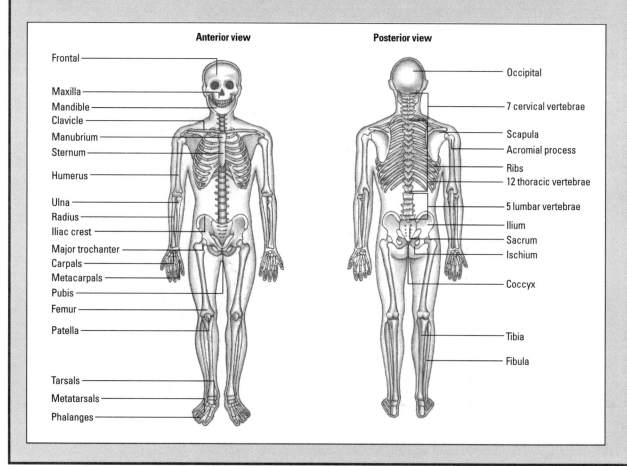

Anterior view

Frontal
Maxilla
Mandible
Clavicle
Manubrium
Sternum
Humerus
Ulna
Radius
Iliac crest
Major trochanter
Carpals
Metacarpals
Pubis
Femur
Patella
Tarsals
Metatarsals
Phalanges

Posterior view

Occipital
7 cervical vertebrae
Scapula
Acromial process
Ribs
12 thoracic vertebrae
5 lumbar vertebrae
Ilium
Sacrum
Ischium
Coccyx
Tibia
Fibula

• flat, such as the shoulder blade (scapula), ribs, and skull
• irregular, such as bones of the vertebrae and jaw (mandible)
• sesamoid, such as the kneecap (patella).
    Bones of the head and trunk — called the axial skeleton — include the:

Just your average long bone

A typical long bone has a diaphysis (main shaft) and an epiphysis (end). The epiphyses are separated from the diaphysis with cartilage at the epiphyseal line. Beneath the epiphyseal articular surface lies the articular cartilage, which cushions the joint.

Getting beneath the bone surface

Each bone consists of an outer layer of dense compact bone containing haversian systems and an inner layer of spongy (cancellous) bone composed of thin plates, called trabeculae, that interlace to form a latticework. Red marrow fills the spaces between the trabeculae of some bones. Cancellous bone doesn't contain haversian systems.

Compact bone is located in the diaphyses of long bones and the outer layers of short, flat, and irregular bones. Cancellous bone fills central regions of the epiphyses and the inner portions of short, flat, and irregular bones. Periosteum — specialized fibrous connective tissue — consists of an outer fibrous layer and an inner bone-forming layer. Endosteum (a membrane that contains osteoblast producing cells) lines the medullary cavity (inner surface of bone, which contains the marrow).

Blood reaches bone by way of arterioles in haversian canals; vessels in Volkmann's canals, which enter bone matrix from the periosteum; and vessels in the bone ends and within the marrow. In children, the periosteum is thicker than in adults and has an increased blood supply to assist new bone formation around the shaft.

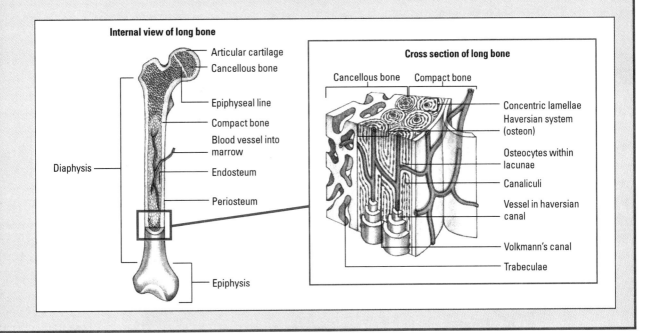

- facial and cranial bones
- hyoid bone (a u-shaped bone at the base of the tongue, beneath the thyroid cartilage)
- vertebrae
- ribs

• breast bone (sternum).

Bones of the extremities — called the appendicular skeleton — include the:

• collarbone (clavicle)
• scapula
• humerus
• radius
• ulna
• hand bones (metacarpals)
• pelvic bone
• femur
• patella
• fibula
• tibia
• foot bones (metatarsals).

Bones at work

Bones perform the following mechanical and physiologic functions:

• protecting internal tissues and organs (for example, the 33 vertebrae that surround and protect the spinal cord)
• stabilizing and supporting the body
• providing a surface for muscle, ligament, and tendon attachment
• moving, through lever action, when contracted
• producing red blood cells in the bone marrow (hematopoiesis)
• storing mineral salts (for example, approximately 99% of the body's calcium).

Cartilage and bone formation

Bones begin as cartilage. Cartilage is a growing fibrous or elastic tissue that hardens to form bone. At 3 months' gestation, cartilage makes up the fetal skeleton. At about 6 months' gestation, the fetal cartilage hardens (ossifies) into bony skeleton. The process whereby cartilage hardens into bone is called endochondral ossification. Some bones — especially those of the wrists and ankles — don't ossify until after a baby's birth.

> The fontanels of the skull are incomplete at birth. These "soft spots" allow for compression during birth.

In endochondral ossification, bone-forming cells produce a collagenous material called osteoid that hardens. Note that endochondral means "occurring within cartilage."

Was that an osteoblast or an osteoclast?

Bone-forming cells called osteoblasts deposit new bone. In other words, osteoblastic activity results in bone formation.

Large cells called osteoclasts reabsorb material from previously formed bones, tearing down old or excess bone structure and allowing osteoblasts to rebuild new bone (a process called resorption). Osteoblastic and osteoclastic activity promotes longitudinal bone growth.

Estrogen levels are linked to calcium uptake and to rebuilding of bone tissue.

It happens during adolescence

Longitudinal bone growth continues until adolescence, when bones stop lengthening. During adolescence, the epiphyseal growth plates located at bone ends close.

A lifelong process

Osteoblasts and osteoclasts are responsible for remodeling — the continuous creation and destruction of bone within the body. When osteoblasts complete their bone-forming function and are located within the mineralized bone matrix, they transform themselves into osteocytes (mature bone cells). Bone renewal continues throughout life, though it slows down with age.

Estrogen plays its part

Researchers have found that estrogen secretion plays a role in calcium uptake and release and helps regulate osteoblastic activity (bone formation). Decreased estrogen levels have been linked to decreased osteoblastic activity.

Let's talk about sex, race, and age

A patient's sex, race, and age also influence bone. They affect:
- bone mass
- bone's structural ability to withstand stress
- bone loss.

Men commonly have denser bones than women and blacks commonly have denser bones than whites. Bone density and structural integrity decrease after age 30 in women and age 45 in men. Thereafter, bone density and strength tend to continually decline at a more or less steady rate.

Everything you ever wanted to know about cartilage

Cartilage is a dense connective tissue consisting of fibers embedded in a strong, gel-like substance. Cartilage is avascular (bloodless) and is not innervated. Cartilage supports, cushions, and shapes body structures. It may be fibrous, hyaline, or elastic:

• Fibrous cartilage forms the symphysis pubis and the intervertebral disks. This type of cartilage provides good cushioning and strength.

• Hyaline cartilage covers the articular bone surfaces (where one or more bones meet at a joint). It also appears in the trachea, bronchi, and nasal septum and covers the entire skeleton of the fetus. This type of cartilage cushions against shock.

• Elastic cartilage is located in the auditory canal, the external ear, and the epiglottis. It provides support with flexibility.

Cartilage cushions and absorbs shock, preventing direct transmission to the bone.

Joints

The union of two or more bones is called a joint. The body contains three major types of joints, classified by how much movement they allow:

☝ synarthrosis joints, which permit no movement; for example, joints between bones in the skull

✌ amphiarthrosis joints, which allow slight movement; for example, joints between the vertebrae

🤟 diarthrosis joints, which permit free movement; for example, the ankle, wrist, knee, hip, and shoulder.

Joints are further classified by shape and by connective structure, such as fibrous, cartilaginous, and synovial.

Don't forget these important structures

In a free-moving joint, a fluid-filled space known as the joint space exists between the bones. The synovial membrane, which lines this cavity, secretes a viscous lubricating substance called synovial fluid, which allows two bones to move against one another. Ligaments, tendons, and muscles help to stabilize the joint.

Bursae (small sacs of synovial fluid) are located at friction points around joints and between tendons, ligaments, and bones. In joints like the shoulder and knee, they act as cushions, easing stress on adjacent structures.

Tendons are bands of fibrous connective tissue that attach muscles to the fibrous membrane that covers the bones (the periosteum). Tendons enable bones to move when skeletal muscles contract.

Ligaments are dense, strong, flexible bands of fibrous connective tissue that tie bones to other bones. Ligaments that connect the joint ends of bones either limit or facilitate movement. They also provide stability.

Joints get additional support from ligaments, tendons, and muscles.

Movement

Skeletal movement results primarily from muscle contractions, although other musculoskeletal structures also play a role. Here's a general description of how body movement takes place:

• Skeletal muscle is loaded with blood vessels and nerves. To contract, it needs an impulse from the nervous system and oxygen and nutrients from the blood.

• When a skeletal muscle contracts, force is applied to the tendon.

• The force pulls one bone toward, away from, or around a second bone, depending on the type of muscle contracted and the type of joint involved.

Usually, one bone moves less than the other. The muscle tendon attachment to the more stationary bone is called the origin. The muscle tendon attachment to the more movable bone is called the insertion site.

Thirteen options

There are 13 angular and circular musculoskeletal movements. They are:

- circumduction — moving in a circular manner
- flexion — bending, decreasing the joint angle
- extension — straightening, increasing the joint angle
- internal rotation — turning toward midline
- external rotation — turning away from midline
- abduction — moving away from midline
- adduction — moving toward midline
- supination — turning upward
- pronation — turning downward
- eversion — turning outward
- inversion — turning inward
- retraction — moving backward
- protraction — moving forward.

Most movement involves groups of muscles rather than one muscle. Most skeletal movement is mechanical; the bones act as levers and the joints act as fulcrums (points of support for movement of the bones).

Musculoskeletal disorders

The musculoskeletal disorders discussed in this chapter are gout and osteoporosis.

Gout

A metabolic disease, gout is marked by red, swollen, and acutely painful joints. Gout may affect any joint but is found mostly in those of the feet, especially the great toe, ankle, and midfoot.

Pathophysiology

Primary gout usually occurs in men over age 30 and in postmenopausal women who take diuretics. It follows an intermittent course. Between attacks, patients may be symptom-free for years.

The underlying cause of primary gout is unknown. In many patients it results from decreased excretion of uric acid by the kidneys. In a few patients, gout is linked to a genetic defect that causes overproduction of uric acid. This is called hyperuricemia.

Secondary gout may develop in the wake of another disease, such as obesity, diabetes mellitus, high blood pressure, leukemia and other blood disorders, bone cancer, and kidney disease. Secondary gout can also follow treatment with certain drugs, such as hydrochlorothiazid or pyrazinamide.

Gout is marked by localized deposits of uric acid salts...

A painful progression

Left untreated, gout progresses in four stages:

During the first stage, the patient develops asymptomatic hyperuricemia (urate levels rise but don't produce symptoms).

...these deposits cause painfully arthritic joints.

The second stage is marked by acute gouty arthritis. During this time, the patient experiences painful swelling and tenderness. Symptoms usually lead the patient to seek medical attention.

The third stage, the interictal stage, may last for months to years. The patient may be asymptomatic or may experience exacerbations.

The fourth stage is the chronic stage. Without treatment, urate pooling may continue for years. Tophi (clusters of urate crystals) may develop in cartilage, synovial membranes, tendons, and soft tissues. This final unremitting stage of the disease is also known as tophaceous gout.

The tale of the tophi

Tophi are clusters of urate crystals generally surrounded by inflamed tissue. This can cause deformity and destruction of hard and soft tissues. In joints, tophi lead to destruction of cartilage and bone as well as other degeneration. (See *A close look at gout tophi*, page 380.)

Tophi form in diverse areas:
- hands
- knees
- feet
- outer sides of the forearms
- pinna of the ear
- Achilles tendon.

Rarely, internal organs such as the kidneys and heart may be affected. Kidney involvement may cause kidney dysfunction.

What's the fuss about hyperuricemia?

When urate crystals are deposited in tissues, a gout attack strikes.

Urates are uric acid salts. They predominate in the plasma, fluid around the cells, and synovial fluid. Hyperuricemia, the hallmark of gout, is a plasma urate concentration greater than 420 μmol/L (7 mg/dl). Hyperuricemia indicates increased total-body urate. Excess urates result from:
• increased urate production
• decreased excretion of uric acid
• a combination of increased production and decreased excretion.

With hyperuricemia, plasma and extracellular fluids are supersaturated with urate. This leads to urate crystal formation. When crystals are deposited in other tissues, a gout attack strikes.

Provoking factors

Many different factors may provoke an acute attack of gout. Examples include:
• stress
• trauma
• infection
• hospitalization
• surgery
• starvation
• weight reduction
• excessive food intake
• use of alcohol and some medications.

Relief in sight

A sudden increase in serum urate may cause new crystals to form. However, a drop in serum and extracellular urate concentrations may cause previously formed crystals to partially dissolve and be excreted.

What to look for

In some patients, serum urate levels increase but produce no symptoms. In symptom-producing gout, the first acute attack strikes suddenly and peaks quickly.

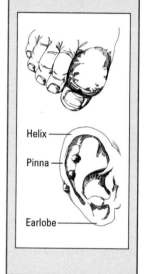

A close look at gout tophi

In advanced gout, urate crystal deposits develop into hard, irregular, yellow-white nodules called tophi. These bumps commonly protrude from the great toe and the pinna.

Helix

Pinna

Earlobe

Although it may involve only one or a few joints, the first acute attack causes extreme pain. Untreated gout attacks usually resolve in 10 to 14 days. Mild acute attacks usually subside quickly but tend to recur at irregular intervals. Severe attacks may last for days or weeks.

Gout patients usually enjoy symptom-free intervals between attacks. Most patients have a second attack between 6 months and 2 years after the first; in some patients, however, the second attack is delayed for 5 to 10 years. Delayed attacks, which may involve several joints, are more common in untreated patients. These attacks tend to last longer and produce more symptoms than initial episodes.

Chronic gout is marked by numerous, persistently painful joints. With tophi development, joints are typically swollen and dusky red or purple, with limited movement. The actual tophi — hard, irregular, yellow-white nodules — may be seen, especially on the ears, hands, and feet.

Late in the chronic stage of gout, the skin over the tophi may ulcerate and release a chalky white exudate or pus. Chronic inflammation and tophi lead to joint degeneration; deformity and disability may develop.

Warmth and extreme tenderness may be felt over the joint. The patient may have a sedentary lifestyle and a history of hypertension and kidney stones. The patient may wake during the night with pain in the great toe or other part of the foot.

Initially moderate pain may grow so intense that eventually the patient can't bear the weight of bed linens or the vibrations of a person walking across the room. He may report chills and a mild fever.

Complications

Potential complications include kidney disorders such as renal calculi, infection that develops with tophi rupture and nerve damage, and circulatory problems, such as atherosclerotic disease, cardiovascular lesions, cerebrovascular accident, coronary thrombosis, and hypertension.

Patients who receive treatment for gout have a good prognosis. (See *Treating gout,* page 382.)

What tests tell you

The following tests aid in the diagnosis of gout:

Battling illness

Treating gout

Treatment for gout varies depending on whether gout is acute or chronic. Dietary restrictions and weight loss may also be a part of care.

Treatment for gout has three goals:

 terminate the acute attack

 reduce uric acid levels

prevent recurrent gout and renal calculi.

Acute attacks
Treatment of an acute attack consists of the following:
• bed rest and elevation of the extremity
• immobilization and protection of the inflamed, painful joints
• local application of cold.

A bed cradle can be used to keep bed linens off sensitive, inflamed joints. Analgesics, such as acetaminophen, relieve the pain of mild attacks. Acute inflammation, however, requires nonsteroidal anti-inflammatory drugs or intramuscular corticotropin.

Colchicine or corticosteroids are occasionally used to treat acute attacks, although they don't affect uric acid levels. The patient should also drink plenty of fluids (at least 2 L a day) to help prevent renal calculi.

Acute gout can attack 24 to 96 hours after surgery; even minor surgery can trigger an attack. Colchicine may be administered before and after surgery to help prevent gout attacks.

Chronic gout
To treat chronic gout, serum uric acid levels are reduced to less than 6.5 mg/dl using various medications, depending on whether the patient over- or underproduces uric acid. If he overproduces uric acid, he may be given allopurinol. If he underproduces uric acid, he may be treated with probenecid or sulfinpyrazone. Serum uric acid levels are monitored regularly, and sodium bicarbonate or other agents may be given to alkalinize the patient's urine.

Adjunctive therapy
Adjunctive therapy emphasizes:
• avoiding alcohol (especially beer and wine)
• avoiding purine-rich foods (which raise urate levels), such as anchovies, liver, sardines, kidneys, sweetbreads, and lentils.

Weight loss
Obese patients should begin a weight-loss program because weight reduction decreases uric acid levels and stress on painful joints as well. To diffuse anxiety and promote coping, the patient should be encouraged to express his concerns about his condition.

• Needle aspiration of synovial fluid (called arthrocentesis) or of tophi for microscopic examination reveals needlelike crystals of sodium urate in the cells and establishes the diagnosis. If test results identify calcium pyrophosphate crystals, the patient probably has pseudogout, a disease similar to gout.
• Blood and urine analysis are used to determine serum and urine uric acid levels.
• X-ray studies initially produce normal results. However, in chronic gout, X-rays show damage to cartilage and bone. Outward displacement of the overhanging margin from the bone contour characterizes gout.

Osteoporosis

In this metabolic bone disorder, the rate of bone resorption accelerates and the rate of bone formation decelerates. The result is decreased bone mass. Bones affected by this disease lose calcium and phosphate and become porous, brittle, and abnormally prone to fracture.

In osteoporosis, bones deteriorate faster than the body can replace them.

Pathophysiology

Osteoporosis may be a primary disorder or occur secondary to an underlying disease.

Primary osteoporosis

Bones lose calcium and phosphate salts and become porous, brittle, and prone to fracture.

The cause of primary osteoporosis is unknown. However, the following are contributing factors:
• mild but prolonged lack of calcium due to poor dietary intake or poor absorption by the intestine secondary to age
• hormonal imbalance due to endocrine dysfunction
• faulty metabolism of protein due to estrogen deficiency
• a sedentary lifestyle.

Primary osteoporosis is classified as one of three types:

☝ postmenopausal osteoporosis (type I), usually affecting women ages 51 to 75; related to the loss of estrogen and its protective effect on bone; characterized by vertebral and wrist fractures

✌ senile osteoporosis (type II), occurring mostly between ages 70 and 85; related to osteoblast or osteoclast shrinkage or decreased physical activity; characterized by fractures of the humerus, tibia, femur, and pelvis

🖖 premenopausal (type III), involving higher estrogen levels that may inhibit bone resorption by affecting the sensitivity of osteoclasts to parathyroid hormone.

Secondary osteoporosis

Secondary osteoporosis may result from:
• prolonged therapy with steroids or heparin
• bone immobilization or disuse (such as paralysis)
• alcoholism
• malnutrition
• rheumatoid arthritis

Osteoporosis may be age-related or linked to an underlying disease.

- liver disease
- malabsorption of calcium
- scurvy
- lactose intolerance
- hyperthyroidism
- osteogenesis imperfecta (an inherited condition that causes brittle bones)
- trauma leading to atrophy in the hands and feet, with recurring attacks (Sudeck's disease).

Trabeculae in trouble

Osteoporosis is characterized by a reduction in the bone matrix and in remineralization, resulting in soft bones that fracture easily. Bone mass is lost because of an imbalance between bone resorption and formation.

Recall that cancellous bone is the inner layer of spongy bone. Cancellous bone is composed of trabeculae, sharp, needlelike structures forming a meshwork of interconnecting spaces. Trabeculae have a larger surface volume than compact bone (the outer layer of dense bone) and therefore are lost more rapidly as bone mass decreases. This loss leads to fractures.

 ## What to look for

Bone fractures are the major complication of osteoporosis. Fractures occur mostly in the vertebrae, the femur, and the distal radius. Signs of redness, warmth, and new sites of pain may indicate new fractures.

The patient is typically postmenopausal or has one of the conditions that cause secondary osteoporosis. She may report that she heard a snapping sound and felt a sudden pain in her lower back when she bent down to lift something. She may say that the pain developed slowly over several years. If the patient has vertebral collapse, she may describe a backache and pain radiating around the trunk. Movement or jarring aggravates the pain.

The patient may have a humped back (dowager's hump); the curvature worsens as repeated vertebral fractures increase spinal curvature. The abdomen eventually protrudes to compensate for the changed center of gravity. The patient often reports a gradual loss of height, decreased exercise tolerance, and trouble breathing. Palpation may reveal muscle spasm. The patient may also have decreased spinal movement, with flexion more limited

Height loss in osteoporosis

Usually, a patient with osteoporosis loses height gradually. Height may be reduced as much as 7″.

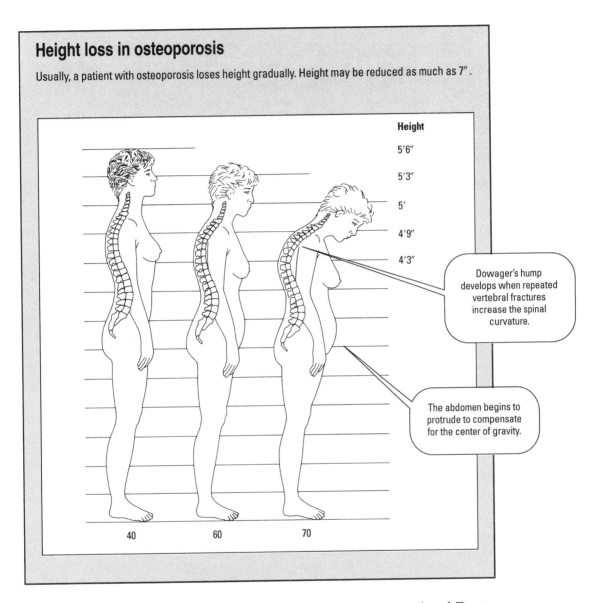

Height

5′6″

5′3″

5′

4′9″

4′3″

Dowager's hump develops when repeated vertebral fractures increase the spinal curvature.

The abdomen begins to protrude to compensate for the center of gravity.

40   60   70

than extension. (See *Height loss in osteoporosis* and *Treating osteoporosis,* page 386.)

What tests tell you

A diagnosis excludes other causes of bone disease, especially those that affect the spine, such as cancer or tumors.

Battling illness

Treating osteoporosis

Treatment focuses on a physical therapy program of gentle exercise and activity and drug therapy to slow the disease's progress.

Goals of treatment
Care seeks to:
• control bone loss
• prevent fractures
• control pain.

How goals are achieved
Measures may include supportive devices, such as a back brace, and, possibly, surgery to correct fractures. Estrogen may be prescribed within 3 years after menopause to decrease the rate of bone resorption. A balanced diet should be rich in nutrients, such as vitamin D, calcium, and protein, that support skeletal metabolism.

Analgesics and heat may be administered to relieve pain. The following drugs may be prescribed:
• Sodium fluoride may be given to stimulate bone formation.
• Calcium and vitamin D supplements may help to support normal bone metabolism.
• Calcitonin may be used to reduce bone resorption and slow the decline in bone mass.
• Etidronate is the first agent proved to increase bone density and restore lost bone.

Keep it safe
Safety precautions include keeping side rails up on the patient's bed and moving the patient gently and carefully at all times.

Be sure to discuss with ancillary hospital personnel how easily an osteoporotic patient's bones can fracture.

• X-ray studies show characteristic degeneration in the lower vertebrae. Loss of bone mineral appears in later disease.
• Serum calcium, phosphorus, and alkaline phosphatase levels remain within normal limits; parathyroid hormone levels may be elevated.
• Bone biopsy allows direct examination of changes in bone cells.

• Computed tomography scan allows accurate assessment of spinal bone loss.
• Radionuclide bone scans display injured or diseased areas as darker portions.
• Dual photon or dual energy X-ray absorptiometry can detect bone loss in a safe, noninvasive test.

Quick quiz

1. Most muscles of the musculoskeletal system are:
  A. striated and involuntary.
  B. smooth and voluntary.
  C. striated and voluntary.

Answer: C. The musculoskeletal system consists mostly of skeletal muscle, which is striated and can be moved at will.

2. Finding urate crystals during an arthrocentesis exam indicates:
  A. gout.
  B. osteoporosis
  C. psuedogout.

Answer: A. Urate crystals are formed in gout and tend to collect in joints. Arthrocentesis is fluid aspiration of a joint.

3. Osteoporosis is characterized by:
  A. crystal deposition and brittleness.
  B. brittleness and swelling of the joints.
  C. porosity and brittleness.

Answer: C. Osteoporosis is a metabolic bone disorder in which bone loses calcium and phosphate and becomes porous, brittle, and abnormally vulnerable to fractures.

4. Which of the following is not a contributing factor for primary osteoporosis:
  A. a sedentary lifestyle.
  B. malnutrition.
  C. lack of calcium due to poor dietary intake.

Answer: C. Malnutrition may be a cause of secondary osteoporosis.

5. The administration of colchicine and anti-inflammatory drugs and the immobilization of affected areas are recommended in the treatment of:

   A. gout.
   B. osteoporosis.
   C. long bone fractures.

Answer: A. Gout is treated by these measures.

6. At what point during a person's life does the process of bone renewal stop?

   A. adolescence
   B. once the person reaches full height
   C. not during a person's lifetime.

Answer: C. Bone renewal continues throughout life, though it slows down with age.

Scoring

☆☆☆   If you answered all six items correctly, take a bow! You're great — no bones about it.

☆☆   If you answered four or five correctly, right on! Jump up and dance a victory jig using any of the 13 angular and circular musculoskeletal movements.

☆   If you answered fewer than four correctly, there's only one thing we can say: You're going to have to bone up.

> That's the last quick quiz. Thank goodness; I'm bone-tired.

Glossary and index

Glossary

acute illness: illness having severe symptoms and a short course

agranulocyte: leukocyte (white blood cell) not made up of granules or grains; includes lymphocytes, monocytes, and plasma cells

allele: one of two or more different genes that occupy a corresponding position (locus) on matched chromosomes; allows for different forms of the same inheritable characteristic

allergen: substance that induces an allergy or a hypersensitivity reaction

anaphylaxis: severe allergic reaction to a foreign substance

androgen: steroid hormone that stimulates male characteristic (The two main androgens are androsterone and testosterone.)

anoxia: absence of oxygen in the tissues

antibody: immunoglobulin molecule that reacts only with the specific antigen that induced its formation in the lymph system

antigen: foreign substance, such as bacteria or toxins, that induces antibody formation

antitoxin: antibody, produced in response to a toxin, that is capable of neutralizing the toxin's effects

atrophy: decrease in size or wasting away of a cell, tissue, organ, or body part

autoimmune disorder: disorder in which the body launches an immunologic response against itself

autosome: any chromosome other than the sex chromosomes (Twenty-two of the human chromosome pairs are autosomes.)

bacteria: one-celled microorganism that breaks down dead tissue, has no true nucleus, and reproduces by cell division

benign: not malignant or recurrent; favorable for recovery

bone marrow: soft organic material filling the cavities of bones

bursa: fluid-filled sac or cavity found in connecting tissue in the vicinity of joints; acts as a cushion

calculus: any abnormal concentration, usually composed of mineral salts, within the body; for example, gallstones and kidney stones

carcinogen: substance that causes cancer

carcinoma: malignant growth made up of epithelial cells that tends to infiltrate surrounding tissues and metastasize

cardiac output: volume of blood ejected from the heart per minute

cartilage: dense connective tissue consisting of fibers embedded in a strong, gel-like substance; supports, cushions, and shapes body structures

cell: smallest living component of an organism; the body's basic building block

chromosome: linear thread in a cell's nucleus that contains DNA; occurring in pairs in humans

chronic illness: illness of long duration that includes remission and exacerbation

cholestasis: stopped or decreased bile flow

clearance: complete removal of a substance by the kidneys from a specific volume of blood per unit of time

coagulant: substance that promotes, accelerates, or permits blood clotting

collagen: main supportive protein of skin, tendon, bone, cartilage, and connective tissue

compensation: the counterbalancing of any defect in structure or function; measure of how long the heart takes to adapt to increased blood in ventricles

complement system: major mediator of inflammatory response; a functionally related system consisting of 20 proteins circulating as functionally inactive molecules

cytoplasm: aqueous mass within a cell that contains organelles, is surrounded by the cell membrane, and excludes the nucleus

cytotoxic: destructive to cells

degeneration: nonlethal cell damage occurring in the cell's cytoplasm while the nucleus remains unaffected

demyelination: destruction of a nerve's myelin sheath, which interferes with normal nerve conduction

deoxyribonucleic acid (DNA): complex protein in the cell's nucleus that carries genetic material and is responsible for cellular reproduction

differentiation: process of cells maturing into specific types

disease: pathologic condition that occurs when the body can't maintain homeostasis

disjunction: separation of chromosomes that occurs during cell division

dominant gene: gene that expresses itself even if it's carried on only one homologous (matched) chromosome

dyscrasia: a condition related to a disease, usually referring to an imbalance of component elements (A blood dyscrasia is a disorder of cellular elements of the blood.)

dysplasia: alteration in size, shape, and organization of adult cells

embolism: sudden obstruction of a blood vessel by foreign substances or a blood clot

endemic: disease of low morbidity that occurs continuously in a particular patient population or geographic area

endochondral ossification: process in which cartilage hardens into bone

endocrine: pertaining to internal hormone secretion by glands (Endocrine glands, including the pineal gland, the islets of Langerhans in the pancreas, the gonads, the thymus, and the adrenal, pituitary, thyroid, and parathyroid glands secrete hormones directly into circulation.)

endogenous: occurring inside the body

erythrocyte: red blood cell; carries oxygen to the tissues and removes carbon dioxide from them

erythropoiesis: production of red blood cells or erythrocytes

estrogen: female sex hormone produced by the ovaries; produces female characteristics

etiology: cause of a disease

exacerbation: increase in the severity of a disease or any of its symptoms

exocrine: external or outward secretion of a gland (Exocrine glands discharge through ducts opening on an external or internal surface of the body; they include the liver, the pancreas, the prostate, and the salivary, sebaceous, sweat, gastric, intestinal, mammary, and lacrimal glands.)

exogenous: occurring outside the body

fungus: nonphotosynthetic microorganism that reproduces asexually by cell division

gland: specialized cell cluster that produces a secretion used in some other body part

glomerulus: a network of twisted capillaries in the nephron, the basic unit of the kidney; brings blood and waste products carried by blood to the nephron

glucagon: hormone released during the fasting state that increases blood glucose concentration

granulocyte: any cell containing granules, especially a granular leukocyte (white blood cell)

hematopoiesis: production of red blood cells in the bone marrow

hemoglobin: iron-containing pigment in red blood cells that carries oxygen from the lungs to the tissues

hemolysis: red blood cell destruction

hemorrhage: escape of blood from a ruptured vessel

hemostasis: complex process whereby platelets, plasma, and coagulation factors interact to control bleeding

heterozygous genes: genes having different alleles at the same site (locus) (If each gene in a chromosome pair produces a different effect, the genes are heterozygous.)

homeostasis: dynamic, steady state of internal balance in the body

homologous genes: gene pairs sharing a corresponding structure and position

homozygous genes: genes that have identical alleles for a given trait (When each gene at a corresponding locus produces the same effect, the genes are homozygous.)

hormone: chemical substance produced in the body that has a specific regulatory effect on the activity of specific cells or organs

host defense system: elaborate network of safeguards that protects the body from infectious organisms and other harmful invaders

hyperplasia: excessive growth of normal cells that causes an increase in the volume of a tissue or organ

hypersensitivity disorder: state in which the body produces an exaggerated immune response to a foreign antigen

hypoxia: reduction of oxygen in body tissues to below normal levels

idiopathic: disease with no known cause

immunocompetence: ability of cells to distinguish antigens from substances that belong to the body and to launch an immune response

immunodeficiency disorder: disorder caused by a deficiency of the immune response due to hypoactivity or decreased numbers of lymphoid cells

immunoglobulin: serum protein synthesized by lymphocytes and plasma cells that has known antibody activity; main component of humoral immune response

immunosurveillance: defense mechanism in which the immune system continuously recognizes cancer cells as foreign and destroys them (An interruption in immunosurveillance can lead to overproduction of cancer cells.)

insulin: hormone secreted into the blood by the islets of Langerhans of the pancreas; promotes the storage of glucose, among other functions

ischemia: decreased blood supply to a body organ or tissue

joint: the site of the union of two or more bones; provides motion and flexibility

karyotype: chromosomal arrangement of the cell nucleus

leaflets: cusps in the heart valves that open and close in response to pressure gradients to let blood flow into the chambers

leukocyte: white blood cell that protects the body against microorganisms causing disease

leukocytosis: increase in the number of leukocytes in the blood; generally caused by infection

leukopenia: reduction in the number of leukocytes in the blood

ligament: band of fibrous tissue that connects bones or cartilage, strengthens joints, provides stability, and limits or facilitates movement

lymphedema: chronic swelling of a body part from accumulation of interstitial fluid secondary to obstruction of lymphatic vessels or lymph nodes

lymph node: structure that filters the lymphatic fluid that drains from body tissues and is later returned to the blood as plasma; removes noxious agents from the blood

lymphocyte: leukocyte produced by lymphoid tissue that participates in immunity

macrophage: highly phagocytic cells that are stimulated by inflammation

malignant: condition that becomes progressively worse and results in death

megakaryocytes: giant bone marrow cells

meiosis: process of cell division by which reproductive cells (gametes) are formed

melanin: dark skin pigment that filters ultraviolet radiation and is produced and dispersed by specialized cells called melanocytes

metastasis: transfer of disease via pathogenic microorganisms or cells from one organ or body part to another not directly connected with it

mitosis: ordinary process of cell division, in which each chromosome with all its genes reproduces itself exactly.

multifactorial disorder: disorder caused by both genetic and environmental factors

myelin: a lipid-like substance surrounding the axon of myelinated nerve fibers that permits normal neurologic conduction

myelitis: inflammation of the spinal cord or bone marrow

necrosis: cell or tissue death

neoplasm: abnormal growth in which cell multiplication is uncontrolled and progressive

nephron: structural and functional unit of the kidney that forms urine

neuritic plaques: areas of nerve inflammation; found on autopsy examination of the brain tissue of people with Alzheimer's disease

neuron: highly specialized conductor cell that receives and transmits electrochemical nerve impulses

nevus: circumscribed, stable malformation of the skin and oral mucosa

nondisjunction: failure of chromosomes to separate properly during cell division; causes an unequal distribution of chromosomes between the two resulting cells

opportunistic infection: infection that strikes people with altered, weakened, immune systems; caused by a microorganism that doesn't ordinarily cause disease but becomes pathogenic under certain conditions

organ: body part, composed of tissues, that performs a specific function

organelle: structure of a cell found in the cytoplasm that performs a specific function; for example, the nucleus, mitochondria, and lysosomes

osmolality: concentration of a solution expressed in terms of osmoles of solute per kilogram of solvent

osmoreceptors: specialized neurons located in the thalamus that are stimulated by increased extracellular fluid osmolality to cause release of antidiuretic hormone, thereby helping to control fluid balance

osteoblasts: bone-forming cells whose activity results in bone formation

osteoclasts: giant, multinuclear cells that reabsorb material from previously formed bones, tear down old or excess bone structure, and allow osteoblasts to rebuild new bone

pancytopenia: abnormal depression of all the cellular elements of blood

parasite: single-celled or multi-celled organism that depends on a host for food and a protective environment

parenchyma: essential or functional elements of an organ as distinguished from its framework

pathogen: disease producing agent or microorganism

pathogenesis: origin and development of a disease

peristalsis: intestinal contractions, or waves, that propel food toward the stomach and into and through the intestine

phagocyte: cell that ingests microorganisms, other cells, and foreign particles

phagocytosis: engulfing of microorganisms, other cells, and foreign particles by a phagocyte

plasma: liquid part of the blood that carries antibodies and nutrients to tissues and carries wastes away from tissues

platelet: disk-shaped structure in blood that plays a crucial role in blood coagulation

polypeptide chains: chains of amino acids linked by a peptide bond that make up hemoglobin

pulmonary alveoli: grapelike clusters found at the ends of the respiratory passages in the lungs; sites for the exchange of carbon dioxide and oxygen

pulsus paradoxus: pulse marked by a drop in systemic blood pressure greater than 15 mm Hg during inspiration

recessive gene: gene that doesn't express itself in the presence of its dominant allele (corresponding gene)

remission: abatement of a disease's symptoms

remyelination: healing of demyelinated nerves

renin: enzyme produced by the kidneys in response to an actual or perceived decline in extracellular fluid volume; an important part of blood pressure regulation

resistance: opposition to airflow in the lung tissue, chest wall, or airways

sepsis: pathologic state resulting from microorganisms or their poisonous products in the bloodstream

stasis: stagnation of the normal flow of fluids, such as blood and urine, or within the intestinal mechanism

stenosis: constriction or narrowing of a passage or orifice

surfactant: lipid-type substance that coats the alveoli, allowing them to expand uniformly during inspiration and preventing them from collapsing during expiration

synovial fluid: viscous, lubricating substance secreted by the synovial membrane, which lines the cavity between the bones of free-moving joints

synovitis: inflammation of the synovial membrane

tendon: fibrous cord of connective tissue that attaches the muscle to bone or cartilage and enables bones to move when skeletal muscles contract

thrombolytic: clot dissolving

thrombosis: obstruction in blood vessels

tissue: large group of individual cells that perform a certain function

tophi: clusters of urate crystals surrounded by inflamed tissue; occur in gout

toxin: a poison, produced by animals, certain plants, or pathogenic bacteria

toxoid: toxin treated to destroy its toxicity without destroying its ability to stimulate antibody production

trabeculae: needlelike, bony structures that form a supportive meshwork of interconnecting spaces filled with bone marrow

translocation: alteration of a chromosome by attachment of a fragment to another chromosome or a different portion of the same chromosome

trisomy 21: aberration in which chromosome 21 has three homologous chromosomes per cell instead of two; another name for Down syndrome

tubercle: a tiny, rounded nodule produced by the tuberculosis bacillus

tubules: small tubes; in the kidney, minute, reabsorptive canals that secrete, collect, and conduct urine

uremic frost: white, flaky deposits of urea on the skin of patients with advanced uremia

urate: salt of uric acid found in urine

urticaria: wheals associated with itching; another name for hives

virus: microscopic, infectious, parasite that contains genetic material and needs a host cell to replicate

V̇/Q̇ ratio: ratio of ventilation (amount of air in the alveoli) to perfusion (amount of blood in the pulmonary capillaries); expresses the effectiveness of gas exchange

X-linked inheritance: inheritance pattern in which single gene disorders are passed through sex chromosomes; varies according to whether a man or woman carries the gene (Because the male has only one X chromosome, a trait determined by a gene on that chromosome is always expressed in a male.)

Index

i refers to an illustration; t refers to a table.

i refers to an illustration; t refers to a table.

i refers to an illustration; t refers to a table.

i refers to an illustration; t refers to a table.

Herpes simplex *(continued)*
  reactivation of, 50
  recurrence of, 49
  signs and symptoms of, 50
  transmission of, 49
  treating, 51
  types of, 48-49
Herpes zoster, 51-53
  cause of, 51
  diagnostic test for, 53
  progression of symptoms in, 51-52
  reactivation of virus in, 51
  treating, 52
Hodgkin's disease, 25-27
  diagnostic tests for, 27
  Epstein-Barr virus and, 25
  Reed-Sternberg cells and, 26i
  signs and symptoms of, 26
  treating, 25
Homeostasis, 8-10
Hyperosmolar hyperglycemic nonketotic syndrome, 144
Hypersensitivity disorders, 73
Hypertension, 227-232
  causes of, 228-230
  complications of, 230, 231i
  diagnostic tests for, 231-232
  treating, 230
  types of, 227-228
Hyperthyroidism, 149-153
  characteristic findings in, 151-153
  diagnostic tests for, 153
  factors that influence development of, 150-151
  treating, 154
  types of, 150
Hypertrophic cardiomyopathy
  complications of, 233
  diagnostic tests for, 233-235
  effects of, 232-233
  signs and symptoms of, 233
  treating, 234
Hypothyroidism, 153-156
  causes of, 153-155
  classifying, 153

Hypothyroidism *(continued)*
  diagnostic tests for, 156
  signs and symptoms of, 155-156
  treating, 156

I-K
Idiopathic thrombocytopenic purpura, 300-302
  diagnostic tests for, 301-302
  progression of, 300
  signs and symptoms of, 301
  treating, 302
Immune disorders, 73-92
Immune system, 69-70
Immunity, types of, 70-72
Immunodeficiency disorders, 73
Immunoglobulins, 71-72
Infection, 45-48
Infectious disorders, 48-67
Infectious mononucleosis, 53-55
  development of, 53-54
  diagnostic tests for, 55
  Epstein-Barr virus as cause of, 53, 54i
  hallmarks of, 53
  signs and symptoms of, 54-55
  treating, 54

L
Leukemia, 27-32
Lung cancer, 32-35
  causes of, 32
  development of, 33-34
  diagnostic tests for, 33t, 34-35
  forms of, 32, 33t
  signs and symptoms of, 33t
  treating, 34
Lupus erythematosus, 87-90
  cause of, 88
  diagnostic tests for, 90
  drugs that activate, 89
  predisposing factors for, 89
  prognosis for, 87-88
  signs and symptoms of, 89-90
  treating, 88
  types of, 87

i refers to an illustration; t refers to a table.

i refers to an illustration; t refers to a table.

i refers to an illustration; t refers to a table.

i refers to an illustration; t refers to a table.